高等医药院校药学专业本科双语教材

基础化学实验
General Chemistry Experiment

主　编：刘　静
副主编：黎红梅

东南大学出版社
·南京·

图书在版编目(CIP)数据

基础化学实验:汉英对照 / 刘静主编. —南京:
东南大学出版社,2010.9(2018.7重印)
 ISBN 978-7-5641-2437-3

Ⅰ.①基… Ⅱ.①刘… Ⅲ.①化学实验－双语教学－
高等学校－教材－汉.英 Ⅳ.①O6-3

中国版本图书馆 CIP 数据核字(2010)第 177283 号

东南大学出版社出版发行
(南京四牌楼2号 邮编:210096)
出版人 江建中
江苏省新华书店经销 常州市武进第三印刷有限公司印刷
开本:787 mm×1092 mm 1/16 印张:18.5 字数:462 千字
2010 年 9 月第 1 版 2018 年 7 月第 4 次印刷
ISBN 978-7-5641-2437-3
印数:9001~12000 定价:32.00 元

(凡因印装质量问题,可直接向我社发行科调换。电话:025-83792328)

前　言

本书为大学一年级的基础化学实验教材。基础化学实验作为学生进入大学所接触到的第一门化学实验课,具有启蒙作用,要解决诸多重要问题,例如,使学生形成科学态度和科学方法,使他们养成良好的习惯,培养他们的基本实验技能和技巧。这些对学生的成长影响深远。因此,本教材注重对学生化学实验基本知识的培养、实验操作技能以及综合运用知识能力的训练,同时注重对学生进行实验研究方法的启蒙和初步训练。

本教材内容分为"基本要求与实验室规则"、"实验基本操作与技能"、"实验数据的处理"、"实验部分"以及"附录"几个部分。在"基本要求与实验室规则"部分,对本课程的学习提出了明确的要求,特别强调了学习的目标和方法以及实验诚信原则;"实验基本操作与技能"不仅介绍了基本仪器及操作,还用较大的篇幅介绍应该如何做规范的实验记录、如何处理与表达实验数据;为了使学生尽早明白"量"的概念、懂得任何测量都存在某种程度的误差,我们在"实验数据的处理"部分介绍了实验误差理论;"实验部分"的安排本着循序渐进的原则,将各个基本操作和训练有机分布在各个类型的实验中,使用时教师可根据实验内容指导学生阅读前几部分的相关内容。本教材中所列实验包括四大板块:无机制备、无机定性分析、化学定量分析、化学常数测定。这些实验既可以独立开设,也可以对其中的一些实验稍加整合作为系列实验进行。为方便学生查找相关信息,在"附录"部分,除了常用的数据表以外,还附有Excel作图指南,此外还有中国药典中常见的无机离子的鉴别。

由于本教材主要用于以化学作为基础课的药学专业学生,"实验部分"有意结合了药物的相关内容,使学生对用实验手段研究实际问题有一些感性认识,以期提高学生的学习兴趣。

我们编写双语教材的主要目的是帮助学生尽早接触专业英语。在编写时参考了大量国外同行使用的教学内容和专业用语,尽可能用流畅、地道的英语表达教学内容,同时尽可能保持中英文的一致性。在教与学中尽量使用英语,但不唯英语,早日学会阅读英文文献、用英语进行学术交流,这才是使用双语教学的目的。当然,我们主要的目标还是通过实验教学使学生掌握实验的方法和技术。

本教材在中国药科大学无机化学教研室多年的实验教学改革和双语实验教学的基础上编写而成。刘静负责本书第三部分、第五部分和部分实验的编写,黎红梅负责第一、第二部分和部分实验的编写。参编人员还有王越、熊晔蓉、何海军、李嘉宾、陈亚东,王越、熊晔蓉老师还参加了本书的校阅工作。由于编者水平有限,书中如有错误和不妥之处,欢迎读者批评指正。

编　者

2010年8月

Contents

Part 1 Laboratory General Information

1.1 Learning Objectives ·· (2)
1.2 General Guidelines ··· (4)
1.3 Honor Principle ·· (6)
1.4 General Safety Rules ·· (6)

Part 2 General Equipment and Lab Techniques

2.1 General Equipment ·· (14)
2.2 Cleaning Laboratory Glassware ··· (20)
2.3 Digital balance and weighing ··· (24)
2.4 Separation of solids from liquids ··· (28)
2.5 Heating ·· (38)
2.6 Measuring and Delivering liquid ·· (40)
2.7 pH Meter and pH Measurement ··· (48)
2.8 Notebooks ··· (54)
2.9 Analysis of Data and Presentation of Results ································· (62)

Part 3 Quantitative Treatment of Data

3.1 Precision of Instrument Readings and Other Raw Data ··················· (70)
3.2 Absolute and Relative Uncertainty ··· (72)
3.3 Accuracy and Precision ··· (74)
3.4 Error ·· (74)
3.5 Average Deviation and Standard Deviation ······································ (76)
3.6 Significant figure ··· (80)
3.7 Statistics Applied to Small Data Sets ·· (84)
3.8 Propagation of Indeterminate Errors ·· (96)

Part 4 Experiments

Experiment 1 Checking and Cleaning Equipment ····································· (100)
Experiment 2 Weighing Exercise and Preparing Solutions ······················· (106)
Experiment 3 Determination of Crystal Water in Barium Chloride ············ (112)

I

目　　录

第一部分　基本要求与实验室规则

1.1　课程教学目标 …………………………………………………………………… (3)
1.2　实验的基本要求 ………………………………………………………………… (5)
1.3　实验的诚信原则 ………………………………………………………………… (7)
1.4　实验室安全 ……………………………………………………………………… (7)

第二部分　实验基本操作与技能

2.1　化学实验的常用仪器 …………………………………………………………… (15)
2.2　仪器的洗涤 ……………………………………………………………………… (21)
2.3　电子天平及称量 ………………………………………………………………… (25)
2.4　固液分离 ………………………………………………………………………… (29)
2.5　加热 ……………………………………………………………………………… (39)
2.6　溶液的量取和转移 ……………………………………………………………… (41)
2.7　pH 计与 pH 值测定 …………………………………………………………… (49)
2.8　实验记录 ………………………………………………………………………… (55)
2.9　实验数据分析与结果表达 ……………………………………………………… (63)

第三部分　实验数据的处理

3.1　仪器读数和其他原始数据的精密度 …………………………………………… (71)
3.2　绝对不确定度和相对不确定度 ………………………………………………… (73)
3.3　准确度与精密度 ………………………………………………………………… (75)
3.4　误差 ……………………………………………………………………………… (75)
3.5　平均偏差和标准偏差 …………………………………………………………… (77)
3.6　有效数字 ………………………………………………………………………… (81)
3.7　有限数据的统计运用 …………………………………………………………… (85)
3.8　不可定误差（偶然误差）的传递 ………………………………………………… (97)

第四部分　实验部分

实验 1　仪器的认领和洗涤 …………………………………………………………… (101)
实验 2　称量练习和溶液的配制 ……………………………………………………… (107)
实验 3　氯化钡结晶水的测定 ………………………………………………………… (113)

Contents

Experiment 4	Preparation of Ferrous Ammonium Sulfate Hexahydrate	(116)
Experiment 5	Preparation of Zinc Gluconate	(120)
Experiment 6	Preparation of Medicinal Sodium Chloride and Examination of Impurities Limitation	(124)
Experiment 7	Synthesis of Potassium Aluminum Sulfate (Alum)	(136)
Experiment 8	Qualitative Analysis of Cations (1)	(144)
Experiment 9	Qualitative Analysis of Cations (2)	(154)
Experiment 10	Designing Scheme for Qualitative Analysis of Cations	(166)
Experiment 11	Calibration of Volumetric Glassware	(172)
Experiment 12	Preparation and Standardization of a Standard Sodium Hydroxide Solution	(178)
Experiment 13	Preparation and Standardization of a Standard Solution of Hydrochloric Acid	(182)
Experiment 14	Determination of Aspirin (Acetylsalicylic Acid) using Back Titration	(186)
Experiment 15	Preparation and Standardization of EDTA Solution	(190)
Experiment 16	Content Assay of Zinc Gluconate	(194)
Experiment 17	Determination of Water Hardness	(198)
Experiment 18	Precipitation Titration: Determination of Chloride by Mohr Method	(202)
Experiment 19	Preparation and Standardization of Perchloric Acid in Non-aqueous Solvent	(206)
Experiment 20	Determination of Sodium Salicylate in Non-aqueous Solvent	(210)
Experiment 21	Preparation and Standardization of Potassium Permanganate Solution	(214)
Experiment 22	Assay of Ferrous Ammonium Sulfate	(218)
Experiment 23	Determination of Medical Hydrogen Peroxide	(222)
Experiment 24	Iodimetric Titration of Ascorbic Acid in Vitamin C Tablets	(226)
Experiment 25	Determination of Dissociation Constant of Weak Acid	(232)
Experiment 26	Buffer Action and Buffer Solution	(240)
Experiment 27	Determining Coordination Number of $[Ag(NH_3)_n]^+$ Complex Ion	(248)
Experiment 28	Determination of Reaction Rate and Activation Energy	(254)
Experiment 29	Determination of K_{sp} of Silver Acetate	(262)

Part 5 Appendix

5.1	Using Excel for Graphing	(268)
5.2	Identification Tests of Common Inorganic Ions	(272)

实验 4　硫酸亚铁铵的制备 …………………………………………………… (117)
实验 5　葡萄糖酸锌的制备 …………………………………………………… (121)
实验 6　药用氯化钠的制备、性质及杂质限度检查 …………………………… (125)
实验 7　硫酸铝钾（明矾）的合成 ……………………………………………… (137)
实验 8　阳离子定性分析（1） …………………………………………………… (145)
实验 9　阳离子定性分析（2） …………………………………………………… (155)
实验 10　阳离子定性分析方案的设计 ………………………………………… (167)
实验 11　容量仪器的校正 ……………………………………………………… (173)
实验 12　NaOH 标准溶液的配制与标定 ……………………………………… (179)
实验 13　盐酸标准溶液的配制与标定 ………………………………………… (183)
实验 14　返滴定法测定阿司匹林（乙酰水杨酸） ……………………………… (187)
实验 15　EDTA 标准溶液的配制和标定 ……………………………………… (191)
实验 16　葡萄糖酸锌含量的测定 ……………………………………………… (195)
实验 17　水的硬度的测定 ……………………………………………………… (199)
实验 18　Mohr 法测定药用氯化钠的含量（沉淀滴定） ………………………… (203)
实验 19　高氯酸标准溶液的配制和标定 ……………………………………… (207)
实验 20　非水滴定法测定药用水杨酸钠的含量 ……………………………… (211)
实验 21　高锰酸钾溶液的配制与标定 ………………………………………… (215)
实验 22　硫酸亚铁铵含量的测定 ……………………………………………… (219)
实验 23　医用双氧水的含量测定 ……………………………………………… (223)
实验 24　碘量法测定维生素 C 的含量 ………………………………………… (227)
实验 25　弱酸电离常数的测定 ………………………………………………… (233)
实验 26　缓冲溶液与缓冲作用 ………………………………………………… (241)
实验 27　银氨配离子配位数的测定 …………………………………………… (249)
实验 28　化学反应速率与活化能的测定 ……………………………………… (255)
实验 29　醋酸银的 K_{sp} 测定 ………………………………………………… (263)

第五部分　附　录

5.1　Excel 作图 …………………………………………………………………… (269)
5.2　药典中常见无机离子的鉴别 ……………………………………………… (273)
5.3　常见无机酸碱的解离常数 ………………………………………………… (280)
5.4　常用基准物质（Common Primary Standards） …………………………… (282)
5.5　常用 pH 缓冲溶液（Common pH Buffer Solution） ……………………… (283)
5.6　常用指示剂 ………………………………………………………………… (284)

基础化学实验

General Chemistry Experiment

Part 1　Laboratory General Information

1.1　Learning Objectives

This course is the first year of general chemistry laboratory. The lab will provide you with instruction in common laboratory manipulations, data collection and interpretation, record keeping, and stoichiometric (chemical) calculations. You will become familiar with some of the chemist's basic laboratory equipment and will learn why and when this equipment is used. This lab should help you connect theory with the real world. You will be concentrating on the correct use of various kinds of laboratory equipment and procedures, but the lab will reinforce chemical principles covered in your lecture. You will be educated for the scientific method of chemistry. This lab helps you establish a "practice first" point of view. You will develop a scientific approach of seeking truth and a scientific method of logical thinking. You will be trained to conduct experiments in the correct, meticulous, tidy way. The objectives of this course are following as:

Technique

Considering the ease and speed of use and the desired precision and accuracy, you should be able to choose the right piece of equipment and follow proper technique to use it. You must be familiar with the following:

　　1. Mass measurement: proper weighing technique and use of the balance.

　　2. Volume measurement: volumetric pipet, Mohr pipet, buret, volumetric flask, graduated cylinder and other measuring equipment. This includes interpolation and estimating between the marks on a scale; using appropriate number of significant figures in reporting measured quantities.

　　3. Temperature measurement: use of thermometer.

　　4. Heating procedures: use of waterbath and hot plate.

　　5. pH measurement: use of pH paper and pH meter.

　　6. Separation of solids from liquids: gravity filtration, vacuum filtration and centrifugation.

　　7. Qualitative analysis: using flame tests, observing and describing the formation and dissolving of precipitates and gases.

　　8. Quantitative techniques: titration, quantitative transferring and weighing.

　　9. Proper cleaning of glassware, mixing of solutions, and use of reagents without causing cross-contamination.

　　10. Laboratory safety: safe use and disposal of chemical reagents and safe use of lab equipment; awareness of lab hazards and safety equipment procedures.

Calculations

　　1. Basic calculations used in this lab: percent by mass, concentration, using a chemical formula to calculate molar mass, conversion from grams to moles or moles to grams using the molar mass, use of concentration units such as molarity, calculation of the simplest formula,

第一部分　基本要求与实验室规则

1.1　课程教学目标

　　本课程是基础化学实验课程之一，是同学们进入大学的第一门实验课程。本课程将引导同学学习如何进行常用实验仪器的操作、如何收集和分析实验数据、如何做实验记录以及如何进行相关化学计算。在本课程的学习中，同学们要熟悉化学实验基本仪器，掌握如何选择和使用这些常用的仪器。化学实验可以帮助大家学会"理论联系实际"。因此，本课程不仅注重于训练同学正确使用仪器和掌握正确的操作技能，同时也注重加深同学们对化学原理的理解。化学实验课程蕴含科学方法的训练，同学们要通过实验逐步树立"实践第一"的观点，养成实事求是的科学态度和科学的逻辑思维方法，逐步培养正确、细致、整洁地进行科学实验的良好习惯。下面是本课程的具体教学目标。

操作技能

　　要想方便、快速地开展实验，同时又能满足一定的精密度和准确度要求，必须做到正确选择和使用实验仪器。同学们必须熟悉和掌握的基本操作有：

1. 质量测量：正确的称量方法和天平的使用。
2. 体积测量：移液管、吸量管、滴定管、容量瓶、量筒等仪器的使用，这些仪器可用于液体体积的测量，但其精度和体积大小各不相同，要求能根据实验具体情况正确选用，定量测定时能根据仪器的精密度正确记录测量值的有效数字。
3. 温度测量：温度计的使用。
4. 加热操作：水浴加热和电炉加热。
5. pH 值测量：pH 试纸和 pH 计的使用。
6. 固液分离：常压过滤、减压过滤和离心分离。
7. 定性分析：焰色反应，观察和描述沉淀的生成和溶解，观察气体的生成。
8. 定量操作：滴定操作、定量转移和称量。
9. 玻璃仪器洗涤、溶液混合，以及在多种试剂使用时能避免溶液的交叉污染。
10. 实验室安全技能：正确使用和处理实验用化学试剂，保证实验设备的安全使用，对实验室可能发生的事故有清醒的认识，对实验设备的安全操作心中有数。

simple dilution calculations using $M_1V_1 = M_2V_2$, use of chemical equations in calculations, calculation of the median and the average from a set of data taken in the lab, calculation of deviation and relative standard deviation.

2. Data recording: proper format and rules.

3. Significant figures should be understood and the appropriate number of digits should be used for recording data, doing calculations, and predicting precision; in addition, mean or average, median, and error analysis will be introduced.

4. Graphing data and using linear regression to predict values.

1.2 General Guidelines

Process of an Experiment

Before lab, write the prelab write-up You must preview the materials entirely and comprehend the experimental purpose, principle, procedures and the cautions, also, complete the assigned prelab problems for the experiment provided in the lab manual. You must turn in your prelab write-up before the lab lecture begins. The prelab write-up is a part of the notebook, which contains sections, such as the experimental objective, reference, the outline or the flowchart of the experiment procedure, and the method to deal with the data. You will be allowed to do the experiment only when you have completed the prelab write-up. Please come to the lab 10 minutes before the lab lecture and prepare the equipment used in the experiment before. Your instructor will check your prelab work. You must complete the assigned prelab problems and notebook preparation before you come to lab lecture or you will not be permitted to continue the lab. In the lab lecture, your instructor will comment this experiment and answer your questions. Then, you can start to do the experiment.

Before leaving the laboratory, check your data You must have your TA (teaching assistant) check that your notebook contains all the data and observations necessary for the successful completion of the lab write-up. He or she will initial your notebook, and if items are missing, your TA or faculty instructor will decide how the missing work is to be completed. Barring major disasters, you will have time to do all the needed work during your regular lab period, especially if you come to lab prepared and stay organized during the lab period.

After the lab, write the experiment report You should look at your results, complete all the calculations and complete either your data sheet or formal report. You have one week to complete and turn in the formal report.

Your formal reports will be examined and graded by your TA to evaluate theory section, your procedure, data, presentation of results, including graphs, answers to questions posed in the manual, a concluding discussion and a discussion of errors and uncertainties, when appropriate. Specific point credit distributions will vary from experiment to experiment, but emphasis will be placed on the accuracy of your final results in those experiments which are quantitative in nature. Grade deductions will be made for late notebooks.

Cleanliness Points

Poor housekeeping can contribute to accidents in the lab and it is your responsibility to keep the lab clean and safe. You will be sharing glassware and equipment with students in other lab sections, so it is common courtesy to clean up your area when you are done. Your teaching assistant (TA) will check to make sure you have cleaned up your equipment and chemicals at the

计算技能

1. 基本计算：质量百分比计算，浓度计算，根据化学式计算摩尔质量，质量和物质的量间换算，能正确使用浓度单位，化简，根据 $M_1V_1=M_2V_2$ 进行溶液稀释的浓度计算，根据化学方程式进行有关计算，平行实验数据的平均值以及中值计算，偏差和相对标准偏差的计算。

2. 数据记录：按正确的格式和规则记录实验数据。

3. 有效数字：正确理解有效数字的意义，并能正确记录数据的有效数字，掌握有效数字的运算法则，能根据有效数字预测实验的精度。另外，要了解平均值、中值以及误差分析等方面的知识。

4. 数据作图和线性回归：通过对实验数据进行作图或线性回归分析，预测实验变量之间的关系。

1.2 实验的基本要求

实验环节

实验前写预习报告 实验前做好充分的预习，弄清实验的目的、原理、步骤以及注意事项，回答实验教材提出的有关实验问题，写预习报告。预习报告是实验记录的一部分，包括实验目的、参考文献、实验大致步骤或实验流程图、数据处理方法。只有写好预习报告的同学才可以进入实验室进行实验。实验时，请提前10分钟到达实验室，并做实验前的仪器准备。实验老师会检查预习情况，没有写好预习报告的同学不允许进实验室做实验。实验课开始，实验老师先对本次实验做讲解，并回答同学们的提问，之后，同学们方可动手实验。

离开实验室时检查实验记录 实验记录要经老师检查数据是否完全、记录是否正确，经签字同意后方可离开实验室。如果有遗漏或错误，则需要继续实验直至完成。为了避免较大错误的出现，事先需要做好准备和筹划。实验时要做到有条不紊，认真操作，以保证能在规定的时间里完成全部实验。

实验后写正式实验报告 实验结束后，要尽快检查实验结果，完成有关计算，然后写正式的实验报告。实验报告要求在实验后一周之内上交。

教师对实验报告的评价要点包括：实验原理和实验过程的表述、实验数据处理、实验数据和结果及表达（包括图表）、回答实验教材所提出的问题情况、根据实验的具体情况对误差和不确定度的讨论以及总结性讨论。不同类型的实验评分标准的侧重点有所不同，如，对于定量实验其重点自然会放在最后实验结果的准确度上。另外，迟交实验报告实验成绩将会受到影响。

实验室卫生要点

混乱是实验事故的诱因，所以保持实验室的整洁和安全是每个人的责任。每位同学都与其他班的多位同学共用同一实验柜和同一套仪器。因此，做完实验后，请务必将实验仪器清洗干净，并整齐摆放在实验柜中。这项工作做得好坏不仅体现了个人的实验素养，而且还

Part 1 Laboratory General Information

end of each lab before you can leave.

1. Make sure your lab area cleaned up. Your lab area is all your work places, including the lab bench, reagent rack, sink, and balance table.

2. Make sure all the equipment you used is cleaned up and your lab drawer is set up correctly.

3. Make sure your glassware drawer is inventoried, with missing items replaced and extra items removed.

If you do not follow the requirements, you will lose points from your lab write up for the week.

1.3 Honor Principle

The principle of academic honesty is at the very heart of experimental science. The following remarks apply to the laboratory component:

1. Use of another student's laboratory data is a violation of the Honor Principle, unless permission is granted by the instructor.

2. When use of another's data is allowed, the source of the data must be indicated with a clear reference in the laboratory notebook.

3. Fabrication of data, alteration of your own data, or fabrication of observations to secure some desired result is a clear violation of the Honor Principle.

4. All laboratory reports must represent your independent calculations and individual conclusions, although comparison of numerical results with those of another student is permitted, with appropriate attribution of the other student's results.

5. Copying of any portion of another student's laboratory report is a serious violation of the Honor Principle.

1.4 General Safety Rules

Safe laboratory practice is based on understanding and respect, not fear. The regulations below are intended to help you work safely with chemical reagents. These guidelines cover ordinary hazards and apply to any laboratory experiments you will encounter. Your instructors will discuss specific safety precautions relevant to each experiment during laboratory lecture. Your laboratory textbook or manual will point out specific hazards and precautions. Before beginning an experiment, be sure you have this information at hand and that you understand it. Do not hesitate to consult with your instructors if you have questions about any experiment or about these regulations.

Safety Rules in the Chem Lab

1. Contact lens don't be worn at all times when in a laboratory.

2. Open flames(burners, matches, *etc.*) are not permitted.

3. Use every precaution to keep all chemicals off your skin and clothing, out of your nose, mouth and eyes, and away from flames. It is strictly forbidden to eat or drink anything(including water) in the laboratory.

4. Long hair and billowy clothing must be confined when in the laboratory. Shoes are mandatory; sandals or open-toed shoes are not allowed, even if socks are worn. The lab cloth must be worn at all times when in a laboratory.

5. All accidents, including contact with chemicals, cuts, burns, or inhalation of fumes must

体现了个人的品行。为了督促同学做好此项工作,教师会在实验结束前做如下检查,凡是没有按要求做的同学,实验成绩将会受到一定的影响。

1. 实验工作区有没有打扫和整理。实验工作区包括你用过的所有区域,如实验台、试剂架、水槽、天平操作台。
2. 实验仪器是否清洗干净,实验仪器有没有全部、正确地放回实验柜中。
3. 实验柜里的仪器是否多余或者缺失,要将多余的拿出来,不够的要补上。

1.3　实验的诚信原则

科学诚信是实验科学的灵魂。在实验课中,我们要遵守如下规则:
1. 没有得到指导老师的同意,不能擅自使用其他同学的实验数据。
2. 使用他人的数据和资料时,需要在实验记录本和实验报告中清楚地注明出处。
3. 不得伪造和修改实验数据或对观察现象进行修改使其符合期望的结果,否则视为严重违反诚信原则的行为。
4. 可以将自己的结果与其他同学的结果进行比较,也可以与其他同学进行讨论,甚至还可以引用他人结果,但必须建立在独立完成计算和得出结论的基础上。实验报告必须自己独立完成。
5. 不得抄袭他人的实验报告,哪怕只有一点,都是严重违反科学诚信的行为。

1.4　实验室安全

实验室安全操作规范的形成不是由于人们对实验的诚惶诚恐,而是基于对实验室安全的充分认识和高度重视。以下规则有助于化学试剂的安全使用,对实验中通常可能出现的一般危险之防范皆适用。至于某实验中的特殊安全要求,指导老师将会在该次实验的讲解中特别指出,实验教材或实验指导对实验中可能发生的具体危害和防范措施也有说明,实验开始前,请务必清楚和理解这些有关的实验安全要求。如果对实验或安全规则有任何疑问,应该毫不犹豫地立即向指导老师咨询。

实验室安全规则

1. 在实验室中不要戴隐形眼镜。
2. 轻易不要使用明火,如打火机、火柴等。
3. 使用药品时要极其小心。任何药品都不要让其接触到皮肤和衣服,要避免药品进入鼻子、嘴巴和眼睛里,药品更要远离火源。严禁在实验室喝水和吃东西。
4. 在实验室时,要将长头发扎起来。不要穿拖鞋之类容易脱落的鞋子,最好不穿露脚指头的鞋子,并尽量穿上袜子,以免药品不慎溅到脚上。在实验室里,要自始至终穿着实验服。
5. 无论什么样的事故发生,如皮肤接触到化学药品、割伤、烧伤、吸入有害气体等,都要

be reported to an instructor immediately. Any treatment beyond emergency first aid will be referred to the student infirmary. Severe emergencies will be referred to the hospital emergency room.

6. It is your responsibility to read and abide by the "Laboratory Safety" section of the lab manual and to keep it with you in the laboratory. Any other safety handouts or special precautions mentioned during lab lecture must be scrupulously observed.

7. Laboratory equipment and work area must be cleaned after finishing work.

Failure to observe laboratory safety rules and procedures may result in injury to you or to fellow students. Accidents in the laboratory are often the result of carelessness or ignorance either by you or by your neighbors. It is in your own best interest to stay alert and to be aware of possible hazards in the laboratory. Do not hesitate to call the attention of the instructors to unsafe practices by your colleagues.

Safety Hazards

The safety precautions outlined below will be worthless unless you understand, and think through the consequences of every operation before you perform it. The common accidents, which often occur simultaneously, are fire, explosion, chemical and thermal burns, cuts from broken glass tubing and thermometers, absorption of toxic, non-corrosive chemicals through the skin, and inhalation of toxic fumes. Less common, but obviously dangerous, is the ingestion of a toxic chemical.

1. Fire. There should never be open flames in the lab. Make it a working rule that water is the only nonflammable liquid you are likely to encounter. Treat all others in the vicinity of a flame as you would gasoline. Specifically, never heat any organic solvent in an open vessel, such as a test tube, Erlenmeyer flask, or beaker, with a flame. Such solvents should be heated in a hood with a steam bath, not a hot plate. Never keep volatile solvents, such as ether, acetone, or benzene in an open beaker or Erlenmeyer flask. The vapors can and will creep along the bench, ignite, and flash back if they reach a flame or spark.

2. Explosion. Never heat a closed system or conduct a reaction in a closed system (unless specifically directed to perform the latter process and then only with frequent venting). Before starting a distillation or a chemical reaction, make sure that the system is vented. The results of an explosion are flying glass and spattered chemicals, usually both hot and corrosive.

3. Chemical and Thermal Burns. Many inorganic chemicals such as the mineral acids and alkalis are corrosive to the skin and eyes. Likewise, many organic chemicals, such as acid halides, phenols, and so forth are corrosive and often toxic. If these are spilled on the desk, in the hood, or on a shelf, call for assistance in cleaning them up. Be careful with hot plates to avoid burns. Always assume that hot plates are HOT.

4. Cuts. The most common laboratory accident is probably the cut received while attempting to force a cork or rubber stopper onto a piece of glass tubing, a thermometer, or the side-arm of a distilling flask. Be sure to make a proper-sized hole, lubricate the cork or stopper (lubrication is essential with a rubber stopper), and use a gentle pressure with rotation on the glass part. Severed nerves and tendons are common results of injuries caused by improper manipulation of glass tubes and thermometers. Always pull rather than push on the glass when possible.

5. Absorption of Chemicals. Keep chemicals off the skin. Many organic substances are not corrosive, do not burn the skin, or seem to have not any serious effects. They are, however,

立即报告老师。非紧急情况送医务室处理,紧急情况需要尽快送医院。

6. 务必认真阅读和遵守实验手册中关于"实验室安全规则"的内容,做实验时,要将这些规则时刻铭记在心。对于指导老师提到的任何安全知识和注意事项都必须小心、认真对待。

7. 实验完毕,实验仪器和操作台必须进行打扫和整理。

不遵守安全规则和操作程序就有可能给你或他人造成伤害。实验事故常常是由不慎和疏忽引起的,为了维护每个同学切身的安全和利益,我们要意识到事故随时都可能发生,并时刻保持警惕;发现有同学进行不安全的操作时,要立即向老师报告。

安全事故的预防

只有事前对安全措施的每个步骤都能做到充分的理解和认真的思考,才能在应用它们时得心应手,否则,临时抱佛脚是无济于事的。化学实验室的一般事故,如火灾、爆炸、化学灼伤和热烫伤、被破碎玻璃管或温度计割伤、吸入毒物、非腐蚀性化学药品的透皮吸收、有毒烟雾的吸入等等,常常会同时发生。有毒化学药品入口的情况虽不常见,但显然也是非常危险的。

1. 防火灾

在实验室里最好永远不要使用明火。记住:水是唯一不会产生火焰的液体,除水之外,所有处于火焰附近的液体都应将其视为汽油处理,要将这种观念当作一条规则。在此,要特别指出的是:永远不要在火焰上使用敞口容器(如试管、锥形瓶、烧杯)加热有机溶剂。加热有机溶剂时,应将有机溶剂放在加盖的容器中,采用水浴或蒸汽浴在通风橱中进行,而不能用电炉直接加热。也不要将挥发性溶剂(如乙醚、丙酮、苯等)存放在开口容器(如烧杯、锥形瓶)中,以免其蒸气浓度到达着火浓度,达到着火浓度的蒸气一旦被点燃就会着火。

2. 防爆炸

不要加热封闭容器,除非特殊需要,也不要在密闭容器中进行化学反应。在进行蒸馏或化学反应之前,一定要确保系统有排气口与外界相连。爆炸时,崩裂乱飞的碎玻璃和飞溅的化学药品会伤到人,而且溅出的化学药品往往是烫人的且具有腐蚀性的,这将会造成更严重的后果。

3. 防化学灼伤和热烫伤

许多无机酸、碱对皮肤和眼睛都具有腐蚀性,同样,许多有机试剂如卤代酸、苯酚等等,不仅具有腐蚀性,而且还具有毒性,一旦将这些试剂洒落,要立即擦净。使用电炉时要小心,避免烫伤。记住:热电炉总是会烫人的。

4. 防割伤

实验室最为常见的事故大概就是割伤了。例如,用力将一根玻璃管、温度计、蒸馏烧瓶的支管插入橡皮塞或软木塞时,很容易发生割伤。这时,需要选择合适的孔径,并对软木塞或橡皮塞的孔洞进行润滑,以轻轻转动方式将玻璃插入塞子,这些措施会大大降低割伤事故发生的概率。玻璃管、温度计使用不当极有可能使手腕神经和肌腱受到严重的损伤,在操作玻璃仪器时,要尽可能采用"拉"的方式代替"推"的方式。

5. 防化学药品的透皮吸收

不要让皮肤接触到化学药品。许多有机物质虽然表面看没有什么危险,既无腐蚀性又

absorbed through the skin, sometimes with dire consequences. Others will give a serious allergic reaction upon repeated exposure, as evidenced by severe dermatitis. Be careful about touching your face or eyes in the lab; make sure your hands are clean first. Gloves will be available in the lab. However, gloves provide only a temporary layer of protection against chemicals on your skin and may be permeable to some chemical reagents, without visible deterioration. If your gloves come in contact with a chemical reagent, remove them, wash your hands, and get a new pair immediately.

6. Inhalation of Chemicals. Keep your nose away from chemicals. Many of the common solvents are extremely toxic if inhaled in any quantity or over a period of time. Do not evaporate excess solvents in the laboratory; use the hood or a suitable distillation apparatus with a condenser. Some compounds, such as acetyl chloride, will severely irritate membranes in your eyes, nose, throat, and lungs, while others, such as benzyl chloride, are severe lachrymators, i. e. they induce eye irritation and tears. When in doubt, use the hood or consult with the laboratory instructor about the use of chemicals required for your work. Specific safety information about chemicals used is included in each experiment write up.

7. Ingestion of Chemicals. The common ways of accidentally ingesting harmful chemicals are: by pipet, from dirty hands, contaminated food or drink and food use of chemicals taken from the laboratory. Below are ways to avoid accidental ingestion of chemical reagents.

(1) Pipets must be fitted with suction bulbs to transfer chemicals. DO NOT USE MOUTH SUCTION.

(2) Wash your hands before handling anything(cigarettes, chewing gum, food) which goes into your mouth. Wash your hands when you leave the laboratory.

(3) Do not eat or drink in the laboratory. Use the water fountains for a drink—not a laboratory faucet.

(4) Never use chemicals(salt, sugar, alcohol, bicarbonate, *etc.*) from the laboratory on food. The source containers may be contaminated or mislabeled.

(5) Never use laboratory glassware as a food or drink container.

(6) Never store food or drink in a laboratory refrigerator or ice machine. Never consume ice from a laboratory ice machine.

What should be done with a safety hazard?

1. Eyes. GET HELP IMMEDIATELY! Chemicals in the eyes must be removed at once by flooding with copious quantities of water. Help the victim. Otherwise, place the victim on the floor, by force if necessary. One person should straddle the victim with knees on the floor, pouring a moderate stream of water from a flask or beaker onto the bridge of the victim's nose so that both eyes are flooded. Another person should squat at the head of the victim and roll back the eyelids of both eyes; use the thumb and forefinger to spread the eyelids open. Use at least several liters of water. When it is reasonably certain that the excess chemicals have been washed away, take the victim to the Emergency Room for immediate medical attention.

2. Chemicals on the skin. All chemicals which come in contact with the body should be considered toxic and washed off completely with soap and water even if they do not appear to be corrosive. Wash off corrosive substances by flooding with tap water, using the safety shower if necessary, stripping off any clothing and shoes that are soaked with chemicals. Except as follows, do not try to neutralize acids with alkali or alkalis with acid.

不会灼烧皮肤,但是它们却有可能透过皮肤,造成可怕的后果。有些物质与皮肤反复接触时,会诱发严重过敏反应,例如,常见的严重皮炎。在实验室里,擦脸、擦眼睛时一定要小心,之前要确认自己的手是干净的。做实验时要戴着手套。但手套也只是皮肤的临时保护层,即使手套没有明显破损,一些化学药品也可能透过手套接触到皮肤。如果手套直接接触过这样的药品,要丢掉它,将手洗净,再换上新的。

6. 防有毒蒸气吸入

鼻子应该远离化学试剂(不要用鼻子直接闻化学试剂)。许多常用有机溶剂的蒸气有很高的毒性,吸入少量或是在其蒸气中呆一段时间,都存在危险。不要在实验室内蒸发过量的有机溶剂,蒸发溶剂应在通风橱中进行,并采用合适的带冷凝装置的蒸馏设备。有些物质,如氯乙酸,对眼结膜、鼻黏膜、喉咙和肺等有强烈的刺激性;有些物质,如氯苯,对眼睛有强烈刺激,具有催泪作用。在对有机溶剂性质不是很清楚的情况下,实验要在通风橱中进行,或向老师咨询。实验中有关化学药品的具体安全要求要写在报告中。

7. 防有毒化学药品入口

有害化学药品入口的可能途径有:用嘴使用移液管移液、弄脏的手、被污染的食物或饮料、将实验室的化学药品当作食物等。因此,可以从以下几方面避免该事故的发生:

(1) 用移液管转移溶液时,只能使用洗耳球,切记不可用嘴吸;

(2) 用手拿入口的东西(香烟、食物、口香糖等)前要洗手,离开实验室时要洗手;

(3) 不在实验室吃东西、喝饮料,特别是不喝实验室的蒸馏水以及实验室水管里流出来的水;

(4) 不要食用实验室的任何化学药品(如盐、糖、酒精、碳酸氢钠等),因为这些物质很可能已被污染或弄错标签;

(5) 不用实验室的玻璃器皿盛放食物和饮料;

(6) 不将食物和饮料放在实验室的冰箱或制冰机里,不食用实验室制冰机制造的冰。

实验事故的处理

1. 眼睛受伤

立即寻求帮助!当化学药品进入眼睛时,要立即用大量的水进行冲洗。先让受伤者躺在地面上,救助者一人膝盖着地,跨骑在伤者身上,将杯子里的水以中等流速倾倒到伤者的鼻梁上,使水流向两边漫过眼睛;另一人则蹲在伤者的头边,用手不断翻动伤者眼睑(用拇指和食指开合眼睑)。清洗需要数升水,当化学药品确实被冲洗完后,将伤者立即送往医院的急救室进行救治。

2. 皮肤接触化学药品

所有与身体接触过的化学药品都应视作有毒物质对待。尽管有些药品没有明显的腐蚀性,当其与身体接触后,也要立即用肥皂和水对身体进行冲洗。如果是腐蚀性的物质,则需要用大量的自来水进行冲洗。如果衣服鞋帽浸染了化学试剂,需要立即脱下,有必要时还需要进行淋浴。除非有明确的要求,一般情况下,不要用碱来中和酸,也不能用酸来中和碱。

3. 吸入有毒气体

室内要通风,到室外呼吸新鲜空气,并立即报告实验指导老师,因为长时间暴露在有害气体中可能诱发头昏和头疼。

3. Inhalation of chemicals. Get fresh air. Report to the laboratory instructor all incidents where prolonged exposure to laboratory fumes has induced faintness or a headache.

4. Ingestion of chemicals. Call the Poison Center for recommended treatment. Vomiting is often dangerous, especially if vomit gets into the lungs; most serious poisoning problems are best treated in the hospital emergency room. If necessary, you can induce vomiting by swallowing as much warm water as possible and rapidly as possible. The addition of one or two teaspoons of table salt per glass of warm water will help. Vomiting should be encouraged by swallowing additional water until the vomited liquid is clear.

5. Cuts. Serious bleeding should be controlled by direct pressure on the wound, preferably with a clean gauze or cloth pad. Minor cuts should be washed with tap water and allowed to bleed briefly. Then press with a clean cotton pad or piece of gauze. Cover with a gauze pad to keep the wound clean. More severe cuts, especially with glass or other foreign objects in the wound, require medical attention.

6. Clothing fires. Call for help and a blanket. When someone's clothing catches fire, the flame and smoke can rise and be inhaled. Unless a shower is immediately available or can be reached without inhalation damage, put the victim prone on the floor, forcibly if necessary, and roll in a blanket; beat the flames out with hands or smother with heavy garments.

7. Thermal burns. Any burn that is extensive or severe, or which involves the eyes or face, should be considered serious. A burn involving the respiratory tract could be critical. Call the ambulance. Minor burns should be treated by washing the burned area with cold tap water. Wrapping the burned area lightly with a clean wet cloth or holding it in an ice bath will help reduce the pain. Do not use ointments or salves unless so instructed by a physician.

8. Faintness. If the victim is conscious, have them sit down and place their head between their knees. Support them to prevent a fall. If they are weak or have fainted, lay them on their back on the floor, raise their feet and legs a little above the level of their head. Call for medical assistance if the victim fails to regain consciousness within a half minute. Insist that they remain quiet, seated or lying down, for a few minutes after recovery unless it is necessary to take them to the hospital for treatment.

Material Safety Data Sheets (MSDS) contain information about substances used in the lab. They contain extensive safety information, for use in laboratories, as well as in industry and manufacturing. Keep in mind that you will always be using small amounts of materials in the lab, and some safety information applies of much larger quantities. For example, if you look up the MSDS for ethanol, the active ingredient of alcoholic beverages, you will learn that it is highly flammable and severely toxic and may cause death if ingested. Something to keep in mind next time someone offers you a beer.

4. 有毒化学药品入口

给医院的中毒治疗中心打电话,遵照医生的要求对中毒者进行处理。呕吐常常比较危险,特别是当呕吐物进入肺部时。严重中毒的救护最好在医院的急救室中进行。如果必要,可以采取催吐措施,方法是让中毒者以尽可能快的速度饮入尽可能多的温水,每杯水中加入一到两汤匙盐效果会更好,让中毒者吞入温水诱导呕吐直至呕吐物为清水为止。

5. 割伤

流血严重时,要用清洁的纱布直接压住伤口控制流血,立即送医院救治。受伤不严重时,用自来水清洗伤口,让伤口自行流一会儿血后,再用干净的纱布按压,然后将伤口包扎起来。严重的割伤,特别是伤口有碎玻璃或其他异物时,在治疗时要引起注意。

6. 衣服着火

立即呼救!如果衣服着火产生烟雾,要防止被烟雾呛着。在没有水龙头的情况下,或无被烟雾窒息危害情况下,最好能找到一条湿的厚毯子或外套裹在身上,然后倒在地面上,用力来回滚动。若火势不大,则可以用厚重物进行拍打。

7. 烫伤

当烫伤很严重时,如大面积烫伤或伤到脸、眼睛等一定要认真对待,特别是伤及呼吸道,要叫救护车。对于不严重的烫伤,可以用冷自来水冲洗烫伤处,然后用干净的湿纱布轻轻包裹,也可以用冰浴帮助减轻疼痛。没有医生的指导不要随便涂抹药膏。

8. 头晕

如果受伤者意识清醒,则让其坐下,将头靠在膝盖休息,扶住受伤者防止其晕倒。如果受伤者虚弱或者晕倒,则让其平躺,脚和腿部略微抬高。倘若受伤者在半分钟内不能苏醒的话,要立即叫医生。使受伤者在恢复后的数分钟之内依然保持安静、坐或躺的姿势,如有必要送医院治疗。

试剂安全手册对化学试剂在实验室以及工业中生产和使用时的安全作了仔细的说明,使用化学试剂前查一下试剂安全手册会很有帮助。安全手册中有些数据是针对药品用量大的情况,因此要时刻牢记:在实验室中,化学药品的用量应当尽量少。例如,乙醇是酒精饮料中的主要成分,当你看到试剂安全手册介绍乙醇可燃、有毒、摄入后可能致命以后,下次当有人请你喝酒的时候你就要当心,记得要少喝一些。

Part 2   General Equipment and Lab Techniques

2.1   General Equipment

Most basic laboratory equipment is made of glass so one can easily see what is happening inside. The basic labwares are listed in Figure 2-1. Considering the ease and speed of use and the desired precision and accuracy, you should be able to choose the right piece of equipment and follow proper technique to use it. Labwares are designed for either experimental procedure such as reaction, volume or mass measurement and so on. The following is the important points you should keep in mind at the first.

Reaction Container

Erlenmeyer flasks and beakers are used for mixing, transporting, and reacting, but not for accurate measurements. The volumes stamped on the sides are approximate and accurate to within about 5%. Beakers are probably most commonly used; their wide mouth and spouts make it very easy to transfer solutions from one beaker to another. The flask (or, to be more precise, the "Erlenmeyer flask") is an excellent choice to run reactions if you do not plan to transfer solutions frequently. The tapered neck of the flask makes it very easy to grab, hold, swirl and manipulate.

Glassware of Volume Measurement

Graduated cylinder, volumetric pipet, Mohr pipet, buret, volumetric flask are measuring equipments. Graduated cylinders are useful for measuring liquid volumes to within about 1%. They are for general purpose use, but not for quantitative analysis. If greater accuracy is needed, use a pipet or volumetric flask. See the sections "uncertainties in lab data and result" and "significant figures" in the lab manual.

The precision of each of the volume stamping glasswares is shown as follows:

Equipment	Precision of Volume Measurement
250 mL beaker	±10 mL
250 mL graduated cylinder	±1 mL
100 mL volumetric flask	±0.08 mL
250 mL volumetric flask	±0.1 mL
25 mL graduated cylinder	±0.2 mL
10 mL pipet (Mohr)	±0.02 mL
25 mL buret	±0.04 mL
5 mL, 10 mL volumetric pipet	±0.01 mL

第二部分　实验基本操作与技能

2.1　化学实验的常用仪器

大多数常见实验仪器是由玻璃制作的，这可以使我们能清楚地看到其中发生的化学变化。常见实验仪器所图 2-1 所示。为了使仪器使用起来既方便又快捷，并能获得实验所需要的精密度和准确度，就必须正确选用仪器和规范进行实验操作。每件实验仪器都有特定的用途，如专门为反应、测定体积或质量等而设计的仪器。要正确选择和使用仪器，下面的要点是应该首先记住的。

反应容器：锥形瓶和烧杯用于试剂的混合、转移和反应，而不能用于体积的准确测量。烧杯和锥形瓶器壁上的刻度值是近似值，误差在 5% 左右。烧杯应该是使用频率最高的仪器，其宽阔的大口使溶液在烧杯间的转移很容易进行。如果不需要频繁转移反应溶液，锥形瓶则是不错的选择，其渐细的瓶颈很容易抓、握，能很方便地进行振摇与操控。

体积测量仪器：量筒、移液管、吸量管、滴定管、容量瓶是液体体积测量仪器。量筒是最常用的液体体积量取的仪器，其精密度通常在 1% 之内，用于一般的体积测量，但不能用于定量分析。如果有更高的准确度要求，则需要移液管或容量瓶等精密度更高的仪器。参看本书"误差与误差分析"和"有效数字"部分。常见的体积测度仪器的精密度如下：

仪　　器	体积测量的精密度
250 mL 烧杯	±10 mL
250 mL 量筒	±1 mL
100 mL 容量瓶	±0.08 mL
250 mL 容量瓶	±0.1 mL
25 mL 量筒	±0.2 mL
10 mL 吸量管(Mohr)	±0.02 mL
25 mL 滴定管	±0.04 mL
5 mL、10 mL 胖肚移液管	±0.01 mL

Part 2 General Equipment and Lab Techniques

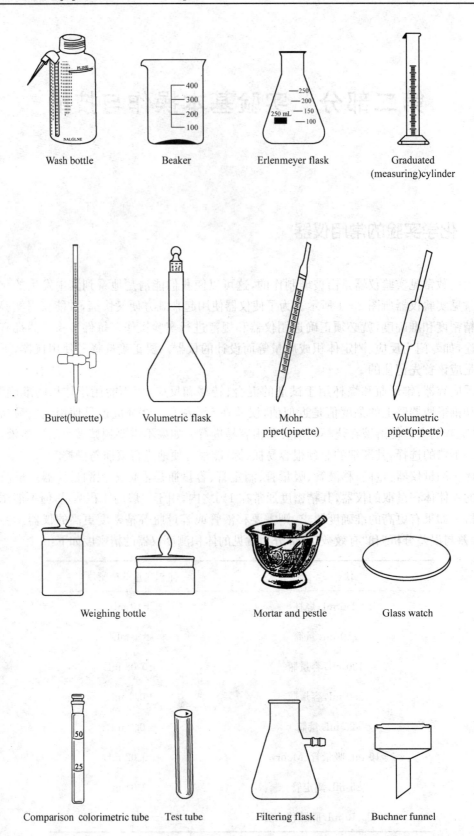

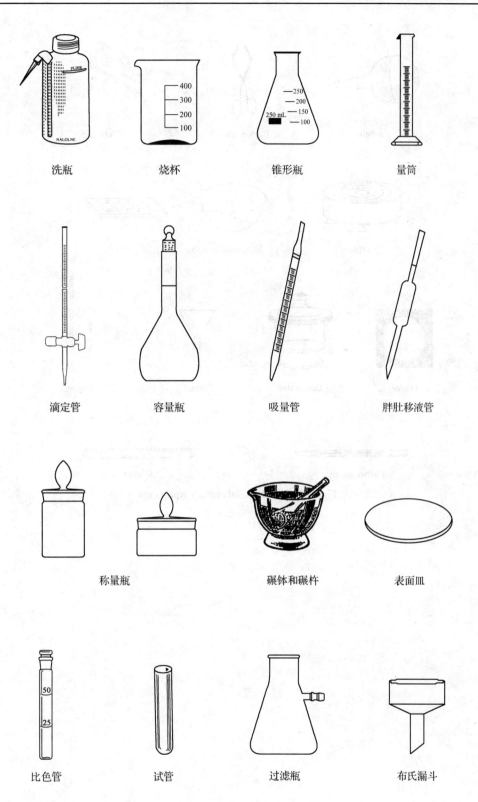

Part 2 General Equipment and Lab Techniques

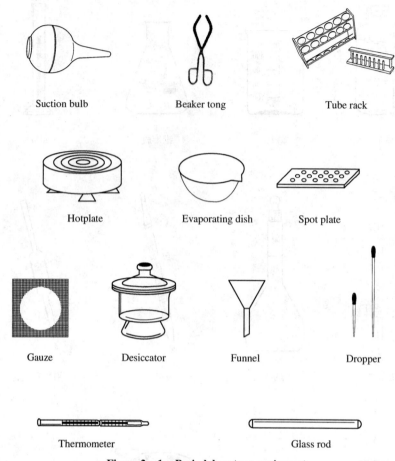

Figure 2-1 Basic laboratory equipment

第二部分 实验基本操作与技能

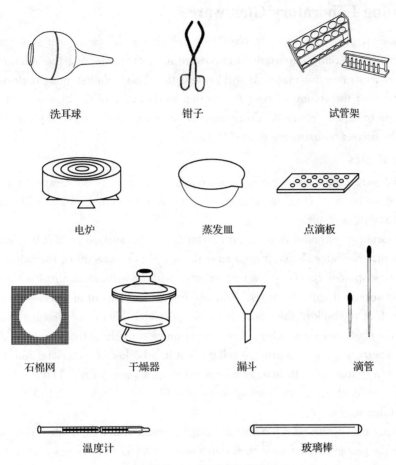

图 2-1 常见实验仪器

2.2　Cleaning Laboratory Glassware

In all instances, glassware must be physically and chemically clean. All glassware must be grease-free. This is especially important in glassware used for measuring the volume of liquids. Grease and other contaminating materials will prevent the glass from becoming uniformly wetted. This in turn will alter the volume of residue adhering to the walls of the glass container and thus affect the volume of liquid delivered. Furthermore, in pipets and burets, the meniscus will be distorted and the correct adjustments cannot be made.

Cleaning Basics

1. Wash labware as quickly as possible after use. If labware is not cleaned immediately, it may become impossible to remove the residue. If a thorough cleaning is not possible immediately, put glassware to soak in water.

2. The criterion of cleanliness is uniform wetting of the surface by distilled water. If the glassware is clean, the water should form a smooth sheath as it runs off of the sides.

3. Usually, soap, detergent, or cleaning powder(with or without an abrasive) may be used. Begin by rinsing very well with tap water. You will find an assortment of brushes near each sink; rinse out as much of the bulk of the contaminants as possible. Then, add cleaning agent, use the brush to clean the glassware thoroughly; if you are cleaning something that cannot be cleaned with a brush(like a pipette), dissolve a little cleaning agent in a beaker of water and run it through as best you can. After cleaning with detergent, rinse the glassware VERY THOROUGHLY with TAP water to get rid of all of the excess soap. Finally, rinse it 2 - 3 TIMES with SMALL amounts of distilled water.

4. Sometimes detergent and tap water are neither required nor desirable. You can rinse the glassware with the proper solvent, and then finish up with a couple of rinses with distilled water, followed by final rinses with deionized water if needed.

How to Wash out Common Lab Chemicals

1. Water Soluble Solutions

(e.g., sodium chloride or sucrose solutions) Rinse with tap water, then rinse 2 - 3 times with distilled water before putting the glassware away.

2. Water Insoluble Solutions

(e.g., solutions in hexane or chloroform) Rinse 2 - 3 times with ethanol or acetone, rinse 2 - 3 times with deionized water, then put the glassware away. In some situations other solvents need to be used for the initial rinse.

3. Concentrated Acids

(e.g., concentrated HCl or H_2SO_4) Carefully rinse the glassware with copious volumes of tap water. Rinse 2 - 3 times with distilled water.

4. Concentrated Bases

(e.g., 6 mol·L^{-1} NaOH or concentrated NH_3) Carefully rinse the glassware with copious volumes of tap water. Rinse 2 - 3 times with distilled water, then put the glassware away.

5. Glassware Used for Organic Chemistry

Use the appropriate solvent to rinse the glassware, e.g., distilled water for water-soluble contents and ethanol for ethanol-soluble contents. Rinse with other solvents as needed, followed by ethanol and finally distilled water if needed. If the glassware requires scrubbing, scrub with a

2.2 仪器的洗涤

在化学实验中,玻璃仪器要做到物理和化学清洁。所谓物理清洁就是没有污垢,仪器看起来洁净透明;所谓化学清洁就是仪器内没有与研究体系之中的物质进行反应的污染物。玻璃仪器不能有油脂,这对于液体体积测量仪器来说尤为重要。油脂和其他污染物会使玻璃表面的润湿性变得不一致,这将导致有残余液附着在玻璃器壁上,从而影响液体量取体积的准确性。此外,如果移液管和滴定管内玻璃表面有油污,则液体弯月面会变得扭曲,且这种扭曲无法进行校正。

玻璃仪器的清洁要领

1. 用过的实验仪器要尽早洗涤。仪器若不能立即清洗,残留污物可能变得很难去掉。如果来不及立即洗涤,要将玻璃仪器浸泡在水里。

2. 以玻璃表面能被蒸馏水均匀润湿为洁净的标准。即,如果一个仪器是洁净的,水在其表面流过时应该能形成均匀的水膜。

3. 通常情况下,可以使用肥皂水、洗涤剂或去污粉作为去污剂。洗涤时,开始要用自来水对仪器进行彻底的冲洗。每个水槽边都有规格不同的毛刷,利用不同的毛刷尽可能将大部分污染物冲走。然后,加入清洁剂,用毛刷将玻璃仪器进行彻底的刷洗。有些仪器不能使用毛刷刷洗(如移液管),可以用水将清洁剂溶解在烧杯里,再设法将清洁剂灌入仪器中。仪器用清洁剂洗过后,要用自来水彻底冲洗以去掉残余的洗涤剂。最后,用少量蒸馏水将仪器润洗 2~3 次。

4. 在很多情况下,洗涤玻璃仪器可以既不需要自来水也不需要洗涤剂,可以用合适的溶剂对仪器进行多次润洗,再用蒸馏水多次润洗,如果需要,再用去离子水润洗。

有针对性的洗涤方法

1. 水溶性物质
(如 NaCl 或蔗糖溶液)先用自来水冲洗,再用去离子水润洗 2~3 次,然后存放起来。

2. 脂溶性物质
(如己烷或氯仿溶液)用乙醇或丙酮润洗 2~3 次,再用去离子水润洗 2~3 次,然后进行存放。在某些情况下,还需要在乙醇或丙酮润洗之前用其他溶剂进行润洗。

3. 浓酸
(如浓 HCl 或 H_2SO_4)先小心地用大量自来水冲洗,再用去离子水润洗 2~3 次。

4. 浓碱
(如 6 mol·L^{-1} NaOH 或浓氨水)先小心地用大量自来水冲洗,再用去离子水润洗 2~3 次。

5. 有机化学实验的玻璃仪器
先选择合适的有机溶剂进行润洗。如水可用于除去水溶性污物,醇可用于除去醇溶性污物。先用必需的有机溶剂润洗,再用乙醇润洗,必要时再用蒸馏水润洗。如果玻璃仪器需

brush using hot soapy water, rinse thoroughly with tap water, followed by rinses with distilled water.

6. Burets

Wash with warm soapy water, rinse thoroughly with tap water, and then rinse 2 – 3 times with distilled water. Be sure the final rinses sheet all the inside wall of the glass.

7. Pipets and Volumetric Flasks

In some cases, you may need to soak the glassware overnight in soapy water. Clean pipets and volumetric flasks using warm soapy water. The glassware may require scrubbing with a brush. Rinse with tap water followed by 2 – 3 rinses with distilled water.

要刮刷,可以在热的肥皂水里用刷子刷洗,再用自来水彻底冲洗后用蒸馏水润洗 2~3 次。

6. 滴定管

先用温热的肥皂水洗涤,再用自来水彻底冲洗,然后用蒸馏水润洗 2~3 次,润洗时要确保整个滴定管内壁都被润洗到。

7. 移液管和容量瓶

有时可以把需要洗涤的仪器放在肥皂水中浸泡过夜。洗涤移液管和容量瓶时可以用温的肥皂水,也可以使用专门的毛刷进行刷洗。洗过后用自来水冲洗,然后再用蒸馏水润洗。

Part 2 General Equipment and Lab Techniques

2.3 Digital balance and weighing

The balance perhaps is the single most critical piece of equipment in chemlab because chemists associate the number of molecules(moles of molecules) with the mass of a substance since there is no instrument capable of counting the exact number of atoms or molecules. Digital balances get easier to use today. The basic operation of these balances is trivial, place your object on the pan, and the mass appears on the display. This section describes not only the methods for weighing but also the basics of maintenance that must be observed by all users and, finally, how to correctly read the display to avoid errors.

Maintenance

Careful use of the balance is critical. For economic reason, a good quality balance easily will cost several thousand yuan, while high end("analytical") balances will cost tens of thousands of dollars. For technical reason, balances are extremely precise instruments. This means that balances are very easily damaged, susceptible to both mechanical and chemical damage.

Some forms of damage are obvious, such as mechanical damage. If you shake or hit the balance, you can quickly and easily damage the mechanical components that do the work in a balance, especially the "knife edge". Modern electronics balances are also designed to work on a level, draft free surface. These high tech devices still rely on the good old-fashioned low-tech "bubble" leveling device. *It is good practice to check the level bubble to ensure that the balance is level before you begin. It is bad practice to move the balance when the balance is used.*

Balances are also very susceptible to corrosion. Always use a container or weighing paper when weighing a chemical; *do not place any chemical directly onto the balance pan*! Waxed weighing paper, small beakers, watch glasses, small vials etc. are all convenient containers for weighing chemicals. If you should inadvertently spill something on the balance, clean it up as soon as possible(a brush and paper towels work well for this, and will be located near the balance). Additional care must be used when weighing liquids. If possible, flasks containing liquids must be sealed with stoppers to prevent spilling or evaporation during weighing.

How to use the balance

1. Check the bubble to *be sure the balance is level* before turning the balance on.

2. Press the button "ON/OFF" and wait a few seconds for the digital display to read "0.0000 g".

3. To weigh an object, slide open one of the draft doors and *place the object in the center of the pan lightly.*

Always use tongs, clamps, or a tissue to handle solid objects or liquid containers. Do not use your fingers, as oil and water from them initially adds a few micrograms to the mass that then partially evaporates, resulting in an error.

4. *Close the door* and wait a few seconds for the digital display to *read a constant mass.*

Lighter objects($<$105 g or 110 g, depending on the scale) are weighed to the 0.0001 g level. Heavier objects($>$105 g or 110 g, depending on the scale) are weighed to the 0.001 g level. All numbers displayed should be recorded as they indicate the sensitivity of the scale(they are all significant figures).

5. Cleaning up and shutting down the balance.

When you are done with the balance, make sure you have properly cleaned up any chemicals that may have spilled on the balance. At the end, the balance can be turned off by lifting up gently

2.3 电子天平及称量

天平是质量测量工具,是化学实验室最重要的仪器。因为没有一种仪器能直接测量分子或原子个数,因此在测定分子或原子物质的量时,只能通过对物质质量的测定来间接换算得之,所以正确使用天平是非常重要的。如今,天平已由机械天平发展到数字天平(电子天平),电子天平操作方便,使称量变得非常简单。本节内容不仅要介绍称量方法,还要介绍天平使用者必须掌握的电子天平基本保养知识,以及如何避免错误操作以获得正确读数的一些注意事项。

天平的保养

使用天平的关键就是小心谨慎。从经济角度看,一架好的天平动辄数千元,而高端天平("分析天平")则要上万元;从技术角度看,天平是非常精密的仪器,需要精细的操作才能获得好的测量结果。这些意味着天平是很娇贵的、易受损的,很容易受到来自机械的和化学的损害。

有些损害是显而易见的,比如机械损害。天平的工作机械元件,特别是天平的"刀口"(平衡支点)很容易损坏,所以天平必须放置在水平固定的台面上。要避免振动,摇晃或敲击极易损坏天平。现代电子天平工作时同样需要安放在水平的、无气流通过的台面上。这种高科技的装置安装了古老而优秀的"低科技"水平仪——"水泡",查看水泡是否处于水平仪的中央可以判断天平是否水平。使用天平之前检查天平是否水平是一种良好的操作习惯,要努力养成。在使用天平的过程中**挪动天平是一种非常糟糕的习惯**,必须努力克服。

天平也很容易遭到腐蚀。在称量化学药品时,一定要使用容器或称量纸,任何化学物质都不能直接放在天平秤盘上进行称量!蜡质的称量纸、小烧杯、表面皿、小瓶等都可以做称量化学药品的容器。如果不慎将药品洒落在天平里,要及时进行清理(可以用刷子或纸巾进行清扫,这些物品应该放在天平旁边)。称量液体物质时要格外小心。如果可能,装液体的容器要塞上塞子,以防称量过程中液体洒落或挥发。

天平的使用

1. 在打开天平之前要检查天平是否水平。
2. 按控制面板上"ON/OFF"键,几秒钟之后,天平的显示器上显示"0.0000 g"。
3. 称量物品时,轻轻地推开活动门,将物品轻轻地放置在秤盘的中央。拿取称量物体和容器时,需用纸包住,或使用钳子、镊子。手不能直接接触称量物,因为手指上的油脂或水会粘在称量物品上,使其称量质量多出几微克,而这些微量的水和油脂后来又会蒸发,给称量结果带来误差。同样的道理,天平的操作台面必须洁净,不能有灰尘和水。
4. 读数时,要将天平门关上,等显示器显示的数字稳定后方可读数。轻的物品(与天平的量程有关,如,<105 g 或 110 g)其称量精密度水平为 0.0001 g,重的物品(与天平的量程有关,如,>105 g 或 110 g)其称量精密度水平为 0.001 g。记录称量数据时要将天平显示的所有数字全部记下,因为这些数字代表测量的灵敏度水平,都是有效数字。

on the control bar.

Methods for weighing

To use the balance, first you must decide what the plan is. There are a couple of ways that the balance can be used.

Direct weighing

(1) Press the key "tare", the readout will read zero.

(2) Place the weighed material on the pan of the balance and close the door. The mass of the material is read directly.

Weighing by taring

(1) These substances must always be weighed using an appropriate weighing container. Place the weighing container on the balance pan and close the doors.

(2) Tare the container by briefly pressing the control bar. The readout will read zero with the container sitting on the pan. This allows the mass of your sample to be read directly.

(3) Add the substance to be weighed. Be careful not to spill chemicals on the balance. If need be, you can remove the container from the weighing chamber while you add the sample provided that no one presses the control bar before you weigh your sample.

(4) With the sample and its container sitting on the pan, close the chamber doors and read the display to find the mass of your sample.

Weighing by difference

(1) Add the substance to a weighing bottle. Stopper the weighing bottle.

(2) Place the weighing bottle on the balance pan and close the doors.

(3) Tare the weighing bottle by briefly pressing the control bar.

(4) Take the bottle out of the chamber. Be carefully to tip up the bottle and spill the appropriate mount of the chemical into a container. Stopper the weighing bottle.

(5) With the weighing bottle sitting on the pan, close the chamber doors and read the display to find the mass of your sample removed out of the weighing bottle.

5. 清理和关闭天平

天平使用完毕,要检查是否有药品洒落在天平里面,采用正确的方法仔细清扫干净,然后按"ON/OFF"键关闭天平。

称量方法

天平的称量方式有好几种,要根据实验的要求进行选用。

直接称量法

(1) 按控制面板的除皮键"tare",天平读数归零;

(2) 将被称量物品放在秤盘上,关上天平门,天平的显示器上直接显示出被称量物的质量。

固定称量法

(1) 先将称量容器(药品必须放在称量容器中称量)放在天平的秤盘上,关上天平门;

(2) 按除皮键"tare",天平读数归零,即除去容器的质量,以便能从天平读数上直接读出样品的质量;

(3) 小心地向称量容器中加入样品,要防止药品洒落在天平里,如若需要,称量容器可以从天平秤盘上拿下来添加样品,但此时不能再按动控制面板上的操作键;

(4) 将加好样品的称量容器放在秤盘上,关上天平门,天平显示器上显示出样品的质量。

减重称量法

(1) 将需要称量的样品置入称量瓶中,并盖好称量瓶的盖子;

(2) 将称量瓶放在天平秤盘上,关上天平门;

(3) 按除皮键"tare",使天平读数归零;

(4) 打开天平门将称量瓶拿出来,小心地从称量瓶里移出合适量的样品到另一个容器中,然后将称量瓶盖子重新盖好;

(5) 将称量瓶重新放在秤盘上,关上天平门,天平的显示器上直接显示出被移出样品的质量。

2.4 Separation of solids from liquids

Gravity Filtration

The reason it is called "gravity filtration" is because we employ gravity to help us out.

Gravity filtration uses a long-stem funnel and filter paper. Filter paper can have pore sizes ranging from small to large to permit slow to fast filtering. Take an appropriate piece of filter paper with respect to the sizes of the particles and the experimental plan. The paper is folded in half(Figure 2-2), then folded in quarters, and the tip of one corner is torn off to allow for a snug fit in the funnel cone. The paper cone is fitted to the funnel so three thicknesses of the paper line one-half of the cone and one thickness lines the opposite half(Figure 2-2). Now place the funnel into a beaker and wet the filter paper completely with the dominated solvent or solvents in the mixture to be filtered. This step adheres the filter paper to the funnel walls preventing solid from escaping. Then, support the funnel with a clamp or ring(if necessary) and place a clean beaker beneath the funnel so the stem rests against the side of the beaker(this prevents splattering).

Before filtering, allow most of the solid in the mixture to settle. Now pour the supernatant liquid through the filter first. This will allow the initial part of the filtration to proceed faster and may prevent clogging of the filter by the solid. To prevent splattering pour the liquid down a glass rod as shown in Figure 2-3.

Vacuum(or Suction) Filtration

In vacuum or suction filtration, a partial vacuum is created below the filter, causing the air pressure on the surface of the liquid to increase the rate of flow through the filter. So a vacuum filtration is usually faster than a simple gravity filtration. A typical apparatus is illustrated in Figure 2-4.

A Buchner funnel is a flat bottomed, porous, circular porcelain bowl with a short stem. A circle of filter paper just large enough to cover the holes in the bottom of the Buchner funnel should be used. A common error is to try to use a piece of filter paper so large that it must be turned up at the edges. If this is done, it is almost impossible to create a vacuum in the suction flask. Not only will the filtration take much longer, but any material that flows over the edge of the filter paper will run down into the suction flask without being filtered. Double pieces of paper should be used to avoid the large pressure tearing the paper.

Filtration is done by connecting the side arm of the suction flask to the source of vacuum, which is almost always the water aspirator. Vacuum pumps are also applicable. The tubing used for all connections to the source of vacuum must be thick-wall tubing. The thin-wall or flexible tubing will collapse on evacuation of the system. In the assembly, the flask should be connected to the aspirator through a trap, as shown in Figure 2-4. The trap prevents water from the aspirator being sucked back into the filter flask.

Turn the aspirator on just a little at first so as to create a gentle vacuum, wet the filter paper with a small portion of the same solvent used in the solution being filtered while making sure that the paper is being pushed down over the holes, and pour the mixture to be filtered onto the center of the paper. Once the mixture has been added, the vacuum can be increased. When using the water aspirator, be sure to release the vacuum at the trap before turning off the water.

2.4 固液分离

常压过滤

常压过滤又叫重力过滤,过滤时依靠重力作用使滤液通过滤纸达到分离目的。常压过滤用到的仪器有长颈漏斗和滤纸。根据滤纸孔径从大到小分不同规格,大孔径滤纸过滤速度快,小孔径滤纸过滤速度慢,可根据沉淀粒径大小以及实验要求选择不同规格的滤纸。过滤时,先将滤纸对折(如图2-2所示),再对折,撕下其中一个小角以便滤纸能与漏斗圆锥体紧密吻合。打开滤纸,使三层滤纸压住漏斗圆锥的半边、单层滤纸的一边压住另外半边(如图2-2所示)。然后,将漏斗放在烧杯里,用需要过滤的溶液或其溶剂润湿滤纸,使滤纸紧贴漏斗壁以防止固体颗粒流出滤纸。如果需要,可以用铁圈或夹子固定漏斗,在漏斗下面放一个洁净的烧杯,漏斗颈紧贴烧杯内壁(防止溶液飞溅)。

过滤之前先让混合液中的固体沉降下来,上清液先过滤。这样做可以防止滤纸堵塞,使先行过滤的溶液过滤起来快一些。为了防止溶液溅出,倾倒液体时要用玻璃棒引流。如图2-3所示。

减压过滤

减压过滤又叫真空过滤。操作过程中,过滤器中产生部分真空,液体表面的大气压加速液体通过滤纸,因此真空过滤要比简单的常压过滤速度快。真空过滤的一般装置如图2-4所示。

布氏漏斗是一种平底多孔、短颈圆形陶瓷漏斗,所用的滤纸是圆形的,滤纸的大小应该刚好盖住布氏漏斗底部的小孔。选择滤纸时,常发生的错误是所选滤纸过大,使得滤纸边在漏斗中翘起。这种错误应当避免,否则,抽滤瓶内就不能有效形成真空,这不仅使过滤时间延长,同时还使固体颗粒从滤纸边缘的缝隙泄漏到抽滤瓶中,达不到过滤的目的。滤纸一般需要两层,以防止由于真空度过大使滤纸抽破。

过滤时,要将抽滤瓶的支管与真空装置连接起来。最简便的真空装置就是水流抽吸器,真空泵也是常见的真空装置。与真空系统相连的所有管线都必须是厚壁橡皮管,薄壁管或软管在管内形成负压时会变扁、变塌,使气体不能有效地抽出。在减压过滤装置中,抽滤瓶与抽吸器间可以由安全瓶连接,如图2-4所示。安装安全瓶可以防止抽吸器的水被倒吸至抽滤瓶中而污染滤液。(打开抽吸器时,安全瓶的活塞要关闭;关闭抽吸器时,一定要先打开安全阀活塞,释放真空,否则会发生倒吸)

将减压装置安置好之后,先将抽吸器稍稍打开形成较弱的真空度,用少量待过滤溶液或相应溶剂润湿滤纸,确保滤纸紧紧盖住漏斗底部的小孔,然后将要过滤的混合物倒在滤纸的中央。混合液被倒入漏斗之后,可以加大抽吸使真空度增加,直至固体被抽干(此时可用一个干净的平顶瓶塞挤压沉淀,帮助抽干)。如果使用水流抽吸器作为真空装置,一定要在将安全瓶中的真空释放之后再关闭水龙头。

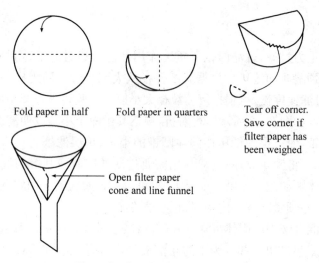

Figure 2-2 The fold of filter paper

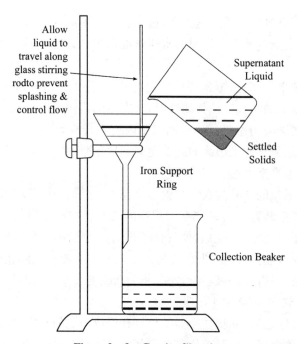

Figure 2-3 Gravity filtration

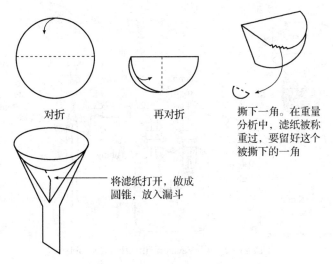

图 2-2 滤纸的折叠

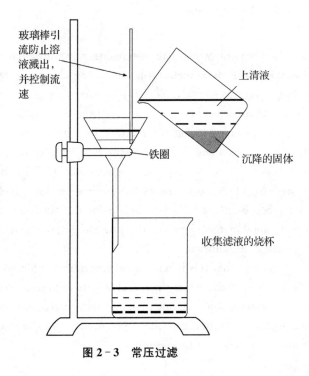

图 2-3 常压过滤

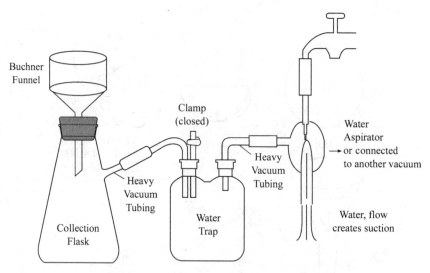

Figure 2 – 4 Vacuum filtration assembly

If the solids are to be washed, turn off and break the vacuum by lifting the Buchner funnel slightly or opening the water tap. Add the wash liquid to the original container, and pour it onto the solids in the Buchner funnel. Very carefully stir the solids and wash them thoroughly, but do not tear the filter paper. Re-apply the vacuum.

Decanting

Decanting is used when you would like to separate a mixture, but it is not necessary to do so with extreme care. Decanting is a method in which one can separate liquids from solids in a mixture rapidly, but relatively sloppily. Typically, decanting gets a desired separation for the solids with a relatively large density or the large size from the mixture.

When decanting, one begins laying the mixture to settle all of the solids to the bottom of the container (if necessary, by centrifuging the mixture). Place a glass rod across the top of the container to help the liquid flow out more easily because it breaks the surface tension which can form without it. See Figure 2 – 5. Carefully pour the supernatant from the mixture into another container. In the procedure, pour slowly and try to avoid agitating the solution.

It must be kept in mind that this is not a good separation technique; it is designed to be fast and crude when this is all that is required. The solid will still have a considerable amount of liquid left on it, and the liquid will have some of the solid in it as well. If you need to get a precise experimental result, you should not be decanting.

Figure 2 – 5 Decanting

Centrifuging

Centrifugation is a process which uses centrifugal force to separate mixtures by density: the more dense material will be on the bottom (typically solids) while the less dense will be on top (typically liquid).

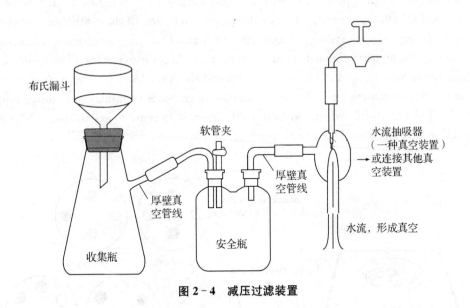

图 2-4 减压过滤装置

固体需要洗涤时,要先解除过滤系统的真空,例如,将布氏漏斗从抽滤瓶口向上稍稍拔起或者打开安全瓶。之后,向原来装过滤混合物的容器中加入洗涤液荡洗器壁,再将洗涤液倒在布氏漏斗中的固体上。搅动固体使其充分洗涤,搅动要小心,防止弄破滤纸。最后重新抽滤。

倾斜过滤

当混合物的固液分离没有严格要求时,可以采用倾斜法过滤。倾斜法过滤是一种快速的固液分离手段,分离效果差强人意。该方法主要用于固体相对密度较大或晶体颗粒较大时的固液分离。

过滤之前,让混合物静置一段时间使固体沉降到容器底部(也可以通过离心使固体沉降)。取一根玻璃棒横放在容器的口上(如图2-5所示),这样可以破坏在没有玻璃棒存在时的表面张力,帮助液体顺利流出。小心将上清液倒入另一个容器中,倾倒的时候要缓慢,避免搅动溶液。

要记住倾斜过滤不是一种好的分离方法,适用于快速粗略的实验过程。分离后的固体中含有大量的液体,而液体中也会混有一些少量固体。如果要得到精确的实验结果,就不能使用倾斜法过滤。

离心分离

离心分离是根据物质密度不同,在离心力作用下对混合物进行分离(不同密度的物质在离心力场中的沉降速度不同):密度大的物质(如固体)沉在底层,密度小的物质(如液体)处在上层。

图 2-5 倾斜过滤

The most important thing about centrifuging is to balance the centrifuge: put a test tube of the same size and design opposite the test tube to be centrifuged. See Figure 2-6. The balancing test tube is filled with tap water as much as the separate mixture. If the centrifuge is not properly balanced, it is very easy to severely damage the centrifuge. Once the centrifuge is balanced, cover the lid. Set up the proper rotational speed and turn it on. Allow it to run for a few minutes. If the centrifuge begins making a lot of noise, turn it off and check the balance. You must stay with the centrifuge during the entire time when it is running, since minor vibrations can cause a centrifuge to "walk" off of the bench. When you turn it off, allow it to come to a stop itself; NEVER put your hand(or anything else) above the centrifuge.

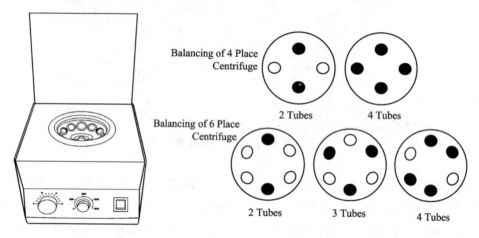

Figure 2-6 Centrifuge(in the left) and balancing of centrifuge(in the right)

Separating Solid & Solution After centrifugation the precipitate should be found packed in the bottom of the tube. Usually the supernatant liquid can be poured off of the precipitate. Sometimes precipitates tend to float on the surface of the solution. If this is the case, use a Beral-type pipet to draw off the supernatant liquid. The supernatant liquid, or centrifugate, is separated from the precipitate by holding the tube at an angle of about 30°(see Figure 2-7) and removing the liquid by slowly drawing it into a capillary syringe. The tip of the syringe is held just below the surface of the liquid. As the pressure on the bulb is slowly released, causing the liquid to rise in the syringe, the capillary is lowered into the tube until all of the liquid is removed. As the capillary approaches the bottom of the tube, the tip must not be allowed to stir up the mixture by touching the precipitate.

Figure 2-7 Separating solid & solution

Washing of the Precipitate The precipitate left in the tube after the removal of a supernatant liquid is still wet with a solution containing the ions of this liquid. The precipitate must be washed, usually with water, to dilute the solution adhering to the precipitate. The wash liquid is added to the precipitate, and the mixture is stirred thoroughly. The mixture is then centrifuged to cause the precipitate to settle again. After centrifugation, the washings are removed by a capillary syringe as described earlier. A precipitate is usually washed at least twice. The first

进行离心分离时,最需要关注的是离心机的平衡:将装有需要离心的混合物的试管置入离心机套筒内,在其对称位置的套筒中放入与其大小、形状相同的试管,用来保持离心机的平衡,如图2-6所示。用于平衡离心机的试管中可加入自来水,使之与待分离混合物的体积相同,以保证每个试管的质量相近。倘若不能正确平衡离心机,则离心机很容易遭到严重的损害。离心机平衡之后,盖上盖板,调整合适的转速,打开电源,让其转动数分钟。如果离心机开始发出较大的噪音,则要立即关掉离心机电源,并检查离心机的平衡情况。在离心机工作的过程中,人不能离开,要守在离心机旁,因为离心机在工作的时候会发生一些振动,很有可能使其"跑"出工作台面。断开电源后,要让离心机自然停下来。在离心机停止转动前,严禁将手或其他物品放在离心机上。

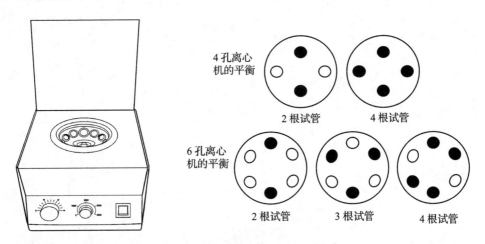

图2-6 离心机(左)与离心机的平衡(右)

固体与溶液分离 离心之后,沉淀沉积在试管的底部。通常上清液可以从试管中直接倒出来,有时候沉淀会漂浮在溶液表面,此时就必须用滴管将上清液吸取出来。将试管倾斜30°拿好,将一根滴管的毛细吸管端插入液面以下,慢慢地将溶液吸入滴管中,这样可以将上清液与沉淀进行分离(见图2-7)。(滴管在插入上清液之前,滴管的乳胶头要先捏住),随着滴管乳胶头内的压力慢慢地释放,液体被不断地吸入滴管中。毛细吸管端口要一直处于液面以下,直至溶液被全部吸出来为止。当毛细吸管端口到达试管底部时,要小心,不要接触到沉淀,以免使混合物被搅动。

图2-7 固体与溶液的分离

洗涤沉淀 取出上清液后,试管中所剩余的沉淀依然被上清液所润湿,含有上清液中所含的离子,因此,沉淀需要用水进行洗涤,以去除附着在沉淀上的这些离子。进行洗涤时,将洗涤液加入沉淀中,充分搅动混合物,然后,混合物再次离心使沉淀重新沉降下来;离心后,如上述步骤,再用滴管将上清液吸出。沉淀通常至少要洗涤两次,第一次洗涤的上清液一般要留下来与最初的上清液合并保存。如果沉淀需要转移至另一个容器中,则要向试管中加入辅助试剂,充分搅拌混合物,将其倒入要转入的容器中;待沉淀沉降下来之后,用其上清液将离心试管中残留的沉淀也转移出来。在定性分析中,没有将沉淀洗涤干净是造成实验错误的主要原因之一。

wash liquid is ordinarily saved and added to the first centrifugate. If the precipitate must be transferred to another container, the reagent to be used is added, the mixture is well stirred, and then it is poured into the other container. After the precipitate has settled, the supernatant liquid may be employed to remove any precipitate remaining in the centrifuge tube. Failure to wash precipitates thoroughly is one of the principal sources of error in qualitative analyses.

Dissolution and Extraction of Precipitates When all or a part of a precipitate is to be dissolved by a reagent, the solvent is added to the precipitate that is in the centrifuge tube and the mixture is stirred. The mixture is then separated by centrifugation. And the operation is repeated using fresh solvent. Often the extraction of a precipitate is more efficient at an elevated temperature.

沉淀的溶解和提取　当要用某种试剂溶解全部或部分沉淀时,在往装有沉淀的试管中加入溶剂之后,混合物要充分搅拌,然后,混合物再经过离心分离。要用新鲜的溶剂重复上述操作。通常,升高温度时沉淀的提取(溶解)效率更高。

2.5 Heating

Hotplate

When using a hotplate, be more careful for preventing from fires and thermal burns. These are tips to help you avoid making mistakes when you use a hotplate:

1. Place the hotplate on an insulation board made of heat-resistive material;
2. Most of the thermal burns occur when you reach across the hot plate in a flurry. Locate the hot plate in an appropriate position on the bench so that you can proceed your operation smoothly;
3. Untie the power cord around the hot plate before turn it on;
4. Turn the hot plate off immediately when the heating is complete;
5. The temperature on the surface of the hot plate exceeds the ignition temperatures of most of organic solvents. All flammable liquid must be in the covered containers and keep them away from the hot plate;
6. Do not put the hot object on the bench made of organic material directly. Place the object on a heat-resistive board or a trivet to prevent the bench being burned;
7. It is not necessary to put a piece of wire gauze on the hot plate, of which the heating cord coils up to make the heat uniform.

When heating the liquid in a container, such as flask, the following items should be given much more attention to:

Choose appropriate size of the container. Do not fill the liquid over 2/3 of the volume of the container; Wipe the outside wall of the container dry before it is heated, since it may crack when the water on the outside wall flows down from the cool top to the hot bottom.

Water Bath

Comparing with a hot plate, a water bath is milder and easily controlled. Most electric thermostatically-controlled water baths are available in modern chemistry laboratories. These equipments have temperature adjustment. Temperature ranges ambient to 100 ℃ with accuracy for your needs. The stainless steel lid with series of tight-fitting rings provides the optional sizes of the holes for the heated containers of various sizes. When using a water bath, fill the appropriate amount of water to immerge the heater in the oven. Be careful to protect the temperature senor and to avoid hitting the senor.

Ice bath

Usually an ice bath keeps the low temperature for chemical reaction. Start with a bath that is ice, and half or less of water; if too much ice melts, pour some of the water out and add more ice, since this can create isotherms that are warmer than you would like (the temperature may be higher 5 - 6 ℃ than 0 ℃). If you need the ice bath colder, add salt.

2.5 加热

电炉加热

使用电炉时要特别小心,谨防火灾和烫伤。为了防止事故发生,使用电炉时要注意以下几点:

1. 电炉要放在耐火材料做的垫板上;
2. 很多烫伤是在手忙脚乱之中将手伸到电炉上时发生的,所以,电炉摆放的位置要合理,这样才能使实验操作有条不紊地进行;
3. 在电炉开始工作前,要先解开缠绕在电炉上的电源线;
4. 电炉随时用随时通电,不用时,要将电炉插头从电源插座上拔下来;
5. 电炉加热温度较高,超过大多数有机溶剂的燃点,因此,使用电炉时要远离易燃的有机溶剂,并且存放这些有机溶剂的容器要加盖;
6. 除水溶液外,凡是从电炉上取下的高温物体,都不能直接放置在有机材料制作的实验台面上,要放在隔热垫板上,以免烫坏实验台;
7. 电炉是"面加热"(加热丝盘成一个平面),因此是均匀加热,不需要在电炉上垫石棉网。

当盛有液体的容器放在电炉上加热时,应当注意:

为了防止加热时液体溢出,要选择适当体积的容器,液体体积不宜超过容器体积的2/3;要先将容器外壁的水珠擦干,防止水珠从冷的容器口流到热的容器底部使其骤冷导致容器破裂。

水浴

与电炉直接加热相比,水浴加热较温和、易于控制。目前实验室中广泛使用的水浴加热装置是电热恒温水浴锅。这类水浴锅有温控装置,能方便设置水温,保证反应在一定温度下(室温～100 ℃)进行。其盖子由一套不同口径的金属圈组成,可以按加热器皿的外径任意选用。使用水浴锅时要检查锅内加水量是否合适,特别要注意水面一定要没过加热管。锅内的温度传感器控制加热的温度,切勿碰撞,以免温度失控。

冰浴

较低的反应温度通常用冰浴来维持。制作冰浴时,冰水混合物中的水量只需要占一半或更少的比例;若冰融化得太多,则要倒出一部分水,再添加一些冰,否则得到的温度要比想要的温度高(可能比 0 ℃高出 5～6 ℃)。倘若需要的温度更低(如小于 0 ℃),则可以在冰水混合物中加入一些盐。

2.6 Measuring and Delivering liquid

Graduated Cylinder

Graduated cylinders are useful for measuring liquid volumes to within about 1%. They are for general purpose use, but not for quantitative analysis. If greater accuracy is needed, use a buret, pipet or volumetric flask.

Volumetric Glassware

The most common types of volumetric glassware are volumetric pipets and volumetric flasks and burets. These containers are calibrated at a specific temperature to deliver or contain VERY PRECISE amounts of liquid. Heating the glassware or using heated solutions distorts the calibrated volume!

Care of Volumetric Glassware

1. Wash glassware with a mild dilute soap solution. Rinse first with tap water, and then deionized(DI) water. If beads of water form on the walls, rewash the glassware. If glassware is to be dried, allow it to drain or use lint-free paper towels. Never dry volumetric glassware in an oven! The heat will distort the glass and change the calibrated volumes. Never dry glassware using air jets! The air system contains oil droplets and fine dust.

2. A piece of volumetric glassware should always be rinsed with a small amount of the solution to be used. This step prevents contamination of the solution from water or other contaminates on the glassware's inside walls and removes the need to dry the glassware.

3. After an experiment is completed, wash the glassware, rinse thoroughly, and return to common storage area in the laboratory.

How to Read the Level of Liquid in Volumetric Glassware

1. A liquid's meniscus is the curvature of the liquid surface in a narrow container. The level of concave(downward curving) liquid surfaces(e. g. water) is read at the bottom of the meniscus. The level of convex (upward curving) liquid surfaces(e. g. mercury) is read from the top of the meniscus.

2. When the shape of the meniscus is difficult to discern (e. g. dark liquids such as purple $KMnO_4$), the liquid level is read from the edge of the liquid.

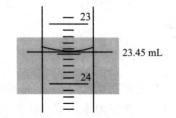

Figure 2-8　Reading volume with a background card

3. Avoid parallax error when reading liquid levels by positioning the eye at the same level as the meniscus. For easier viewing, a white card marked with a dark line can be held behind the glassware to contrast the line between glassware markings and the liquid level. See Figure 2-8.

Volumetric Flasks

When using the volumetric flask, check whether the glass stopper or plastic cap on the top is tight-fitting to the mouth. The calibration mark is a single gradation line on the neck indicates the exact volume the flask will contain at a specified temperature marked on the flask(usually 20 ℃). Prevent warming flask contents(and volume distortions) by handling the flask by the neck instead of the body.

Volumetric flasks are used to make solutions of known concentration by the dissolution of a

2.6 溶液的量取和转移

量筒

量筒是常用的量液仪器,误差约为 1‰,用于一般的体积测量,但不能用于定量分析。如果有更高的精度要求,需要使用滴定管、移液管或容量瓶。

容量仪器

常见的容量仪器有移液管、容量瓶和滴定管。在指定温度下,这些仪器"量出"或"量入"液体的体积经过了校准后,非常精密。容量仪器不能加热或装热溶液,否则会使校准的体积发生改变!

使用容量仪器的注意事项:

1. 应该使用温和的稀洗涤剂溶液洗涤容量仪器,之后,用自来水冲洗,再用蒸馏水润洗。如果有水珠挂壁,则需要重新洗涤。干燥容量仪器时,可以使水沥干,或用无屑纸巾擦干,但绝对不可用烘箱烘干!加热会造成玻璃膨胀,改变已经校准的体积。也不要用吹风机来干燥容量仪器,因为空气中可能会有一些小油滴和细小的灰尘,污染仪器。

2. 使用时,应该用少量待用溶液润洗移液管或滴定管,可以防止装入的溶液被容量容器内壁残留的水或其他的污染物所污染;同时,这样可以免去容器干燥的步骤(即这样操作后,容器就不需要再进行干燥)。

3. 实验完成后,尽快清洗仪器。仪器在彻底冲洗后,要放置在实验室固定的存放位置。

容量仪器的读数

1. 液面会在容器的窄部位形成弯月面。如果形成的弯月面是凹面,例如水,读数时要读凹面的最低处;如果形成的弯月面是凸面,例如汞,则读凸面的最高处。

2. 有时有色溶液的弯月面很模糊,例如紫红色的 $KMnO_4$ 溶液,则读数时要读液面的边沿位置。

3. 读数时要避免视差错误,眼睛与弯月面要处于同一水平面。为了方便读数,可以在仪器后面衬一张画有深色直线的白纸,使直线与液面、刻度线相切。如图 2-8 所示。

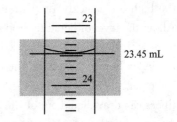

图 2-8 用背景卡帮助体积读数

容量瓶

使用容量瓶时,要检查容量瓶的玻璃塞或塑料盖是否与其口部契合严密。容量瓶的颈部刻有一条细线,指示在一定温度下(通常是 20 ℃)校准过的精确体积。手拿容量瓶时要拿住其颈部而不是下部,这样可以防止手使容量瓶中液体变热,造成液体体积不准。

容量瓶可用来配制已知浓度的溶液:将已知质量的固体溶于体积一定的容量瓶中,或将浓度已知的溶液进行定量稀释。使用前,要先清洗容量瓶,并用溶剂进行润洗。基础化学实验室常用的容量瓶规格有 10.00、25.00、50.00、100.00 和 250.0 mL,有时我们省略了小数点后面的零,但是在进行计算时千万不能忘记这些零也是有效数字,因为这些零指示了仪器

known mass of solid or the dilution of a more concentrated solution. Before use, always wash the flask and then prerinse with the solvent. Some frequently used volumes in general Chemistry lab are 10.00, 25.00, 50.00, 100.00, and 250.0 mL flasks. At times the zeros to the right of the decimal point are omitted. However, these zeros must always be considered in calculations, as they indicate the accuracy of the volume measurement (i.e., they are significant figures).

Volumetric Pipet

A volumetric pipet is an elongated glass bulb with two narrow glass stems at the top and bottom of the bulb (Figure 2-9). The pipet is used "to deliver" a single, fixed volume of liquid at a specific temperature (usually 20.0 ℃) from one container to another. Some frequently used volumes in General Chemistry lab are 1.00, 5.00, 10.00, 20.00 and 25.00 mL pipets. Like the volumetric flasks, the zeros to the right of the decimal point are sometimes omitted but are significant figures. The bottom tip is tapered to deliver a fine stream of liquid and is easily clogged. A single calibration mark on the top stem marks the volume contained at a specific temperature. Above the calibration mark, stem is open so a suction bulb can be attached to draw liquid into the pipet.

Mainpoints of usage of a volumetric pipet: Never pipet by mouth! Always use a pipet bulb to provide the necessary suction. First rinse the pipet with a small amount of the solution to remove any water film from the inside walls. To fill the pipet, compress the rubber pipet bulb and fit it over the top stem of the pipet. Insert the pipet tip into the liquid and slowly release the pressure on the pipet bulb. Allow the liquid level to rise above the calibration mark but do not permit liquid to enter the rubber pipet bulb. Remove the bulb and quickly fit your index finger over the stem. Allow the level of liquid to drop until the meniscus is exactly level with the calibration mark by adjusting the pressure of the index finger. Touch off the hanging drop from the tip of the pipet. Tissues are not recommended because the paper fibers can draw liquid out of the pipet tip by capillary action. To transfer the liquid to another container, simply release finger pressure on the pipet stem and allow the liquid to drain freely. When finished, touch off the drop of liquid hanging at the tip into the transferred liquid (it is part of the delivered volume) but do not blow out any liquid remaining inside the tip of the pipet (if no "blowing" stamped on the pipet)! See Figure 2-9. The pipet has been calibrated to contain this last drop of liquid.

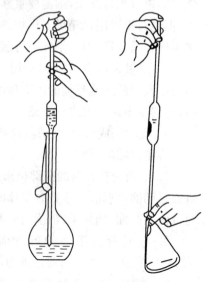

A. To transfer liquid   B. To draw liquid

Figure 2-9 **Usage of volumetric pipet**

Caution: Never use a pipet to transfer heated solutions or incompletely dissolved solids. The heat will distort the calibrated volume and solids can "condense out" on the colder glass walls of the pipet, clogging the narrow stem and tip.

Buret

Like the pipette, the buret is a precision instrument for measuring volumes of liquid. However, the buret is different in two major differences. First, the volume it measures is variable. Secondly, even though the buret is often read to four significant figures as well, it is not

的精密度。

移液管

移液管被设计成中间有胖肚的细长玻璃管(如图2-9所示)。移液管用于在特定温度下(通常为20 ℃),将一份固定体积的液体从一个容器移取到另一容器中。基础化学实验常用的移液管规格有1.00、5.00、10.00、20.00、25.00 mL,与其他容量仪器一样,在表示移液管时可以将小数点后的零省略,但要记住这些零也是有效数字。移液管的下端逐渐变细,可使转移液体时液流变小,但也很容易被堵塞。移液管的上端有经过校准的体积刻度线。移液管刻度线上面的开口可以插入洗耳球,将被转移的液体吸入。

操作要领:绝对不能用嘴吸取液体!只能用洗耳球来吸取。在转移液体之前,先要用少量溶液润洗移液管,以去除移液管内壁可能残留的少量水。装液时,先将洗耳球捏扁,插入移液管的上口中,然后将移液管下端插入液体,慢慢松开洗耳球,液体则慢慢进入移液管内。使液面超过体积刻度线,但要小心不能使液体进入到洗耳球内,移去洗耳球,迅速用食指堵住管口。然后,调节食指压管口的松紧,让液面慢慢降低至弯月面与刻度线相切,堵住管口。将移液管与盛液体的容器器壁相靠,使移液管下端悬挂的液滴落下。一般不建议使用纸巾擦,因为移液管尖端内的一部分液体很可能在纸纤维的毛细作用下被吸出去。将移液管内的液体转移至另一个容器时,只需要将手指松开,让液体自由流出移液管。当液体不再流出时,将移液管下端出口在容器内壁停靠一会儿,使下端悬挂的液体流到容器内(这一滴液体也属于被转移的体积)。见图2-9所示。如果移液管上没有刻"吹"字,其尖端内残留的液体不属于要转移的部分,不能将其吹出,因为在对移液管体积校准时已经考虑这一滴的体积。

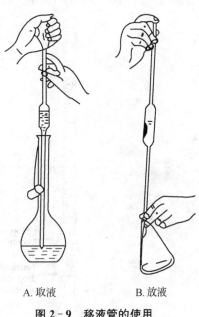

A. 取液　　B. 放液

图2-9　移液管的使用

注意事项:绝对不能用移液管来移取热溶液,也不能用来移取有固体没有完全溶解的液体;热溶液会使校准的体积发生改变,而固体容易在冷的玻璃器壁上沉积,会堵塞移液管的尖端出口。

滴定管

滴定管跟移液管一样,同属于精密的液体体积测量仪器,与移液管主要有两点不同之处:首先,量取的液体体积可变;第二,虽然滴定管读数也有四位有效数字,但其精密度低于移液管,这主要是因为人眼的误差,使得最后一位有效数字的不确定度较大。

滴定管呈上端开口、下端带旋塞的长玻璃管形状(如图2-10所示),旋塞可以控制滴定管内的液体从滴定管尖端流出。滴定管的刻度是从上端开始的,这是因为滴定管是测量从量器中流出的液体体积(量出式),而不是测量量器中装入的液体体积(量入式,如量筒就是测量量器中装入的液体体积)。滴定管的最小精密刻度为0.1 mL,可以估读到百分之一毫

quite as accurate as the pipette. Because of human error, there tends to be larger variance in the last significant figure.

A buret is a long glass cylinder open at one end and fitted at the opposite end with a stopcock valve (Figure 2-10). The stopcock valve controls the flow of liquid from the cylinder through the buret tip. The numbering of the markings begins at the top (open end) of the buret. This is because the buret is designed to show how much volume has been delivered, rather than how much it contains (like a graduated cylinder, for example). The buret is etched with calibration markings at 0.10 mL intervals allowing the estimation of variable liquid levels to the hundredths of a milliliter (0.01 mL). Record all numbers from markings and one number estimated between markings (significant figures) to indicate the sensitivity of the volume measurement. Refer to Figure 2-10 and note the correct way to read the liquid level in a buret.

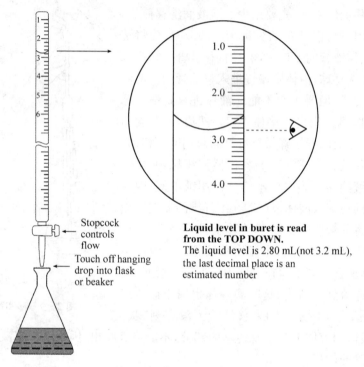

Figure 2-10 Buret

Before use, a buret should always be cleaned and prerinsed with the solution to be delivered. To fill the buret, take the buret out of the clamp. Check the valve to be sure it is closed. Pour the reagent from a beaker into the top of the buret. A funnel may be used if necessary. As you are filling the buret, at some point, pause and look at the tip to be sure the liquid is not pouring out of the bottom. Fill the buret slightly above the "zero" mark.

The buret is still not quite ready to use, because the tip of the buret is probably still filled with air. Put either beaker under the buret tip and open the buret tip fully to expel the air and fill the tip with reagent. It helps to "flick" the buret tip to dislodge any stuck bubbles. When no air remaining, close the stopcock, and check to be sure that the volume in the buret is now BELOW the zero mark. Do not waste time trying to get the volume EXACTLY to zero; it does not matter what volume you actually have, as long as you can read it (which is why it must be below zero). Remove any drop hanging from the buret tip by touching the tip to the side of the beaker or wiping

升(0.01 mL)。记录滴定管读数时,要将刻度上所能读到的数值以及最后估读的数值(这些都是有效数字)全部记录下来,即要将体积测量的灵敏度记录清楚。参照图 2-10。注意读数时视线要与液面保持在同一水平面。

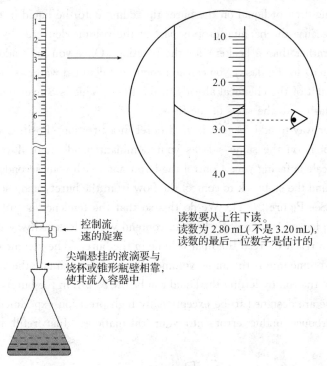

图 2-10 滴定管

滴定管在使用之前要清洗干净。在装溶液之前,滴定管要先用待装的溶液进行润洗。装液时,将滴定管从滴定夹上取下,关好旋塞,将试剂从容器中(如烧杯)倒入滴定管中。为方便起见,也可以使用漏斗(漏斗必须干净干燥)。倒溶液时,要时不时停下来观察溶液有没有洒落。装液一般略高于"0"刻度线。

此时滴定管还不能使用,因为滴定管尖端可能会有气泡。另取一个烧杯(或废液杯)放在滴定管出口的下面,将旋塞完全打开,溶液流速较快可以使气泡冲出来,使溶液充满滴定管的尖端部分。轻轻敲击滴定管有助于驱逐附着在管壁的气泡。当气泡赶走后,关好旋塞,并检查溶液体积,确保溶液的体积在零刻度线以下。体积不必一定要调整到"0.00 mL"刻度线上(可以略低),这是因为滴定管是量出容器,开始的刻度在哪里没有关系,只要能读出来就可以(这也是为什么开始的刻度必须不高于零刻度线的原因)。然后,将滴定管的尖端在烧杯内壁上停靠一下,以去除滴定管尖端悬着的一滴溶液,或者用吸水纸将其拭去。

读取并记录最初的体积读数,体积要估读到 0.01 mL(有效数位要比仪器最小刻度多一位)。从滴定管中移取溶液时,将旋塞打开一定程度,当定量溶液慢慢流入容器后,将旋塞关紧,再将滴定管尖端悬挂的一滴溶液"靠"入容器中,这个步骤很重要,因为这一滴溶液是移取溶液的一部分。如果进行滴定,最好使用锥形瓶而不是烧杯。当滴定剂被滴入锥形瓶后,如有必要可以用少量蒸馏水冲洗锥形瓶内壁,使溅落在锥形瓶内壁上的溶液冲下去。

the tip with a tissue.

Record the starting value; remember to estimate the volume to the nearest 0.01 mL (one more significant digit than the graduations on the buret). To transfer the liquid, open the stopcock valve until the desired amount of liquid is slowly drained into a container. Close the valve and touch off the hanging drop of liquid on the buret tip adding it to the liquid in the container. This step is important because the hanging drop is part of the volume delivered by the buret. Use an Erlenmeyer flask (rather than a beaker) for the titration. Once you have added the titrant (the chemical to be titrated) to the flask, you can add additional distilled water as needed since this will not change the amount of the chemical already in the flask. This is convenient for washing down droplets as they splash onto the sides of the flask.

Note the proper way to add reagent from a buret in a titration. Position the buret such that the stopcock control is on the same side as your dominant hand (note if you are right or left handed), and the scale is facing you. Control the buret and reach your secondary hand around the barrel of the buret and the stopcock to control the flow from the buret, and use your primary hand to swirl the flask. See Figure 2 - 11. We do this so that the tendency is to pull the stopcock in tighter, rather than looser, so we don't have to contend with a leak halfway through a titration. Once the endpoint is reached, read the final volume in the buret. The volume of the titrant added to the flask is this volume minus the initial volume. If you underestimate the volume in the buret, do not try to "save" the run by letting the liquid run below 25 mL in the buret, or by adding liquid mid-titration. These are designed to be exceptionally high precision experiments; either of these techniques will introduce undue error into your calculations. Just refill the buret, and do another run.

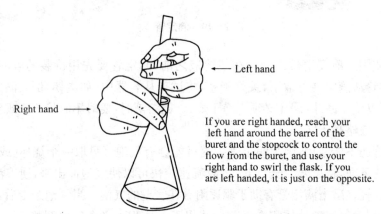

Figure 2 - 11 A right handed man doing titration

下面介绍滴定操作的技巧：首先，将滴定管旋塞的控制边置于右手的一边（注意如果你是左撇子，则放在左边），并将滴定管刻度面对自己。要用左手来控制滴定管，握住旋塞、用手指控制旋塞，通过控制旋塞来控制液流的大小，而右手握住锥形瓶进行旋摇。参看图 2-11。这样操作能使活塞在使用过程中塞得越来越紧，而不是越来越松，可以防止滴定过程中由于旋塞松动造成溶液从滴定管中泄漏出来。到达滴定终点时，读出最后的体积，加入锥形瓶中的滴定剂体积等于最后与最初体积读数之差。如果对滴定所需溶液的体积估计不足，滴定管溶液滴定至最大刻度（例如，25 mL）还没有到达滴定终点，此时，试图记录溶液在最大刻度以下的体积已经没有意义；同样，在中途向滴定管中补充溶液也是没有意义的。这是因为滴定操作是用于精密实验的，这样的操作会为最后的计算带来不应有的误差。因此，每一次滴定开始，都要将滴定管重新装满。

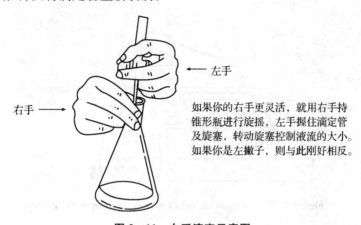

如果你的右手更灵活，就用右手持锥形瓶进行旋摇，左手握住滴定管及旋塞，转动旋塞控制液流的大小。如果你是左撇子，则与此刚好相反。

图 2-11　右手滴定示意图

2.7 pH Meter and pH Measurement

pH Meter

pH meter is a device for pH measurement. pH meters are in most cases the best way to check pH of the solution, as they are much more precise than indicators and pH stripes (papers). Using properly calibrated pH meter with a good electrode one may measure pH with ±0.01 unit accuracy without any problem. A pH meter is nothing else but a precise voltameter connected to the pH electrode-kind of ion selective electrode. See Figure 2-12. Voltage produced by the pH electrode is proportional to the value of pH. The voltameter display of a pH meter is scaled in such a way that the displayed result of measurement is just the pH of the solution.

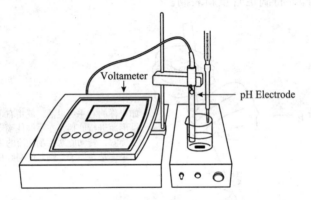

Figure 2-12 Measuring pH of a solution using a pH meter

Construction of a pH Electrode

The pH electrodes used in pH meters consist of a H^+ selective electrode and a reference electrode. Most often used H^+ selective electrodes are glass electrodes. Typical model is made of glass tube ended with small glass bubble. See Figure 2-13, the part in the middle with a bubble is a glass electrode. Inside of the electrode is usually filled with buffered solution of chlorides in which silver wire covered with silver chloride is immersed. The internal solution is a pH reference solution, its pH varies, for example, to be 1.0 (0.1 mol·L^{-1} HCl) or 7.0 (different buffers used by different producers). Active part of the electrode is the glass bubble. The bubble is made to be as thin as possible. Surface of the glass is protonated by both the internal and external solution till equilibrium is achieved. Both sides of the glass are charged by the adsorbed protons, this charge is responsible for potential difference. This potential is called potential of glass electrode. When a glass electrode is immersed in a solution of which the pH wants to be measured, the potential of glass electrode depends on the pH of the external solution since the pH of internal solution is constant. The potential is described by the Nernst equation and is directly proportional to the pH of the solution out sides of the glass.

$$E_{glass} = E_{glass}^{\ominus} + \frac{2.303RT}{F} pH$$

The potential of glass electrode can be derived from the cell potential of a chemical cell which is formed with the glass electrode and a reference electrode: $E_{cell} = E_{glass} - E_{ref}$. Often the reference electrode used in pH meter is an electrode which has a stable and well-known electrode potential, so, the cell potential depends on the pH of the solution which is to be measured. The pH of the

2.7 pH 计与 pH 值测定

pH 计

pH 计是测定溶液 pH 值的仪器。pH 计的精密度比 pH 试纸和指示剂要高出很多,在大多数情况下,使用 pH 计是测定溶液 pH 值的最好方法。电极工作状态良好、经过校正的 pH 计测量精度可以达到 ±0.01 pH 单位。pH 计事实上是一种连接着 pH 电极(H^+ 离子选择电极)的精密电位计,如图 2-12。pH 电极的电位与溶液的 pH 值成正比,因此,pH 电极的电位可以换算成溶液的 pH 值,直接显示在 pH 计的读数上。

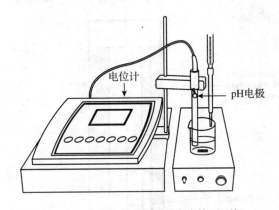

图 2-12 使用 pH 计测定溶液的 pH 值

pH 电极的结构

pH 计使用的电极由 H^+ 离子选择电极和参比电极两部分组成。绝大多数 pH 计的 H^+ 离子选择电极为玻璃电极,典型的玻璃电极由一个玻璃管组成,玻璃管的末端被吹成玻璃泡。如图 2-13 所示,中间的部分即为玻璃电极。在玻璃电极中,玻璃管内灌注含 Cl^- 的缓冲溶液,该缓冲溶液被称为内参比溶液,并插入一根镀有 AgCl 的银丝,其中镀有 AgCl 部分浸入溶液中。不同制造商生产的玻璃电极内参比溶液的 pH 值不尽相同,从 1.0(含 0.1 mol·L^{-1} HCl)到 7.0 不等。玻璃电极的活性部分是玻璃泡,玻璃泡要求做得尽可能薄,膜内外玻璃表面分别被内外水溶液所质子化,并建立质子平衡。由于内外溶液的 pH 溶液不同,则两侧玻璃表面对质子的吸附情况不同,造成玻璃膜两侧电荷不同,从而在玻璃膜两侧形成电势差,这个电势差叫做"玻璃电极的电极电势"。当将玻璃电极放入待测溶液中,由于内参比溶液的 pH 值保持不变,则玻璃电极的电极电势则只随待测溶液(玻璃膜外侧溶液)的 pH 值发生变化,其数值可以用 Nernst 方程描述:

$$E_{玻} = E_{玻}^{\ominus} + \frac{2.303RT}{F}\text{pH}$$

将玻璃电极与参比电极组成原电池,测定电池电动势就可以测定玻璃电极的电极电势: $E_{池} = E_{玻} - E_{参比}$。通常 pH 计所选用的参比电极是电极电势值稳定的电极,因此,电池电动

solution can be obtained from the cell potential.

$$E_{cell} = \left(E^{\ominus}_{glass} + \frac{2.303RT}{F}\text{pH}\right) - E_{ref}$$

The majority of pH electrodes available commercially are combination electrodes that have both glass H⁺ sensitive electrode and additional reference electrode conveniently placed in one housing, such as the pH meter shown in Figure 2-13. For some specific applications separate pH electrodes and reference electrodes are still used since they allow higher precision needed sometimes for research purposes. In most cases combination electrodes are precise enough and much more convenient to use.

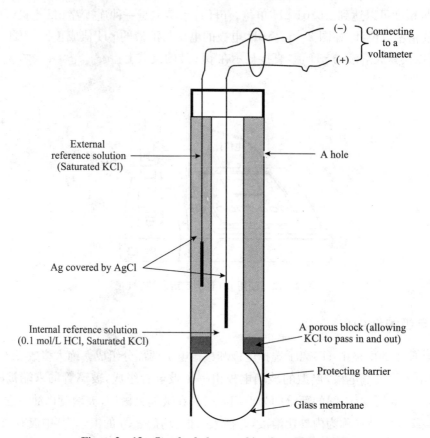

Figure 2-13 Standard glass combination pH electrode

When the pH meter works, it needs a closed circuit through the internal and external solutions and the pH meter. However, for correct and stable results of measurements reference electrode must be isolated from the solution so that they will not cross contaminate. It is not an easy task to connect and isolate two solutions at the same time. Connection is made through a small hole in the electrode body. This hole is blocked by porous membrane, or ceramic wick.

Measuring pH by a pH meter

The steps using a pH meter are: setting up the temperature → calibrating the pH meter → measuring the pH of a solution. Calibrating the acidometer is the most important step.

General Procedure in Calibrating a pH Meter:

1. Choose the mode to "pH". Reset temperature to the temperature of the solution.

势只与待测溶液的 pH 值有关,通过测定玻璃电极与参比电极所构成电池的电池电动势,可以测定溶液的 pH 值:

$$E_{池} = \left(E_{玻}^{\ominus} + \frac{2.303RT}{F}\text{pH}\right) - E_{参}$$

现在,大多数商用 pH 电极为复合电极:将 H^+ 离子选择电极(如玻璃电极)与参比电极合二为一,置入同一根管子中,如图 2-13 所示的 pH 复合电极就是由玻璃电极做 H^+ 选择电极、Ag/AgCl 做参比电极而组成的。除某些实验需要使用特定的分开的电极,以获得研究所需要的特殊实验精度,在大多数情况下使用复合电极,复合电极不仅方便操作,而且精密度也足以满足测量要求。

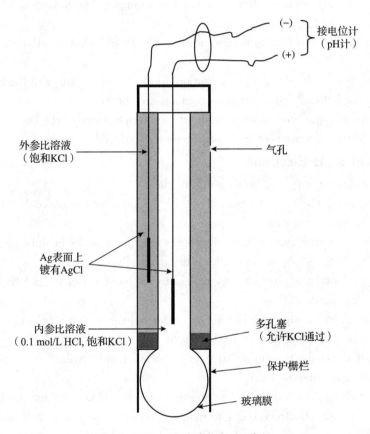

图 2-13 标准玻璃复合 pH 电极

pH 计在测量溶液 pH 值时,需要构成电流回路,即内外溶液与 pH 计要形成回路。但是,另一方面,要使测定结果稳定可靠,参比电极的参比溶液又必须与待测溶液分离,以避免二者的交叉污染。然而,做到参比电极的参比溶液与待测溶液既相互联通又彼此分离是不易的。参比溶液与待测溶液的连接通过电极底部的小孔实现,在小孔中塞入多孔膜材料或者陶瓷芯(可以允许离子通过)。

用 pH 计测量 pH 值

使用 pH 计测定溶液 pH 值时,首先要将电极与电位计相连,如图 2-13 所示。测定分

2. Remove bottle with storage solution, rinse electrode, blot it dry.

3. Immerse the electrode into the standard buffer solution 1, for example, the buffer with pH 4.00.

4. Adjust the meter to the reading to equal to the pH of the buffer, for example, 4.00 with Cal 1 knob.

5. Remove the buffer 1, rinse the electrode, blot it dry.

6. Immerse the electrode into the standard buffer solution 2, for example, the buffer with pH 10.00.

7. Adjust the meter to read the pH of the buffer 2 with Cal 2 knob.

8. Remove the buffer 2, rinse electrode, blot it dry.

9. Measure the pH of the buffer 1 again. For example, the pH should read 4. If not, readjust Cal 1 knob.

10. Return to the buffer 2. For example, the pH should read 10. If not, readjust Cal 2 knob.

11. Repeat the steps 9 - 10 of standardization using the Cal 1 knob with the buffer 1 and the Cal 2 knob with the buffer 2 until consistent readings are obtained.

12. When the pH meter is calibrated well, do not turn any one of the buttons on the meter. You can carry out the measurement within 24 hours without calibration anymore.

Maintain of a pH Electrode

1. Handle electrode with care because it is fragile!

2. Keep electrode always immersed. Use the solution recommended by manufacturer or neutral solution of KCl (3 - 4 $mol \cdot L^{-1}$).

3. Fill electrode with correct filling solution (as recommended by manufacturer, usually KCl solution, 3 $mol \cdot L^{-1}$ to saturated) to not let it dry internally.

4. If dried incidentally or after storing, the electrode should be soaked in a KCl solution for at least 24 hours before used.

5. If you are using the electrode in solution containing substances able to clog the junction or stick to the glass bubble, clean the electrode as soon as possible after use.

6. Don't put electrode in solutions that can dissolve glass—hydrofluoric acid (or acidified fluroide solution), concentrated alkalis.

7. Don't put electrode into dehydrating solution such as ethanol, sulfuric acid, *etc*.

8. Don't rub or wipe electrode bulb, to reduce chance of error due to polarization.

9. Don't use organic solvents for cleaning of the electrode with epoxy body.

以下几个步骤:设定测量温度→校正 pH 计→测量。其中校正 pH 计是最关键的步骤。

校正 pH 计的一般过程:

1. 将 pH-mV 旋钮置于 pH 挡。将温度补偿器旋钮旋至溶液的温度值。
2. 取下电极保护帽(内有存储电极的缓冲溶液),蒸馏水冲洗后,吸干电极上面的水。
3. 将电极插入定位用的标准缓冲溶液 1 中,如 pH=4.00 的缓冲液。
4. 调节"定位"旋钮使读数与标准缓冲溶液 1 的 pH 相同,如 4.00。
5. 将标准缓冲溶液 1 拿走,电极用蒸馏水冲洗,吸干电极。
6. 将电极插入另一个标准缓冲溶液 2 中,如 pH=10.00 的缓冲液。
7. 调节"斜率"旋钮使读数与标准缓冲溶液 2 的 pH 相同。
8. 将标准缓冲溶液 2 拿走,电极用蒸馏水冲洗,滤纸条吸干。
9. 将电极插入缓冲溶液 1,测定缓冲溶液 1 的 pH 值。若 pH 值不对,则需要重新"定位"。
10. 之后,将洗净的电极插入缓冲溶液 2 中,如读数不对,则需要重新调节"斜率"。
11. 重复 9~10 步骤,直至对这两个缓冲溶液的 pH 测定时,读数与其 pH 值相等。
12. pH 计经过校正后,不要再动旋钮,可以直接进行溶液的 pH 值测定。一般开机 24 时内不需要重新校正。

pH 电极的维护

1. 安装电极及测量过程中要注意保护玻璃电极,以免损坏。
2. 电极不用时应该浸泡在电极制造商推荐的 KCl 溶液或中性 KCl 溶液($3\sim4$ mol·L^{-1})中。
3. 避免电极内管溶液的干涸,要正确补充内充液(要根据制造商的推荐进行配制:3 mol·L^{-1}~饱和 KCl)。
4. 如果电极偶尔或经过长时间储存发生干涸,在使用前要在 KCl 溶液中浸泡 24 h。
5. 如果测定溶液有可能造成电极多孔塞堵塞或能在玻璃泡表面附着(如蛋白质溶液),使用过后要尽快对电极进行清洗。
6. 不要将电极置入氢氟酸、浓碱等对玻璃有腐蚀作用的溶液中。
7. 不要将电极置入无水乙醇、浓硫酸等脱水性溶液中。
8. 不要拭擦玻璃电极的玻璃泡,否则将可能导致电极的极化,带来误差。
9. 不要使用有机溶剂清洗电极,因为电极的外套是有机树脂,可能会溶解在有机溶剂中。

2.8 Notebooks

Proper use of a laboratory notebook

One of the most useful skills you will acquire in the laboratory is the proper use of a laboratory notebook. Notebooks, or other formally kept records, are an essential tool in many careers, ranging from that of the research scientist to that of the practicing physician. The effort invested in developing good habits of notebook use will be amply repaid for students who pursue a future in the basic or applied sciences. Experience has indicated that skillful notebook use is developed by most students only through continued special effort—it does not come naturally. Some of the main principles of sound notebook use are outlined below.

1. The laboratory notebook is a permanent, documented, and primary record of laboratory observations. Therefore, your notebook will be a bound journal with pages that should be numbered in advance and never torn out. A notebook will be supplied to you before the first laboratory period.

2. All notebook entries must be in ink and clearly dated. No entry is ever erased or obliterated by pen or "white out". Changes are made by drawing a single line through an entry in such a way that it can still be read and placing the new entry nearby. If it is a primary datum that is changed, a brief explanation of the change should be entered (e. g. "balance drifted" or "reading error"). No explanation is necessary if a calculation or discussion is changed; the section to be deleted is simply removed by drawing a neat "×" through it.

3. In view of the fact that a notebook is a primary record, data are not copied into it from other sources (such as this manual or a lab partner's notebook, in a joint experiment) without clear acknowledgment of the source.

4. Observations are never collected on note pads, filter paper, or other temporary paper for later transfer into a notebook. If you are caught using the "scrap of paper" technique, your improperly recorded data may be confiscated by your TA or instructor at any time.

5. Your notebook should be your primary source of information. Everything you do in the laboratory should be included in your notebook, from procedure to calculations.

Contents in Notebooks

1. Prelab write-up that shows that you were prepared for lab *before* beginning the experiment.

2. Data and associated graphs and calculations that *quantitatively* gauge how successful your laboratory technique was.

3. Enough explanatory information so that someone else with your knowledge of chemistry could, *from your notebook alone*, enter the lab and repeat your work.

4. Your discussion and answers to questions raised from time to time in the laboratory manual itself.

A laboratory notebook should be legible, and data in it should be readily accessible, clearly labeled with units, and grouped in a logical way.

The formal pre-lab and in-lab is described as shown as follows:

1. Objective　State the purpose of the experiment along with a brief statement of basic principles involved and the method to be used.

2. Reference　Cite the source for the experiment. It will suffice to reference the page

2.8 实验记录

实验记录的规范

掌握实验记录的规范性是同学们在实验课中需要养成的最重要技能之一。无论是专门从事科学研究工作的科学家，还是从事生产实践工作的技术人员，在工作中都会将实验记录或其他正式保存的记录当作必不可少的工具。养成规范做实验记录的良好习惯，对以后无论从事基础研究工作还是应用实践工作都大有裨益。实践证明，对于绝大多数学生而言，良好的实验记录技能不可能自然天成，只有通过持续不懈的专门训练才能掌握。以下是掌握良好实验记录技能的几点重要原则：

1. 实验记录具有永久性、档案性、原始性，是实验的第一手材料，因此，实验记录本要标记页码，装订成册，且不能缺页（永远不能撕页）。实验记录本应该在实验开始之前就准备好。

2. 所有的实验记录都必须用墨水笔书写，并注明日期。记录不能擦去、涂改或者"涂白"。需要修改时，可以在要修改部位上面轻轻画上一道线，使修改前的内容还能清楚地看到，然后在附近写上替代的内容。如果被修改的内容为原始数据，则要在旁边简述修改的原因（如"天平漂移"或"读数错误"）。如果是计算或者讨论需要修改，则不需要注明原因。如果删除某一部分时，只需要在删除内容之上画一个清晰的"×"即可。

3. 鉴于实验记录的原始性，任何其他来源的数据（如来自实验讲义、实验合作者的实验数据等）都必须清楚注明来源。

4. 实验记录不能记在便笺、滤纸上，甚至，临时记在某张纸上之后再誊写到记录本上也是不允许的。如果有同学使用"纸片"做实验记录，一旦被老师发现，则"纸片"将被立即没收。

5. 实验记录是实验的第一手材料，所以，从实验过程到计算的每一个环节的记录都应该尽可能详尽。

实验记录的主要内容

1. 实验前的准备（预习报告）。
2. 实验数据以及相关的图表和计算。
3. 充分的解释。实验记录要具有可读性，要有充分的解释，要保证别人不仅能读懂，而且还能根据记录重复该实验。
4. 讨论和相关问题的回答。

实验记录一定要条分缕析，数据要带单位，按一定的逻辑顺序组织起来，使人清楚明白。实验记录的一般格式为：

1. 实验目的
说明实验的目的，并简要说明实验原理以及实验方法。
2. 参考文献

numbers of the Lab Manual from which the procedure comes.

3. Prelab Procedure Flowchart or Outline   Before coming to lab, summarize the procedure you will perform in a flowchart or outline.

4. Calculations or Analysis Flowchart or Outline   Before coming to lab, write in your notebook a description of the calculations that you will perform to analyze your data. Note that when no calculations are required, this section is omitted.

5. Procedure, Data, and Results   Qualitative observations and quantitative data are best entered in a running commentary. This commentary should be recorded in the lab, as the experiment proceeds. High prose standards are not expected. If repeated measurements are made using the same procedure, a table provides the best presentation. If the experimental work is done jointly it must be noted and reported independently. Your notebook must list your co-workers and identify who did what. You may write this commentary on the same page as your Prelab Procedure, in an adjacent column. If you use this technique, prelab and in-lab writing must be clearly distinguished. Write your calculations clearly and include a brief explanation for each step. Remember to include units. If the same calculation is done repeatedly, write one sample calculation in your notebook and report the results of other calculations in a table.

6. Discussion   A discussion of the experiment should include qualitative and quantitative comments on your results. Calculations of precision, accuracy, and possible explanations of any obvious errors may be appropriate. It is often helpful to collect your results in tabular form. Questions posed in the description of the experiment in the Manual should also be answered here. An example discussion for the density experiment is shown in the sample write up in the lab manual.

Note that steps 1 through 4 will be done *before* coming to lab lecture for the week's experiment. Note, too, that a logical tabular form for your data will often be the clearest presentation, but that you should construct these tables in the lab when you obtain the data.

For Example

Density of an Unknown Liquid

Objective
To determine the identity of an unknown liquid by measuring its density.

Reference
"Dartmouth College, Chemistry 3/5 Lab Manual", pp. 25 – 37.

Prelab Procedure Outline
1. Obtain sample of unknown liquid. Note physical properties such as odor, viscosity, and color.
2. Record the mass of clean, dry 10.00 mL volumetric flask and stopper.
3. Carefully transfer unknown liquid to volumetric flask. Fill to mark exactly. Stopper flask.
4. Measure mass of filled flask.
5. Empty and dry flask.
6. Repeat steps 2 – 5 for additional liquid sample.

引用要清楚。如实验步骤来源于实验教材,要注明引用文献的书名、出版社以及页码。

3. 预习的实验流程图或概要

实验前,将要进行的实验步骤加以总结和整理,形成概要或流程图。

4. 数据计算和分析流程图或概要

如果实验结果需要计算和分析,实验前需要在记录本上将数据处理方法用流程图或概要的方式列出。如果实验中不需要处理数据,这一项可以省去。

5. 实验过程、实验数据和结果

定性观察和定量数据的记录最好根据实验进程,采取记流水账的方式,散文式的华丽描述是不可取的。相同操作的重复测定结果最好以表格的形式给出。如果实验工作是与他人共同完成的,应该在实验记录中注明,实验报告则要独立完成。实验记录中必须列出合作者的名单,并写清谁做了什么工作。你可以采取另加一栏的方法,在预习报告之实验过程所在的页面中记录实验观察和数据。如果能掌握这种做记录方法,预习报告和实验记录一定非常清晰,一目了然。数据计算时要写清楚计算的步骤以及简单的说明,记住一定要带单位。如果是相同步骤的重复计算,则写出一个样本的计算过程,然后将其他样本的计算结果用列表表示出来。

6. 讨论

讨论应当包括对实验结果的定性和定量评价,例如,可以讨论结果的准确度、精密度以及出现明显错误的可能原因。尽量用表格的形式表达实验结果,这样会让人一目了然。实验教材上有关该实验的问题也可在讨论部分作答。如何进行讨论可以参见下面一个简单实验的例子:通过测定液体密度对未知液体进行鉴定。

注意:实验记录中的 1 到 4 项要在实验前完成;另外,记录实验数据之表格的设计要富有逻辑性,这样会使数据的表达显得非常清晰,但填写这些表格是应该在实验中完成的,即在获得实验数据后要立即填写在表格中。

范例

未知液体密度的测定

实验目的

通过测定液体密度对未知液体进行鉴别

参考文献

"Dartmouth College, Chemistry 3/5 Lab Manual", pp. 25～37.

预计的实验大致过程

1. 拿到未知液体样品。观察液体的物理性质,如气味、黏度、颜色等。
2. 记录 10.00 mL 干燥洁净的容量瓶及其塞子的质量。
3. 小心将未知液体转移至容量瓶中,使液体刚好装至容量瓶的刻度线。盖好瓶塞。
4. 测定装满液体容量瓶的质量。
5. 将容量瓶倒空并干燥。
6. 重复步骤 2～5,取另外的液体样品进行测定。

or Prelab Procedure Flowchart

Obtain unknown sample in beaker→note odor, color, viscosity
↓
Weigh clean, dry 10 mL volumetric flask and stopper→record
↓
Transfer liquid to flask→fill to mark, use dropper, stopper flask
↓
Weigh filled flask→record
↓
Empty, dry flask. Repeat for additional liquids

Sample Calculations:
If flask is 10.0 g and flask+liquid is 15.0 g
Mass liquid=15.0 g—10.0 g=5.0 g
Density liquid=5.0 g/10.00 mL=0.50 g/mL

Prelab Analysis Outline
1. Subtract mass of empty flask from mass of filled flask to calculate mass of liquid.

$$m_{liquid} = m_{flask\ \&\ liquid} - m_{flask}$$

2. Divide liquid mass by liquid volume to calculate density.

$$D = m_{liquid}/10.00\ mL$$

3. Compare to known densities to identify liquid.

Procedure, Data and Results
1. Washed two 10.0 mL volumetric flasks and dried inside and out.
2. Labeled A and B with pencil.
3. Weighed flasks A and B on analytical balance.
   Flask A 11.2571 g; Flask B 11.3921 g
4. Filled flasks to mark with unknown liquid 1.
5. Weighed on analytical balance.
6. Emptied and dried flasks A and B.
7. Weighed empty flasks.
8. Filled to mark with unknown liquid 2.
9. Weighed on analytical balance.

Unknown liquid 1

	Flask A	Flask B
Mass of empty flask	11.2571 g	11.3921 g
Mass of flask & liquid	19.2571 g	19.6124 g
Mass of liquid	8.0000 g	8.2203 g
Density of liquid	0.8000 g/mL	0.8220 g/mL

Average density of liquid 1: 0.8110 g/mL
Probable Identity: methanol

或:预计的实验流程图

用烧杯取用未知液体→观察气味、颜色、黏度
↓
洁净、干燥的10 mL容量瓶带塞的称重→记录
↓
将液体转移至容量瓶中→用滴管调至刻度,盖上塞子
↓
装满液体容量瓶的称重→记录
↓
倒空,干燥容量瓶,重新装入另外的液体

样品计算

如果容量瓶重为10.0 g,容量瓶+液体为15.0 g
液体的质量=15.0 g−10.0 g=5.0 g
液体的密度=5.0 g/10.00 mL=0.50 g/mL

预计的分析概要

1. 用装满液体容量瓶的质量减去空容量瓶的质量计算液体的质量。

$$m_{液} = m_{瓶+液} - m_{瓶}$$

2. 用液体质量除以液体体积计算液体的密度。

$$D = m_{液}/10.00 \text{ mL}$$

3. 与已知的液体密度相比较对液体进行鉴别。

实验过程、数据与结果

1. 洗净两个10.00 mL容量瓶,并使其内外干燥。
2. 用铅笔将其标记为A和B。
3. 在分析天平上对A、B两个容量瓶进行称重。
   瓶A 11.2571 g,瓶B 11.3921 g
4. 将未知液体1装满至容量瓶的刻度。
5. 在分析天平上进行称重。
6. 倒空并干燥容量瓶A、B。
7. 空瓶称重。
8. 将未知液体2装满至容量瓶的刻度。
9. 在分析天平上称重。

未知液体1

	瓶A	瓶B
空瓶的质量	11.2571 g	11.3921 g
瓶+液体的质量	19.2571 g	19.6124 g
液体的质量	8.0000 g	8.2203 g
液体的密度	0.8000 g/mL	0.8220 g/mL

液体1的平均密度:0.8110 g/mL
可能的物质:甲醇

Unknown liquid 2

	Flask A	Flask B
Mass of empty flask	11.2575 g	11.3918 g
Mass of flask & liquid	21.3308 g	21.5086 g
Mass of liquid	10.0733 g	10.1168 g
Density of liquid	1.0073 g/mL	1.0117 g/mL

Average density of liquid 2: 1.010 g/mL
Probable Identity: water

Discussion

The first liquid unknown was determined to be methanol from its experimental density of 0.8110 g/mL. This compares to the literature value of 0.7914 g/mL (CRC Handbook, 66th ed., pp. C−351). Methanol is colorless, slightly less viscous than water, and has a slightly sweet odor. The second liquid was determined to be water from its experimental density of 1.010 g/mL. The literature value for the density of water at 25 ℃ is 0.997 g/mL (CRC Handbook, 66th ed., pp. F−10). The water sample looked and smelled like water.

This method of determining densities was accurate to within 2% for methanol and to within 1% for water. Limitations on the precision of the determination were the ability of the experimenter to correctly fill the volumetric flask, the accuracy of the flask itself, and the accuracy of the balance. The balance is accurate to 0.0001 g out of 20 g (0.0005%) and the flask is accurate to 0.02 mL out of 10 mL (0.2%). Since both of these values are less than the discrepancy between the experimental and literature density values in this experiment, they were not the limiting factors in the precision of the results. Possible error by the experimenter that would account for the experimental precision being less than expected are not filling the flask to the mark properly or having droplets of liquid in the neck of the flask or on the outside of the flask. Handling the flask with bare hands could have introduced an error in the mass by adding fingerprints. Finally, the densities of liquids vary slightly with temperature and this could account for the discrepancy between the experimental results and literature values. If the temperature in the lab was not 25 ℃, the true values of the densities of the liquids may have differed from the literature values for 25 ℃.

未知液体 2

	瓶 A	瓶 B
空瓶的质量	11.2575 g	11.3918 g
瓶+液体的质量	21.3308 g	21.5086 g
液体的质量	10.0733 g	10.1168 g
液体的密度	1.0073 g/mL	1.0117 g/mL

液体 2 的平均密度：1.010 g/mL
可能的物质：水

讨论

第一种未知液体密度测定值为 0.8110 g/mL，被鉴定为甲醇，这与其文献值 0.7914 g/mL（CRC Handbook，66th ed.，pp. C−351）相当。甲醇无色，黏度较水略低，气味微甜。第二种液体密度测定值为 1.010 g/mL，被鉴定为水。水在 25 ℃时，密度的文献值为 0.997 g/mL（CRC Handbook，66th ed.，pp. F−10）。从外观和气味上看，该被测样品与水相同。

本密度测定方法的准确度对于甲醇而言相对误差为 2%，对于水而言为 1%。测定精密度受操作者使用容量瓶和天平的技能以及容量瓶和天平精密度的限制。天平测量的不确定值为 0.0001 g，样品质量为 20 g，其称量相对误差为 0.0005 %；容量瓶体积不确定值为 0.02 mL，测量体积为 10 mL，其体积相对误差为 0.2%。这些值都小于实验值与文献值间的差异，因此，仪器不是影响本次测定结果精密度的主要因素。实验误差源于操作者，如操作者向容量瓶装液时没有恰好装到刻度的位置，或者将液滴滴落到容量瓶外面，或者拿容量瓶的手不洁净使得容量瓶的质量没有称准。另外，液体的密度与温度有关，实验温度与文献值的温度不同也是二者不同的原因。如果实验温度不是 25 ℃，液体在此温度下的真实密度也就与 25 ℃的文献值不同。

2.9 Analysis of Data and Presentation of Results

Using significant Figures

It is important to take data and report answers such that both the one doing the experiment and the reader of the reported results know how precise the final answer is. The simplest way of expressing this precision is by using the concept of significant figures where a significant figure is any digit that contributes to the precision of an experimentally measured number or to a number calculated from experimentally measured numbers. Please refer to the chemistry textbook for a discussion of the use of significant figures.

Analysis of Data and Presentation of Results

You should have performed multiple trials of your experiment. Think about the best way to summarize your data. At first, take some time to carefully review all of the data you have collected from your experiment. Often, you will need to perform calculations on your raw data in order to get the results from which you will generate a conclusion. Use charts and graphs to help you analyze the data and patterns. Also, charts and graphs should be the best way to present the results.

Tabling

1. Make sure to clearly label all tables and graphs.

2. Be sure to label the rows and columns. Don't forget to include the units of measurement, such as grams, centimeters, liters, *etc*.

3. Pay careful attention because you may need to convert some of your units to do your calculation correctly. All of the units for a measurement should be of the same scale, for example, keep L with L and mL with mL, do not mix L with mL!

4. A spreadsheet program such as Microsoft Excel may be a good way to perform such calculations, and then later the spreadsheet can be used to display the results.

For example, the data and result is presented as in Table 2-1 for standardization of the solution of $KMnO_4$ against the primary standard substance $Na_2C_2O_4$ and Table 2-2 for determination of ferrous ammonium sulfate hexahydrate (FAS) against the solution of $KMnO_4$ of which the concentration has been standardized.

Table 2-1 Standardization of the $KMnO_4$ solution

No.	1	2	3
$W_{Na_2C_2O_4}/g$	0.1516	0.1495	0.1514
V_{KMnO_4}/mL	22.61	22.45	22.63
$c_{KMnO_4}/mol \cdot L^{-1}$	0.02001	0.01988	0.01997
$\bar{c}_{KMnO_4}/mol \cdot L^{-1}$	0.01995		
RSD/%	0.3		

2.9　实验数据分析与结果表达

正确使用有效数字

表达实验数据和结果时,要求人们(无论实验者本人还是读者)能够从已表示出来的实验数据和结果中看出最终结果的精密度,这一点是非常重要的。表示实验精密度的最简单办法是运用有效数字,即用有效数字来记录实验测量和计算结果的精密度。有关有效数字的概念以及应该如何使用的知识,请仔细阅读课本相关内容。

实验数据分析与结果表达

在一个实验中,可能需要进行多次测试,获得系列实验数据,这些实验数据需要以一种最优化的方式收集整理起来。首先,要花些时间仔细检查和分析这些从实验中收集到的数据(看看是否存在异常数据,分析实验数据存在的某种规律等)。通常,需要对原始数据进行相关的计算分析(例如,分析不同物理量之间的关系)之后才能得出相关结论。列表和作图是分析实验数据必不可少的手段,并且也是实验结果表达的有效方式。

列表法

1. 表格和图表一定要有标题。
2. (在每行或每列的第一栏)要标明每行、每列所代表的物理量的名称和单位。不要忘记被测量的物理量所带的单位,如克、厘米、升等等。[如表 1 中"$W_{Na_2C_2O_4}/g$"一栏表示 $Na_2C_2O_4$ 质量($W_{Na_2C_2O_4}$)的单位为"g",即:物理量除以单位等于表格中的纯数字]
3. 计算时,一定要注意单位的换算,应该统一单位。例如:如果体积单位采用"L"时,所有的体积就都统一用"L"表示;如果采用"mL"时,所有的体积都应该统一用"mL"表示。
4. 可以采用诸如 Excel 等表格程序来进行列表,这类软件能提供计算程序,并且将结果显示出来。

如:用基准物质 $Na_2C_2O_4$ 标定 $KMnO_4$ 溶液,再用经过标定的 $KMnO_4$ 溶液对六水合硫酸亚铁铵(FAS)含量进行测定,实验结果由表 2-1 和表 2-2 给出:

表 2-1　$KMnO_4$ 溶液的标定

No.	1	2	3
$W_{Na_2C_2O_4}/g$	0.1516	0.1495	0.1514
V_{KMnO_4}/mL	22.61	22.45	22.63
$c_{KMnO_4}/mol \cdot L^{-1}$	0.02001	0.01988	0.01997
$\bar{c}_{KMnO_4}/mol \cdot L^{-1}$		0.01995	
RSD/%		0.3	

Table 2-2 Determination of the purity of FAS

No.	1	2	3
W_{FAS}/g	0.8078	0.7969	0.7958
V_{KMnO_4}/mL	18.59	18.25	18.35
$w(FAS)/\%$	90.45	90.50	90.62
$\overline{w}(FAS)/\%$		90.52	
RSD/%		0.1	

Graphing

Mathematical relationships between variables are determined by graphing experimental data. For example, the linear relationship between the concentration ratio $\dfrac{c_{A^-}}{c_{HA}}$ and the pH leads to Henderson equation ($pH = pK_a + \lg \dfrac{c_{A^-}}{c_{HA}}$); Once the mathematical relationship is known, experimental quantities can often be calculated from the slope or y-intercept of a graph. For another example, reaction enthalpy ΔH can be determined from the slope of a plot of $\ln K$ vs. $1/T$ ($\ln K = -\dfrac{\Delta H}{R} \cdot \dfrac{1}{T} + c$). Furthermore, extrapolation of graphical data trends can be used to find information about conditions difficult to achieve in the laboratory, such as high or low temperatures and pressures.

The procedure outline of plotting is as follows: chose the variables→plot the coordination→draw the data points→look for the trend of the data→analysis the relationship between the variables.

Plotting Variables A line graph consists of two axes at right angles. The horizontal x-axis, or *abscissa*, is typically chosen to represent the independent variable which is intentionally manipulated by the experimenter. The vertical y-axis, or *ordinate*, is chosen to represent the dependent variable which changes as the independent variable is manipulated.

Example: A student wishes to measure the relationship between volume and temperature of a gas. If the student purposely increases the temperature and measures the resultant changes in gas volume, he should graph the values for T on the x-axis and the resulting values for V on the y-axis. If the student reverses the experiment and purposely increases the volume of the gas and measures the resultant temperature, the variables are then reversed: V is plotted on the x-axis and T is plotted on the y-axis.

Selecting a Title and Labeling the Axes The title must describe the variables measured, chemicals used, and special conditions of the experiment. Examples: "Titration of HAc solution of 20.00 mL using NaOH solution with concentration of 0.1001 mol · L^{-1}" or "pH vs. $\lg \dfrac{c(Ac^-)}{c(HAc)}$". The axes must be labeled with the name (or symbol) of the variable. Units (if any) should be included. Examples: Temperature (℃), or V(mL).

Choosing an Appropriate Scale The scale of the graph should be chosen so the data completely fills the graph and is not restricted to one small region or corner. Examples are shown in Figure 2-14. Note that the scale does not have to begin with zero. If the graph is made on

表 2-2　FAS 纯度的测定

No.	1	2	3
W_{FAS}/g	0.8078	0.7969	0.7958
V_{KMnO_4}/mL	18.59	18.25	18.35
$w(FAS)/\%$	90.45	90.50	90.62
$\overline{w}(FAS)/\%$		90.52	
RSD/%		0.1	

作图法

实验变量间的数学关系可以通过作图来确定。例如,pH 与浓度比 $\dfrac{c_{A^-}}{c_{HA}}$ 之间呈线性关系,这种关系就是 Henderson 方程($pH=pK_a+\lg\dfrac{c_{A^-}}{c_{HA}}$)。一旦变量间的数学关系确定下来,实验所要测定的物理量则可以根据直线斜率或 y 轴截距得到。再如,测定反应焓变 ΔH 时,可以用 $\ln K$ 对 $1/T$ 作图得到一条直线,ΔH 可以根据其直线斜率进行求算($\ln K=-\dfrac{\Delta H}{R}\cdot\dfrac{1}{T}+c$)。另外,作图法的另一个用途是通过测量数据间的关系,作图外推至测量范围之外,得到一些诸如超高(低)温或超高(低)压条件下不能由实验直接测定的数据。

作图的大致过程为:选择变量→画出坐标→标出实验数据点→观察数据趋势→分析变量间的数学关系。

选择变量　平面上的线图有两个相互垂直的坐标轴:水平的 x 轴,即横坐标,一般作为独立变量,是实验者有意控制的物理量;垂直的 y 轴,即纵坐标,作为因变量,其变化随独立变量的变化而发生变化。

例如,某学生要测定气体体积 V 与温度 T 之间的关系。如果该生设计的实验是通过升高温度(改变温度)使体系体积发生改变,即测量温度变化引起对应体积的变化,他就应该以 T 为 x 轴、V 为 y 轴;反过来,若这个学生设计的实验是通过增加气体体积来改变温度,即测量体积的改变引起对应温度的改变,则自变量与因变量相反:V 为 x 轴、T 为 y 轴。

写出作图的标题,并标记坐标轴　标题要注明测定的变量、使用的化学试剂、特殊的实验条件。例如:"0.1001 mol·L^{-1} NaOH 溶液滴定 20.00 mL HAc 的滴定曲线"、"pH 与 $\lg\dfrac{c(Ac^-)}{c(HAc)}$ 的关系"等。坐标轴一定要标记变量的名称(或符号),若变量有单位,则还必须注明变量的单位。例如:温度 $T(℃)$、体积 $V(mL)$。

选择合适的坐标轴比例尺　选择的坐标比例尺应以数据点能填满作图的全部空间为标准,而不是将数据点局限在作图的某一区域或一个角落里,如图 2-14 所示。注意:坐标原点可以不从"0"开始。如果是在坐标纸上手动作图,还要求能将数据的有效数字在图上全部表示出来。

graph paper by hand, significant figures of the data should be completely shown in the graph.

Drawing Data Points Data points can be represented either as a circle, dot, or cross. The size of the circle or bar is often used to represent the deviation or error in the data point. Recorded data should be included on the plot as a separate table. (Figure 2 – 14.)

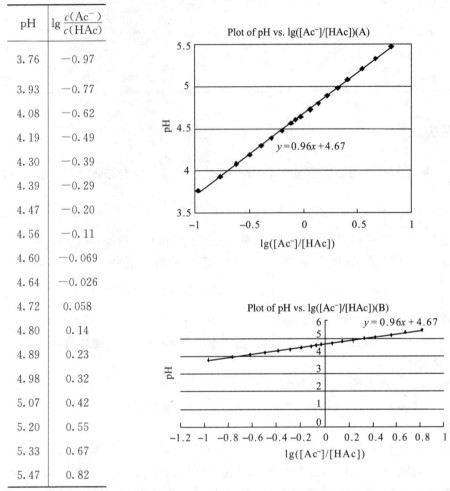

pH	$\lg \dfrac{c(\text{Ac}^-)}{c(\text{HAc})}$
3.76	−0.97
3.93	−0.77
4.08	−0.62
4.19	−0.49
4.30	−0.39
4.39	−0.29
4.47	−0.20
4.56	−0.11
4.60	−0.069
4.64	−0.026
4.72	0.058
4.80	0.14
4.89	0.23
4.98	0.32
5.07	0.42
5.20	0.55
5.33	0.67
5.47	0.82

Graph A and B are from the same data set, but use different scales for the dependent and independent variables. Graph A represents the data range most clearly whereas Graph B seizes the data in the top corner. Note that each graph contains a title and the axes are labeled.

Figure 2 – 14   Plotting with an appropriate scale

Data points from other experimenters should be represented with different symbols. Include a legend citing the source of the additional data points. Example: • My data, ◆ Zhang San's data, † Li Si's data.

Drawing Lines or Curves Draw a curve or line when connecting data points. "Dot-to-dot" connections should be avoided. If the data points do not exactly fit on the line, a best-fit line in which an equal number of data points lie an equal distance above and below the line should be drawn. See the example as shown in Figure 2 – 15. The best fit line "smoothes" the data and helps eliminate random error. The line should be smooth and sharp, which affect the accuracy of the experimental result. It is encouraged to use a computer graphing program to generate the line and its equation.

pH	lg $\dfrac{c(\text{Ac}^-)}{c(\text{HAc})}$
3.76	−0.97
3.93	−0.77
4.08	−0.62
4.19	−0.49
4.30	−0.39
4.39	−0.29
4.47	−0.20
4.56	−0.11
4.60	−0.069
4.64	−0.026
4.72	0.058
4.80	0.14
4.89	0.23
4.98	0.32
5.07	0.42
5.20	0.55
5.33	0.67
5.47	0.82

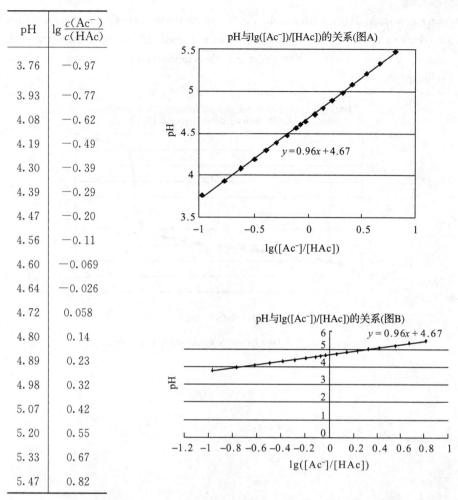

图 A 与 B 使用同一组数据作图，A 图选择的坐标尺度合适，数据点填满图中全部空间，数据点与计算结果清晰明了；B 图坐标选择不合理，使得数据点仅占据图之一隅，计算结果显示的位置也需要调整。

图 2-14　选择合适坐标尺度作图

描绘数据点　数据点可以用诸如圆圈、点或十字叉等符号(○、●、×、△、⊙)表示。符号的大小常常代表数据点的偏差或误差。数据记录分别包括图和数据表(如图 2-14 所示)。他人的数据要用不同的符号表示，并且使用图例说明数据的来源(将其与自己的数据区别开来)。例如：•代表自己的数据，◆代表张三的数据，†代表李四的数据。

绘制直线或曲线　将数据点连接起来画出一条曲线或直线。在绘制直线或曲线时，要避免"点对点"的连接。如果数据点不能完全落在同一条直线上，那么，拟合出来的最优直线应该能使数据点均匀分布在直线的上下两侧，且两侧数据点到直线的距离相等。最优拟合直线或曲线具有"平滑"数据、消除偶然误差的作用，要求所画出来的线光滑均匀、细而清晰(如图 2-15 所示)。连线的好坏会直接影响到实验结果的准确性，有条件的鼓励用计算机作图。

计算机作图　所有的图都可以采用计算机绘制。注意：计算机作图时，一定要正确选择坐标比例尺，要做好标记，注明其他相关信息。记住，好的结果源于好的数据。但是要保证好的数据能给出好的计算结果，就必须学会正确使用绘图软件、进行正确作图。

Computer generated Plots All graphs can be computer generated. Caution: A computer plot must be scaled correctly, have labels, and all other expected information. Remember that good data is necessary for good results. You must use the computer graphing program correctly and obtain a useful graph.

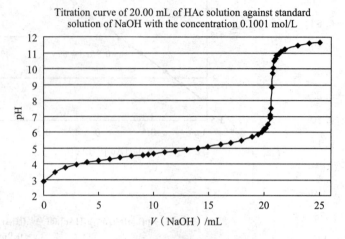

Figure 2 - 15　A best fit curve being smooth and sharp

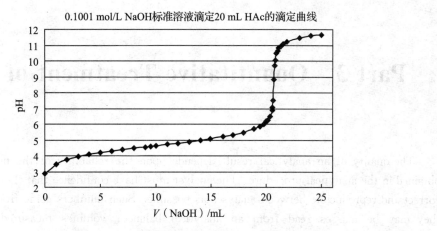

图 2-15　光滑均匀、细而清晰的拟合曲线

Part 3　Quantitative Treatment of Data

　　The quality of an analytical result depends upon the reliability of the numbers used and obtained in the analytical procedure. The analyst must have confidence that these numbers are as correct and reproducible between analyses as possible. Such numbers come from many sources: they may be masses read from an analytical balance; volumes measured from a buret; fundamental constants; or the results of calculations using numbers representing masses, volumes, electrical potentials, light intensities, and other parameters. This section deals with the handling of such numbers in quantitative analysis. First, here are some fundamental things you should realize about uncertainty:

- Every measurement has an uncertainty associated with it, unless it is an exact, counted integer, such as the number of trials performed.
- Every calculated result also has an uncertainty, related to the uncertainty in the measured data used to calculate it. This uncertainty should be reported either as an explicit value or as an implicit uncertainty, by using the appropriate number of significant figures.
- The numerical value of a $\pm$ uncertainty value tells you the range of the result. For example, a result reported as 1.23 ± 0.05 means that the experimenter has some degree of confidence that the true value falls is between 1.18 and 1.28.

3.1　Precision of Instrument Readings and Other Raw Data

　　The first step in determining the uncertainty in calculated results is to estimate the precision of the raw data used in the calculation. Consider three of weighing on a balance in your laboratory:

　　1st weighing of object　　　6.3302 g
　　2nd weighing of object　　　6.3301 g
　　3rd weighing of object　　　6.3303 g

$$\text{The average, or mean, weight of the object} = \frac{\text{sum of values}}{\text{number of values}}$$

$$= \frac{6.3302 \text{ g} + 6.3301 \text{ g} + 6.3303 \text{ g}}{3}$$

$$= 6.3302 \text{ g}$$

　　In this example, the precision or reproducibility of the measurement is ± 0.0001 g. All three measurements may be included in the statement that the object has a mass of (6.3302 ± 0.0001) g. The balance allows direct reading to fourth decimal places, and since the precision is roughly 0.0001 g, or an uncertainty of ± 1 in the last digit, the balance has the necessary sensitivity for this measurement.

　　In the above example, we have little knowledge of the accuracy of the stated mass, (6.3302 ± 0.0001) g. The accuracy of the weighing depends on the accuracy of the internal calibration weights in the balance as well as on other instrumental calibration factors. The stated accuracy of

第三部分　实验数据的处理

分析结果的准确与否取决于在分析过程中使用和得到的数据是否可靠。分析工作者必须对实验数据有把握,保证实验数据是正确的和可再现的。实验数据来自多个方面,可能是从分析天平所得的质量,从滴定管读数获得的体积,也可能是实验常数,或者是由已知的质量、体积、电极电势、光强度或其他参数进行计算的结果。本章节就定量分析中的这类数据进行讨论。首先,我们要熟悉一些关于不确定值的基本概念。

- 每个测定都会有一定的不确定值,除非如实验序号这样的整数。
- 每个计算结果也会有不确定值,这个不确定值与计算所用的实验数据有关。不确定值的表示,既可以将其数值的大小直接报告出来,也可以用合适的有效数字间接反映出来。
- "数值±不确定值"可以反映实验结果的范围。例如 1.23 ± 0.05 表示实验者在某种程度上有把握认为真值在 1.18 和 1.28 范围内。

3.1　仪器读数和其他原始数据的精密度

确定计算结果的不确定值的第一步就是估计所用的原始数据的精密度,考察下面对同一物体的三次称量的数据:6.3302 g,6.3301 g,6.3303 g。

$$\begin{aligned}
被称物的平均值 &= \frac{数值之和}{数值个数} \\
&= \frac{6.3302\ \text{g}+6.3301\ \text{g}+6.3303\ \text{g}}{3} \\
&= 6.3302\ \text{g}
\end{aligned}$$

该例中,测量值的精密度或其再现性是 ±0.0001 g,所有的三个测定值都在(6.3302 ± 0.0001) g 内,天平允许直接读数到小数点后第四位,因此其精密度就大约是 0.0001 g,即最后一位数±1 就是不确定值,这是这个测定中分析天平所必须具备的灵敏度。

该例还没有涉及测量结果(6.3302 ± 0.0001) g 的准确度的知识。称量的准确度取决于天平内部校正砝码以及其他的仪器校正因素。分析天平可以准确到 ±0.0001 g,这可以用校正模式进行检查。

在实验室中还有两种我们常常使用的仪器,这两种仪器的精密度与它们的刻度有关,但又不是其刻度值直接显示的那样。25 毫升的滴定管每 0.1 毫升都有一个刻度,因此也许你会怀疑滴定管的读数是否会精确到 ±0.05 mL,实际上刻度是在一段很长的距离上,在

our analytical balances is ±0.0001 g and this is checked, the balance is put in the calibration mode.

The precision of two other pieces of apparatus that you will often use is somewhat less obvious from a consideration of the scale markings on these instruments.

The 25 milliliter burets used are marked (graduated) in steps of 0.1 mL. Thus you might suspect that readings from a buret will be precise to ±0.05 mL. Actually since the scale markings are quite widely spaced, the space between 0.1 mL marks can be mentally divided into 10 equal spaces and the buret reading estimated to the nearest 0.01 mL. In fact, since every volume needs reading twice, the precision of a buret reading by the average student is probably on the order of ±0.02 mL. Nevertheless, buret readings estimated to the nearest 0.01 mL will be recorded as raw data in your notebook. Similarly, readings of your Celsius (centigrade) scale thermometer can be estimated to the nearest 0.1 ℃ even though the scale divisions are in full degrees.

Every measurement that you make in the lab should be accompanied by a reasonable estimate of its precision or uncertainty.

3.2 Absolute and Relative Uncertainty

Precision can be expressed in two different ways: absolute uncertainty and relative uncertainty.

Absolute precision refers to the actual uncertainty in a quantity. For the example of the three weighings, with an average of (6.3302 ± 0.0001) g, the absolute uncertainty is 0.0001 g.

Suppose that a laboratory buret were drained from an initial volume of 0.55 mL to 22.15 mL. It would *not be correct* to express the volume of solution delivered as 21.6 mL, because the last digit is not uncertain. With a good buret and a careful analyst, the uncertainty in reading a buret is ±0.01 mL. Therefore, the result should be expressed as 21.60 mL, where the last digit is uncertain. It would not be correct to express the volume delivered as 21.602 mL because both of the last two digits are uncertain, and only one uncertain digit should be retained.

The 0.01 mL uncertainty in reading an analytical-quality 25 mL buret is the absolute uncertainty of the measurement. If a number has units, milliliters in this case, the absolute uncertainty has the same units.

The absolute uncertainty in reading a volume from a properly calibrated 25 mL buret is 0.01 mL throughout the measurement range of the buret. Thus an accurately observed 4 mL volume reading on a 25 mL buret should be regarded as having the same 0.01 mL absolute uncertainty as a 20 mL volume reading on the same buret.

The relative uncertainty of a number is defined as:

$$\text{Relative uncertainty} = \frac{\text{Absolute uncertainty of a number}}{\text{Value of the number}} \quad (3-1)$$

Relative uncertainty expresses the uncertainty as a fraction of the quantity of interest. Other ways of expressing relative uncertainty are in percent, parts per thousand, and parts per million. For our example of an object weighing (6.3302 ± 0.0001) g, the relative uncertainty is 0.0001 g/6.3302 g which is equal to 2×10^{-5}. This relative uncertainty can also be expressed as 2×10^{-3} percent, or 2 parts in 100000, or 20 parts per million. Relative uncertainty is a good way to obtain a qualitative idea of the precision of your data and results. The relative uncertainty for a small number is greater than that of an appreciably larger number when both have the same absolute uncertainties, as shown by the following examples.

0.1 mL 间可以分为 10 等分,因此滴定管的读数可以近似到 0.01 mL。事实上,由于每个数据要读两次,那么滴定管读数的精确度就取值±0.02 mL,然而,作为记录本上的原始数据,滴定管读数仍近似到 0.01 mL。同样,温度计的读数可以近似到 0.1 ℃。

总之,我们在实验室中的每一个测定都有一个合理的估值,这个估值由该测定的精密度或不确定值来决定。

3.2 绝对不确定度和相对不确定度

精密度可以用两种不同的方法表示:绝对不确定度和相对不确定度。

绝对精密度表示一个数值的实际不确定值。例如,三个称量值的平均值是(6.3302±0.0001) g,其绝对不确定值就是 0.0001 g。

假如将一支滴定管从初始体积 0.55 mL 放出液体至 22.15 mL,那么放出的溶液的体积用 21.6 mL 表达是不正确的,因为最后一位数 6 是确定值。一支好的滴定管,再加上一位细心的分析师,滴定管的不确定值可以达到±0.01 mL,因此,上述结果应该表示为 21.60 mL,其最后一位数是不确定值,同理,所放出的液体体积若表示为 21.602 mL 也是不正确的,因为它的最后两位数值都是不确定值,而正确的表达只能保留一位不确定值。

25 mL 的滴定管有 0.01 mL 的不确定值,这个数值就叫做测量的绝对不确定值。如果绝对不确定值有单位的话,其单位与原数值单位相同。该例中滴定管体积的单位是毫升,则滴定管体积的绝对不确定值的单位也是毫升。

一支经过校正的 25 mL 的滴定管的绝对不确定值是 0.01 mL,在整个的测定体积范围内这支滴定管的绝对不确定值都是 0.01 mL,因此,只要准确地读出从这支 25 mL 滴定管中放出溶液的体积,无论是 4 mL,还是 20 mL,其绝对不确定值都是 0.01 mL。

相对不确定值可定义为:

$$\text{相对不确定值} = \frac{\text{一个数的绝对不确定值}}{\text{该数的数值}} \tag{3-1}$$

相对不确定值将一个量的不确定值表示为一个分数,此外,还可以用百分数、千分数或百万分数来表示相对不确定值。例如,一个物体重(6.3302±0.0001) g,其相对不确定值是 $0.0001/6.3302 = 2 \times 10^{-5}$,也可以表示为 2×10^{-3}% 或十万分之二或百万分之二十。相对不确定值是一种好的表达方式,它可以表达实验数据和实验结果的精密度的优劣。显然,当绝对不确定值相同时,一个小的数值的相对不确定值要比一个大的数值的相对不确定值大,如下例:

例 1 已知 3.00 mL 体积的绝对不确定值是 0.02 mL,30.00 mL 体积的绝对不确定值是 0.02 mL,这两个体积的相对不确定值是多少?

解:对 3.00 mL 的体积:

$$\text{相对不确定值} = \frac{0.02 \text{ mL}}{3.00 \text{ mL}} = \frac{1}{150} = 0.007 = 0.7\%$$

对 30.00 mL 的体积:

$$\text{相对不确定值} = \frac{0.02 \text{ mL}}{30.00 \text{ mL}} = \frac{1}{1500} = 0.0007 = 0.07\%$$

Example 1 A volume of 3.00 mL is known with an absolute uncertainty of 0.02 mL, and a volume of 30.00 mL is known with an absolute uncertainty of 0.02 mL, what are the relative uncertainties of the two volumes?

Solution for 3.00 mL

$$\text{Relative uncertainty} = \frac{0.02 \text{ mL}}{3.00 \text{ mL}} = \frac{1}{150} = 0.007 = 0.7\%$$

for 30.00 mL

$$\text{Relative uncertainty} = \frac{0.02 \text{ mL}}{30.00 \text{ mL}} = \frac{1}{1500} = 0.0007 = 0.07\%$$

The larger volume has a much smaller relative uncertainty. This is the reason that it is best to measure volumes that are close to the maximum capacity of the buret.

3.3 Accuracy and Precision

Two terms are commonly associated with any discussion of error: "precision" and "accuracy". Precision refers to the reproducibility of a measurement while accuracy is a measure of the closeness to true value. The concepts of precision and accuracy are demonstrated by the series of targets below. If the center of the target is the "true value", then A is neither precise nor accurate. Target B is very precise (reproducible) but not accurate. The average of target C's marks give an accurate result but precision is poor. Target D demonstrates both precision and accuracy-which is the goal in lab.

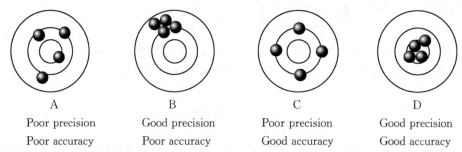

A	B	C	D
Poor precision	Good precision	Poor precision	Good precision
Poor accuracy	Poor accuracy	Good accuracy	Good accuracy

All experiments, no matter how meticulously planned and executed, have some degree of error or uncertainty. In general chemistry lab, you should learn how to identify, correct, or evaluate sources of error in an experiment and how to express the accuracy and precision of measurements when collecting data or reporting results.

3.4 Error

The error of an observation is the difference between the observation and the actual or true value of the quantity observed. Returning to our target analogy, error is how far away a given shot is from the bull's eye. Since the true value, or bull's eye position, is not generally known, the exact error is also unknowable. Errors are often classified into two types: systematic and random.

Systematic Errors (Determinate Errors)

Systematic errors may be caused by fundamental flaws in either the equipment, the observer, or the use of the equipment. For example, a balance may always read 0.0001 g too light because it was zeroed incorrectly. In a similar vein, an experimenter may consistently overshoot the endpoint

较大的体积具有较小的相对不确定值,因此在滴定时我们要尽量使滴定体积接近滴定管最大容量。

3.3 准确度与精密度

在讨论误差时,有两个相关联的术语:准确度和精密度,精密度表示测定结果的再现性,而准确度是测定结果与真实值接近程度的一种度量,精密度和准确度的概念可以用下列一系列的靶图来说明。设靶心是"真实值",那么靶 A 的精密度和准确度都不好,靶 B 精密度好但准确度不好,靶 C 准确度好(四个点均落在靶心周围,其重心在靶心上),但精密度差,靶 D 的精密度和准确度都好,这正是实验者所追求的目标。

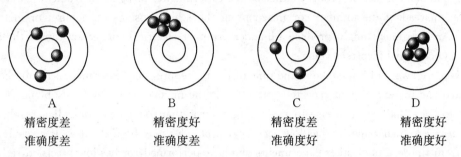

A	B	C	D
精密度差	精密度好	精密度差	精密度好
准确度差	准确度差	准确度好	准确度好

任何实验,无论计划得多缜密,做得多精细,都会有一定的误差。对基础化学实验来说,要学会鉴定、校正和评价误差,还应学会表达实验数据和结果的准确度和精密度。

3.4 误差

误差是实验测得值和实际值即真实值之间的差值。例如,在打靶中,误差可表示为射中的位置和靶心之间的距离,由于我们不可能知道绝对准确的靶心的位置(因为每一个测量和计算都存在某种程度的不确定值),我们也就无法知道绝对误差。误差可分为两类:系统误差和偶然误差。

系统误差(可定误差)

系统误差既可能是由仪器本身的原因、观察不当原因引起,也可能是由操作不当所造成。例如,分析天平由于没有调准零点而总是多读 0.0001 g;类似的,实验人员由于戴的有色眼镜而总是不能正确判断指示剂颜色的变化,造成滴定终点的推迟。有系统误差的实验结果可能精密度好,但准确度差,即使重复多次实验,系统误差通常也不能除去,系统误差对实验结果的影响比较恒定,往往不易发现。由于系统误差的大小值是确定的,可以通过对仪器进行校正、与其他实验方法进行对照等方法对其进行校正。

偶然误差(不可定误差)

偶然误差由一些随机的偶然因素所造成。造成偶然误差的原因常常是实验过程中的微

of a titration because he or she is wearing tinted glasses and cannot see the first color change of the indicator. Systematic errors can result in high precision, but poor accuracy, and usually do not average out, even if the observations are repeated many times. Furthermore, they are frequently difficult to discover.

Random Errors (Indeterminate Errors)

Random errors vary in a completely nonreproducible way from measurement to measurement. However, random errors can be treated statistically, making it possible to relate the precision of a calculated result to the precision with which each of the experimental variables (weight, volume, *etc.*) is known.

Erratic Error

A final type of experimental error is called erratic error or a blunder. These errors are the result of a mistake in the procedure, either by the experimenter or by an instrument. An example would be misreading the numbers or miscounting the scale divisions on a buret or instrument display. An instrument might produce a blunder if a poor electrical connection causes the display to read an occasional incorrect value.

If you are aware of a mistake at the time of the procedure, the experimental result should be discounted and the experiment repeated correctly. If the mistake is not noticed, blunders can be difficult to trace and can give rise to much larger error than random errors. If a result differs widely from a known value, or has low accuracy, a blunder may be the cause. If a result differs widely from the results of other experiments you have performed, or has low precision, a blunder may also be to blame. The best way to detect erratic error or blunders is to repeat all measurements at least once and to compare to known values, if they are available. There are rigorous statistical tests to determine when a result or datum can be discarded because of wide discrepancy with other data in the set.

3.5 Average Deviation and Standard Deviation

In most cases the true value is not known; Whenever several analyses are performed on the same sample and there are no systematic errors, however, experimental result's mean value and the real value are close, it is always possible to express the precision mathematically. The deviation d_i from the mean of a particular result is given by

$$d_i = x_i - \bar{x} \tag{3-2}$$

Average Deviation

The average deviation $\bar{d}$ from the mean is calculated by summing the absolute values of all the individual deviations and dividing by the number of results as given by the equation

$$\bar{d} = \frac{\sum |d_i|}{n} \tag{3-3}$$

The relation average deviation is, of course,

$$\text{Relative average deviation} = \frac{\bar{d}}{x} \tag{3-4}$$

And multiplying this value by 100 and 1000 gives the relative average deviation in percent and parts per thousand, respectively.

小变化，比如质量、体积等，偶然误差是完全不可再现的，是引起分析结果精密度不高的主要原因。偶然误差可以通过统计学方法进行处理，计算结果的精密度与已知的实验测量值，如质量、体积等变量的精密度有关。在同一实验条件下，可以通过增加测量的次数，使偶然误差的平均值趋近于零。

过失误差

最后一种实验误差叫做过失误差或由粗心造成的误差。这种误差由实验过程中的错误操作或过失所产生，它既可以由仪器产生，也可以由操作人员引起。例如，读错数据、算错滴定管的刻度或者是仪器显示错误等。如果仪器接触不良而导致偶然读出的不正确的数据就是仪器所产生的过失误差。

如果我们在实验的过程中发现了影响实验结果的错误时，实验结果将会不可靠，应该重新正确地做实验。但是，如果我们没有注意到错误，失误可能不容易发现，这就可能造成比偶然误差大得多的误差。如果一个实验结果与已知值有较大的出入，即有很低的准确度，则可能产生了过失，如果一个实验结果和其他的实验结果相差很大，即其精密度很低，也可能产生过失。检查偶然或过失误差的最佳方法就是将所有的测量至少重复一次，并将其与已知的数值进行比较(如果有已知结果的话)。当一套数据中有一个数据与其他的数据有很大的差异时，有严格的统计检验法来确定这个结果或这个数据是否要弃去。

3.5 平均偏差和标准偏差

大多数情况下，真实值无法测得(这似乎无法对实验结果的准确度做出判断)。但是，在不存在系统误差时，当对同一个样品进行多次实验，其实验结果的平均值和真实值是很接近的。因此，实验结果的精密度总是可以用数学方法来表示。特定结果的"偏差"可根据平均值进行计算：

$$d_i = x_i - \bar{x} \tag{3-2}$$

平均偏差

来自于平均值的平均偏差 $\bar{d}$ 是用所有单个结果偏差的绝对值相加，再除以实验的测定次数：

$$\bar{d} = \frac{\sum |d_i|}{n} \tag{3-3}$$

当然，
$$\text{相对平均偏差} = \frac{\bar{d}}{\bar{x}} \tag{3-4}$$

该数值乘以 100 和 1000，就分别得到以百分数和千分数表示的相对平均偏差。

> **例 2** 一种酸的物质的量浓度四次的测定值分别是：0.1113，1.1092，1.1120 和 0.1105 mol·L^{-1}，计算结果的平均值、每次结果与平均值的偏差和以千分数表示的相对平均偏差。

Example 2 Four determinations of the amount-of-substance concentration of an acid solution gave values of 0.1113, 1.1092, 1.1120, and 0.1105 mol·L^{-1}, Calculate the mean, the deviation from the mean of each result, and the relative average deviation in parts per thousand.

Solution
$$\bar{x} = \frac{0.1113 + 0.1092 + 0.1120 + 0.1105}{4} = 0.1108$$

x_i	d_i	$|d_i|$
0.1113	0.0005	0.0005
0.1092	−0.0016	0.0016
0.1120	0.0012	0.0012
0.1105	−0.0003	0.0003
	$\Sigma|d_i| =$	0.0036

$$\text{Relative average deviation} = \frac{0.0036/4}{0.1108} \times 1000‰ = 8.1‰$$

Standard Deviation

Standard Deviation is a statistically significant measure of precision for a large set of values. When the true value μ is known, the standard deviation is given the symbol σ and is expressed by the formula

$$\sigma = \sqrt{\frac{\Sigma(x_i - \mu)^2}{n-1}} \tag{3-5}$$

where n and x_i are as defined earlier in this chapter. When the true value is not known, the mean value $\bar{x}$ is substituted for μ, giving s, the standard deviation from the mean, usually simply called standard deviation.

$$s = \sqrt{\frac{\Sigma(x_i - \bar{x})^2}{n-1}} = \sqrt{\frac{\Sigma d_i^2}{n-1}} \tag{3-6}$$

An exactly equivalent form of the above equation that makes calculation easier is

$$s = \sqrt{\frac{\Sigma x_i^2 - (\Sigma x_i)^2/n}{n-1}} \tag{3-7}$$

The formula for relative standard deviation is

$$\text{Relative standard deviation (RSD)} = \frac{s}{\bar{x}} \tag{3-8}$$

This fraction multiplied by 100 gives the relative standard deviation in percent.

Example 3 From the four amount-of-substance concentrations cited, calculate s and the relative standard deviation in percent.

Solution   The value of s can be calculated by substitution into Equation 3−6.

$$s = \sqrt{\frac{d_1^2 + d_2^2 + d_3^2 + d_4^2}{4-1}}$$

$$= \sqrt{\frac{(0.0005)^2 + (0.0016)^2 + (0.0012)^2 + (0.0003)^2}{3}}$$

$$= 1.2 \times 10^{-3}$$

$$\text{RSD} = \frac{1.2 \times 10^{-3}}{0.1108} \times 100\% = 1.1\%$$

解：$\bar{x} = \dfrac{0.1113 + 0.1092 + 0.1120 + 0.1105}{4} = 0.1108$

x_i	d_i	$|d_i|$
0.1113	0.0005	0.0005
0.1092	−0.0016	0.0016
0.1120	0.0012	0.0012
0.1105	−0.0003	0.0003
		$\sum|d_i| = 0.0036$

相对平均偏差 $= \dfrac{0.0036/4}{0.1108} \times 1000‰ = 8.1‰$

标准偏差

标准偏差是对含有大量数据的数据组的精密度的统计描述。已知结果的真实值是 μ，标准误差用符号 σ 表示，σ 可用下式计算：

$$\sigma = \sqrt{\dfrac{\sum(x_i - \mu)^2}{n-1}} \tag{3-5}$$

这里，n 和 x_i 表示测量次数和第 i 次测量结果，当我们不知道真实值时，结果的真实值 μ 用测量平均值 $\bar{x}$ 代替，平均值的标准偏差通常就叫做标准偏差，用符号 s 表示：

$$s = \sqrt{\dfrac{\sum(x_i - \bar{x})^2}{n-1}} = \sqrt{\dfrac{\sum d_i^2}{n-1}} \tag{3-6}$$

式(3-7)与式(3-6)等价，而且更容易计算：

$$s = \sqrt{\dfrac{\sum x_i^2 - (\sum x_i)^2/n}{n-1}} \tag{3-7}$$

相对标准偏差的计算公式是：

$$\text{相对标准偏差(RSD)} = \dfrac{s}{\bar{x}} \tag{3-8}$$

所得分数乘以 100 就得到了以百分数表示的相对标准偏差。

例 3 用例 2 提供的四个物质的量浓度数据，计算 s 和以百分数表示的相对标准偏差。

解：s 的数值可以用公式(3-6)计算，

$$s = \sqrt{\dfrac{d_1^2 + d_2^2 + d_3^2 + d_4^2}{4-1}}$$

$$= \sqrt{\dfrac{(0.0005)^2 + (0.0016)^2 + (0.0012)^2 + (0.0003)^2}{3}}$$

$$= 1.2 \times 10^{-3}$$

相对标准偏差(RSD) $= \dfrac{1.2 \times 10^{-3}}{0.1108} \times 100\% = 1.1\%$

3.6 Significant figure

Concept of Significant Figure

In general, results of observations should be reported in such a way that the last digit given is the only one whose value is uncertain due to random errors. The digits that constitute the result, excluding leading zeros, are then termed significant figure. A brief description of how to use significant figures in calculations is included in the examples, below.

Rule 1 In expressing a number, retain no digits beyond the first uncertain one.

For example the common laboratory analytical balance weighs to the nearest 0.0001 g, Therefore, masses measured on such a balance expressed to the correct number of significant figures are those such as 46.2158 g, 3.8023 g, 0.6714 g, 12.3807 g. Each of these masses has an uncertainty of 0.0001 g; the last digit is the only uncertain one.

Zeros as Significant Digitals

Zeros can cause considerable confusion in the expression of significant figures. Zeros to the left of the first nonzero digit in a number are not significant. Thus, in the number 0.00237, The 237 portion is significant whereas the three zeros are not; therefore, the number has three significant digits. Zeros to the right of any nonzero digit should be considered significant. Thus, in the number 0.084100 there are five significant digits, see Table 3-1.

Table 3-1 Exponential notation and significant digits

Number	Number of significant digits
6.2×10^3	2
6.20×10^3	3
1.2097×10^{-5}	5
9×10^6	1
3.030×10^6	4

Exact Numbers (integers or pure numbers)

Confusion can arise with single-digit numbers that are meant to be exact, but are not followed by any zeros. For example, suppose that a 50 mL pipet was used to withdraw exactly one-fifth of the acid from a 250 mL volumetric flask, and that chemical analysis showed 3.274 mmol of acid in the 50 mL portion. The total amount-of-substance of acid in the 250 mL volumetric flask is given by

$$\text{Total amount-of-substance of acid} = 5 \times 3.274 \text{ mmol} = 16.37 \text{ mmol}$$

Here the 5 is assumed to be 5.000 and does not limit the number of significant figures in the answer.

Rounding off Numbers

Rule 2 In rounding off numbers if the residue (digit to be dropped) is greater than 5, increase the digit to the immediate left of the residue by 1. If the residue is less than 5, leave the last digit to be retained unchanged.

3.6 有效数字

有效数字的概念

一般来说,由于偶然误差的存在,报告测量结果时,其数值只有最后一位数字是不确定的(可疑值)。这种除起首零外,表示测量结果的数字叫有效数字。下面介绍有效数字的使用及其计算。

规则 1 表达一个数值时,只保留一位不确定值,即只有一位可疑数字。

例如,一般实验室的分析天平可精确到 0.0001 g,因此,用天平称量时,正确的有效数字应该表示如下:46.2158 g、3.8023 g、0.6714 g、12.3807 g,每一个质量都有一个不确定值 0.0001 g,最后的一个数是仅有的一个不确定值。

有效数字中的"0"

在表达有效数字时,0 会引起很大的混乱,如果 0 在非 0 数字的左边,就不是有效数字,因此,在数字 0.00237 中,237 部分是有效数字,而 3 个 0 则不是有效数字,故 0.00237 有 3 位有效数字。如果 0 在任何非 0 数字的右边,则认为 0 是有效的,因而 0.084100 有 5 位有效数字。在用指数表达时也要避免混淆,见表 3-1。

表 3-1 指数和有效数字

数 字	有效数字位数
6.2×10^3	2
6.20×10^3	3
1.2097×10^{-5}	5
9×10^6	1
3.030×10^6	4

精确数(整数或自然数)

一个后面没有 0 的精确整数可以引起混乱。例如,假如用一支 50 mL 的移液管从 250 mL 的容量瓶中准确地取出了五分之一的溶液,化学分析表明在这 50 mL 溶液里有 3.274 mmol 的酸,那么在 250 mL 的容量瓶中酸的总物质的量为:

$$5 \times 3.274 \text{ mmol} = 16.37 \text{ mmol}$$

这里假设 5 是 5.000,5 并没有制约答案中的有效数字。

有效数字的修约

规则 2 在修约数字时,如果舍去的数字大于 5,那么舍去这位数字时要向前进位;如果要舍去的数字小于 5,那么不需要向前进位。

For example, the number 32.147 correctly rounded off by dropping the last digit is 32.15.

$$32.147 \rightarrow 32.15$$
↑ Residue greater than 5

For example, The number 7362 correctly rounded off by dropping the last digit is 7.36×10^3.

$$7362 \rightarrow 7.36 \times 10^3$$
↑ Residue less than 5

It would be incorrect to express the rounded number as 7 360 because that could be taken to mean that the last digit, 0, is significant, which is not.

Rule 3 In rounding off numbers, if the residue is exactly 5, leave an even digit to the left of the residue unchanged and raise an odd digit in the same position by 1.

For example, The number 4.865 rounded by dropping the last digit is 4.86.

The number 17035 properly rounded by dropping the last digit is 1.704×10^4.

Rounding Off the Results of Calculations

Rule 4 The result of the addition or subtraction of two or more numbers should retain only as many digits to the right of the decimal point as the number having the fewest such digits. The product or quotient from multiplication or division should have a relative uncertainty approaching as closely as possible the relative uncertainty of the factor with the greatest relative uncertainty.

For example,   + 2.7098
　　　　　　　 −43.213
　　　　　　　 +50.74 ← number with the fewest digits to the right of the decimal point
　　　　　　　 − 0.175 32
　　　　　　　─────────
　　　　　　　 10.061 58 → 10.06
　　　　　　　　　　↑ Result should retain only two digits to the right of the decimal point

For example, $0.0121 \times 25.64 \times 1.05782 = 0.328$

Rounding Logarithms

Rule 5 The logarithm of a number should contain as many significant figures in its mantissa as there are in the number.

For example, $\log(2.71 \times 10^4) = 4.\underline{433}$ ← Mantissa with three significant digits
　　　　　　　　　　　　　　　　　↑
　　　　　　　　　　　　　　Characteristic

$\log(9475) = 3.\underline{976\ 6}$ ← Mantissa with four significant digits
↑
Number with four significant digits

For example, Converse pH = 4.75 into concentration of hydrogen ion H^+, $c(H^+)$

$\log[c(H^+)] = -4.75 = \underline{-5} + \underline{0.25}$ ← Mantissa with two significant digits
　　　　　　　　　　　　↑
　　　　　　　　　Characteristic

例如，将数 32.147 修约，舍弃最后一位数字成为 32.15。

$$32.147 \rightarrow 32.15$$
↑要舍去的数字(大于5)

例如，将数 7362 修约，舍弃最后一位数字成为 7.36×10^3。

$$7362 \rightarrow 7.36\times10^3$$
↑要舍去的数字(小于5)

如果将数字表示为 7360 则是错误的，因为那样就意味着最后一个 0 也是有效数字，但事实上最后一个 0 不是有效数字。

规则3 在修约数字时，如果要修约的数字恰好是 **5**，那么若进位后末位数为偶数时 **5** 则进位，进位后末位数为奇数时 **5** 则舍弃。

例如，将数 4.865 修约最后一位数字成为 4.86。

将数 17035 修约最后一位数字成为 1.704×10^4。

把规则 2 和 3 综合起来，就是"四舍六入五留双"。

有效数值运算结果的修约

规则4 两位或两位以上数字的加减法，计算出来的和或差是由小数点后位数最少（即绝对误差最大的）的数来决定。当进行乘除法运算时，计算结果的积或商则以有效数字最少（即相对误差最大）的数来决定。

例如，　　$+\ 2.7098$
　　　　　-43.213
　　　　　$+50.74\leftarrow$ 该数是小数点后具有最少位数的数
　　　　　$-\ \ 0.17532$
　　　　　―――――――
　　　　　$10.06158 \rightarrow 10.06$
　　　　　　　　　↑结果在小数点后仅保留两位数字

例如，$0.0121\times25.64\times1.05782=0.328$

对数的修约

规则5 对数应该保留该数小数部分数字的位数作为有效数字。

例如，$\log(2.71\times10^4)=4.\underline{433}\leftarrow$ 有 3 位有效数字的尾数
　　　　　　　　　↑
　　　　　　　指数部分
$\log(9475)=3.\underline{9766}\leftarrow$ 有 4 位有效数字的尾数
↑
有 4 位有效数字的真数

$c(H^+) = 1.8 \times 10^{-5}$

↑ Number with two significant digits follows the mantissa (0.25)

These rules of rounding off the result of calculations are based on the ways of accumulation of indeterminate errors in arithmetic operation. It will be discussed in Section 3.8.

3.7 Statistics Applied to Small Data Sets

It should now be apparent that σ and μ are important quantities associated with any measurement and scientists would like to know their values. Because we deal with finite data sets, we can only acquire s and $\bar{x}$, which are estimates of the standard deviation and true value, respectively. However, the reliability of these estimates can be calculated from certain statistical equations.

Confidence Intervals

How "good" the mean is as an estimate of the true value may be answered statistically by defining an interval about the mean, within which one expects to find the true value. The size of the interval is determined in part by the statistical probability (called confidence level) of being correct. If you wish a 100% probability of being correct, the interval would have to be infinitely large to protect against a very large random error, even though such an error is not very likely. Such an interval has no value to the scientist. On the other hand, if we are willing to settle for a finite probability of being correct, we can have a finite confidence interval.

In the absence of determinate error, the confidence interval about the mean, within which the true value lies, is calculated from the equation

$$\mu = \bar{x} \pm \frac{ts}{\sqrt{n}} \tag{3-9}$$

where t is a statistical constant that depends both on the confidence level and the number of measurements involved, the values of t for three different confidence levels are given in Table 3-2.

Table 3-2 Values of t for calculating confidence intervals

Number of observation (n)	Degrees of freedom ($n-1$)	Confidence level		
		90%	95%	99%
2	1	6.31	12.7	63.7
3	2	2.92	4.30	9.92
3	3	2.92	4.30	9.92
4	3	2.35	3.18	5.84
5	4	2.13	2.776	4.60
6	5	2.02	2.57	4.03
7	6	1.94	2.43	3.71
8	7	1.90	2.36	3.50
9	8	1.86	2.31	3.36
10	9	1.83	2.26	3.25
11	10	1.81	2.23	3.17
12	11	1.80	2.20	3.11

例如，将 pH=4.75 转化为氢离子的浓度 $c(H^+)$：

$$\log[c(H^+)] = -4.75 = \underline{-5+0.25} \leftarrow \text{尾数(有 2 位有效数字)}$$
$$\uparrow \text{指数部分}$$

$$c(H^+) = 1.8 \times 10^{-5}$$
$$\uparrow \text{有 2 位有效数位，与对数尾数(0.25)的位数相同}$$

有效数字的运算法则是由测量不可定误差在计算过程中的传递决定的，相关部分的内容详见"3.8 不可定误差的传递"。

3.7 有限数据的统计运用

显然，σ 和 μ 是与测量有关的重要的量，科学家们希望知道其数值。在处理有限的数据组时，我们仅仅可以得到 s 和 $\bar{x}$，s 和 $\bar{x}$ 分别是标准偏差和真值的估计值，然而，这些估计的可靠性可以用统计学的公式来评价。

置信区间

评价一个平均值用于替代真实值是否可靠，可以用统计的方法来回答，这个方法就是，定义一个平均值的区间，并假设真值就在这个区间里。这个区间的大小由真值能出现在该区间的统计概率（即置信水平）确定。如果你希望正确的可能性是100%，这个范围就必须非常大，以防止漏掉一个很大的偶然误差，即使这样的误差不是很可能出现，这样大一个区间对科学家是没什么价值的。另一方面，如果我们希望真实值以某特定的概率出现在该区间，则需要特定的置信区间。

在没有可定（系统）误差时，真值就落在平均值置信区间，平均值置信区间可由下面的方程来计算：

$$\mu = \bar{x} \pm \frac{ts}{\sqrt{n}} \tag{3-9}$$

式中 t 是统计常数，由置信水平和测定次数决定，三种不同置信水平的 t 值见表3-2。

表3-2 用于计算置信区间的 t 值

测定次数 n	自由度 $n-1$	置 信 水 平		
		90%	95%	99%
2	1	6.31	12.7	63.7
3	2	2.92	4.30	9.92
3	3	2.92	4.30	9.92
4	3	2.35	3.18	5.84
5	4	2.13	2.776	4.60
6	5	2.02	2.57	4.03
7	6	1.94	2.43	3.71

续表

Number of observation (n)	Degrees of freedom ($n-1$)	Confidence level		
		90%	95%	99%
13	12	1.78	2.18	3.06
14	13	1.77	2.16	3.01
15	14	1.76	2.14	2.98
20	19	1.729	2.093	2.861
∞	∞	1.645	1.960	2.576

Example 4 Calculator obtained the following results for replicate determinations of calcium in limestone: 14.35%, 14.41%, 14.40%, 14.32%, and 14.37%. Calculate the confidence interval at the 95% confidence level.

Solution The confidence interval is calculated from Equation (3-9)

$$\bar{x} = \frac{\sum x_i}{n} = \frac{14.35\% + 14.41\% + 14.40\% + 14.32\% + 14.37\%}{5} = 14.37\%$$

$$s = \sqrt{\frac{\sum (x_i - \bar{x})^2}{n-1}}$$

$$= \sqrt{\frac{(0.02\%)^2 + (0.04\%)^2 + (0.03\%)^2 + (0.05\%)^2 + (0.00\%)^2}{5-1}}$$

$$= 0.037\%$$

Using $t = 2.776$, for $n = 5$ from Table 3-2 gives us

$$\bar{x} \pm \frac{ts}{\sqrt{n}} = 14.37\% \pm \frac{(2.776)(0.037\%)}{\sqrt{5}} = 14.37\% \pm 0.05\%$$

Comparing values: Tests of significance

Chemists and other scientists often find themselves posing questions about the significance or validity of data that can not be answered with a simple yes or no. The posing and answering of such questions is called *hypothesis testing*. A common procedure called the *null hypothesis* is to state that there is no significant difference between two numbers or sets of numbers, or that a variable exerts no significant effect on the result, and then statistically prove or disprove the statement. It is important to recognize that the proof will be qualified by a confidence level indication the degree of certainty of the proof. Three common questions that can be answered using the null hypothesis approach are:

1. Is the mean of a data set significantly different from the true value?
2. Are the means of two different data sets significantly different?
3. Are the precisions of two different data sets significantly different?

Each of these questions is examined below.

Comparing a mean with a true value A common way to test a method, procedure, instrument, or laboratory technician as a source of determinate error is to perform the determination on a sample whose "true value" is known. How the true value gets known is not important here, but it could be a certified value of a standard sample or the mean of a very large

续表

测定次数 n	自由度 n−1	置信水平 90%	置信水平 95%	置信水平 99%
8	7	1.90	2.36	3.50
9	8	1.86	2.31	3.36
10	9	1.83	2.26	3.25
11	10	1.81	2.23	3.17
12	11	1.80	2.20	3.11
13	12	1.78	2.18	3.06
14	13	1.77	2.16	3.01
15	14	1.76	2.14	2.98
20	19	1.729	2.093	2.861
∞	∞	1.645	1.960	2.576

例4 计算得到了测定石灰石中钙的结果：14.35%、14.41%、14.40%、14.32%和14.37%，计算在置信水平为95%时的置信区间。

解：置信区间由公式(3-9)计算：

$$\bar{x} = \frac{\sum x_i}{n} = \frac{14.35\% + 14.41\% + 14.40\% + 14.32\% + 14.37\%}{5} = 14.37\%$$

$$s = \sqrt{\frac{\sum(x_i - \bar{x})^2}{n-1}} = \sqrt{\frac{(0.02\%)^2 + (0.04\%)^2 + (0.03\%)^2 + (0.05\%)^2 + (0.00\%)^2}{5-1}}$$
$$= 0.037\%$$

由表 3-2 得知当 $n=5$ 时，$t=2.776$。

$$\bar{x} \pm \frac{ts}{\sqrt{n}} = 14.37\% \pm \frac{(2.776)(0.037\%)}{\sqrt{5}} = 14.37\% \pm 0.05\%$$

结果表明：有95%的把握（置信水平）使实验数据在14.37%±0.05%（置信区间）

比较数值：显著性检验

化学家和其他科学家常常要面临这样的问题：实验数据是否有效和有意义，这些问题不可以简单地回答为"是"或"不是"。提出并回答这些问题的方法就叫"假设检验"或"零假设"。一种叫做"否定假设"的一般过程是：先假设两个数据或两组数据之间没有显著性差异，或者实验变量对实验结果没有显著性影响，然后从统计学上证明或反证这种假设。用于证明或反证"假设"的论据要满足一定的置信水平，即，论据要有一定的可信度，认识到这一点是非常重要的。

下面三种常见问题可以用"否定假设"的方法来分析回答：

1. 一组数字的平均值与真值有无明显的差异？
2. 两组不同数据的平均值有无明显不同？

number of careful measurements on the same sample. The test is based on a comparison of the actual difference between the mean and the true value with the largest difference that could be expected as a result of indeterminate error. The latter value is given by $\pm ts/\sqrt{n}$. Thus if $(\bar{x}-\mu) > \pm ts/\sqrt{n}$, the hypothesis that no significant difference exists is rejected.

Example 5 A new procedure for determining trace amounts of zinc in vegetables was evaluated by using it to determine the zinc content of an standard sample, with the following results: 0.083, 0.088, 0.087, and 0.086 ppm Zn. The certified value of the standard sample is 0.082 ppm Zn. Is there a significant difference between the mean of the results and the certified or "true" value?

Solution The first step is to calculate the standard deviation needed to compute the confidence interval, $\pm \dfrac{ts}{\sqrt{n}}$:

$$\bar{x} = \frac{\sum x_i}{n} = \frac{0.083 \text{ ppm} + 0.088 \text{ ppm} + 0.087 \text{ ppm} + 0.086 \text{ ppm}}{4} = 0.086 \text{ ppm}$$

$$s = \sqrt{\frac{\sum(x_i - \bar{x})^2}{n-1}}$$

$$= \sqrt{\frac{(0.003 \text{ ppm})^2 + (0.002 \text{ ppm})^2 + (0.001 \text{ ppm})^2 + (0.000 \text{ ppm})^2}{4-1}}$$

$$= 0.0022 \text{ ppm}$$

From Table 3-2, we find $t = 3.18$ for $n = 4$ at the 95% confidence level; thus

$$\frac{ts}{\sqrt{n}} = \frac{(3.18)(0.0022 \text{ ppm})}{\sqrt{4}} = 0.003 \text{ ppm}$$

Also

$$(\bar{x} - \mu) = 0.086 \text{ ppm} - 0.082 \text{ ppm} = 0.004 \text{ ppm}$$

Since $(\bar{x} - \mu) > \pm ts/\sqrt{n}$, the hypothesis is rejected and the conclusion is that a significant difference does exist.

Comparing two means There are many occasions when chemists wish to determine if two independently obtained results are essentially the same. For example, an oil refiner that purchases a large amount of high-grade crude oil as it is being loaded aboard a tanker may wish to have it analyzed at the time of purchase and again upon its arrival at the refinery to ensure that the oil being delivered is indeed the same oil that was purchased. If n_1 replicate determinations were made during loading and n_2 determinations on arrival, we may write

$$\mu_1 = \bar{x}_1 \pm \frac{ts_1}{\sqrt{n_1}} \tag{3-10}$$

and

$$\mu_2 = \bar{x}_2 \pm \frac{ts_2}{\sqrt{n_2}} \tag{3-11}$$

If we assume that $\mu_1 = \mu_2$ and $\sigma_1 = \sigma_2$, Equations (3-10) and (3-11) can be combined to give

$$\bar{x}_1 - \bar{x}_2 = \pm ts_p \sqrt{\frac{n_1 + n_2}{n_1 n_2}} \tag{3-12}$$

where s_p is the pooled standard deviation of the two data sets and is calculated from the equation

3. 两组不同数据的精密度之间有无明显差异？

这些问题可以用下面的方法检查。

平均值和真值的比较　检验一个实验方法、实验过程、实验仪器或实验技术是否产生系统误差，一般的方法是：对一个已知"真值"的样品进行测定，这里，真值是怎样得到的并不重要，但是，它却是权威机构对一个标准品的测定值，或是对同一个样品经过大量的仔细测定得到的平均值。检验基于用一个偶然误差带来的最大的误差和真值$(\bar{x}-\mu)$之间的差异进行比较。其中，偶然误差的最大值为$\pm ts/\sqrt{n}$，因此，如果$(\bar{x}-\mu)>\pm ts/\sqrt{n}$，则真值与平均值之间就不存在显著性差异。

例5　通过对锌标准品中的锌含量进行测定，来评价测定蔬菜中痕量锌含量的新方法是否可靠。新实验方法所测的结果是：0.083,0.088,0.087 和 0.086 ppm Zn，已知标准品的含量是 0.082 ppm Zn。测定结果的平均值与已知的"真值"之间有显著性差异吗？

解： 计算标准偏差先计算置信区间，$\pm\dfrac{ts}{\sqrt{n}}$：

$$\bar{x}=\frac{\sum x_i}{n}=\frac{0.083\text{ ppm}+0.088\text{ ppm}+0.087\text{ ppm}+0.086\text{ ppm}}{4}=0.086\text{ ppm}$$

$$s=\sqrt{\frac{\sum(x_i-\bar{x})^2}{n-1}}$$

$$=\sqrt{\frac{(0.003\text{ ppm})^2+(0.002\text{ ppm})^2+(0.001\text{ ppm})^2+(0.000\text{ ppm})^2}{4-1}}$$

$$=0.0022\text{ ppm}$$

从表 3-2 中可以查到在置信水平为 95%，$n=4$ 时，$t=3.18$，于是

$$\frac{ts}{\sqrt{n}}=\frac{(3.18)(0.0022\text{ ppm})}{\sqrt{4}}=0.003\text{ ppm}$$

同时

$$(\bar{x}-\mu)=0.086\text{ ppm}-0.082\text{ ppm}=0.004\text{ ppm}$$

因为$(\bar{x}-\mu)>\pm\dfrac{ts}{\sqrt{n}}$，所以，假设不成立，新测定方法不可靠。

比较两组数据的平均值　在许多情况下，化学家希望确定两个独立得到的实验结果在本质上是否是相同的。例如，一个炼油商购买很多高级原油，这些被装到油轮上的油在交货时就分析了成分，他希望到达炼油厂的油与所购买的油的成分是相同的。如果 n_1 是装载时的测定方法，n_2 是到达时的测定方法，则有

$$\mu_1=\bar{x}_1\pm\frac{ts_1}{\sqrt{n_1}} \qquad (3-10)$$

和

$$\mu_2=\bar{x}_2\pm\frac{ts_2}{\sqrt{n_2}} \qquad (3-11)$$

Part 3 Quantitative Treatment of Data

$$s_p = \sqrt{\frac{(n_1-1)s_1^2 + (n_2-1)s_2^2}{n_1+n_2-2}} \quad (3-13)$$

The value for t is based on n_1+n_2-2 degrees of freedom. If

$$(\bar{x}_1 - \bar{x}_2) > \pm t s_p \sqrt{\frac{n_1+n_2}{n_1 n_2}} \quad (3-14)$$

The null hypothesis that no significant difference exists is rejected.

It is not necessary to assume that σ_1 equals σ_2, as this can be tested using the standard deviations, s_1 and s_2, as shown in the next section.

> **Example 6** A ship of copper ore from Chile was purchased by a local metal refiner. The analysis certificate made out while the ship was being loaded, showed that $w(Cu) = 14.66\%$ with a standard deviation of 0.07% for 5 measurements. When the ore arrived at the refinery, it was analyzed with the following results: $w(Cu) = 14.58\%$, 14.61%, 14.69% and 14.64%. Should the refiner accept the ore?
>
> **Solution** To compare the two means, use Equation (3-14). To calculate the pooled standard deviation, we must first calculate s_2.
>
> $$\bar{x}_2 = \frac{14.58\% + 14.61\% + 14.69\% + 14.64\%}{4} = 14.63\%$$
>
> Then
>
> $$s_2 = \sqrt{\frac{(0.05\%)^2 + (0.02\%)^2 + (0.06\%)^2 + (0.01\%)^2}{4-1}} = 0.04\%$$
>
> Using the value of t from Table 3-2 for 7 degree of freedom ($n_1+n_2-2 = 5+4-2$) at the 95% confidence level gives us:
>
> $$\pm t s \sqrt{\frac{n_1+n_2}{n_1 n_2}} = (2.36)(0.04\%)\sqrt{\frac{5+4}{(5)(4)}} = 0.06\%$$
>
> $$\bar{x}_1 - \bar{x}_2 = 14.66\% - 14.63\% = 0.03\%$$
>
> Since $(\bar{x}_1 - \bar{x}_2)$ is not greater than $\pm t s_p \sqrt{(n_1+n_2)/n_1 n_2}$, the hypothesis that on significant difference between means exists is accepted and the refiner should accept the shipment of ore.

Comparing two precisions: The F test It is sometimes desired to determine if the standard deviation, s_1, from one data set is significantly different from the standard deviation, s_2, from another data set. The test is made by comparing the square of each standard deviation.

$$F_c = \frac{s_1^2}{s_2^2} \quad (3-15)$$

If the calculated value, F_c, exceeds a tabulated, statistical value, F_t, the hypothesis of no difference is rejected. An abbreviated compilation of F_t values is given in Table 3-3.

Table 3-3 Values of F_t for comparing variances at the 95% confidence level

Number of observations in denominator	Number of observations in numerator					
	3	4	5	6	7	10	∞
3	19.00	19.16	19.25	19.30	19.33	19.38	19.50
4	9.55	9.28	9.12	9.01	8.94	8.81	8.53

如果假设 $\mu_1=\mu_2$ 和 $\sigma_1=\sigma_2$，联立方程(3-10)和(3-11)：

$$\bar{x}_1-\bar{x}_2=\pm ts_p\sqrt{\frac{n_1+n_2}{n_1n_2}} \quad (3-12)$$

s_p 是两组实验数据的普尔标准偏差，可由下面的方程求出：

$$s_p=\sqrt{\frac{(n_1-1)s_1^2+(n_2-1)s_2^2}{n_1+n_2-2}} \quad (3-13)$$

t 值是以 n_1+n_2-2 为自由度求出的，如果

$$(\bar{x}_1-\bar{x}_2)>\pm ts_p\sqrt{\frac{n_1+n_2}{n_1n_2}} \quad (3-14)$$

则存在显著性差异。即认为二者之间不存在显著性差异的零假设是不成立的。

下例中不需要假设 $\sigma_1=\sigma_2$，直接用标准偏差 s_1 和 s_2 检验。

例6 一家本地的金属提炼厂购买了一船来自于智利的铜矿石，分析报告单是装船时商家做出的，其结果是：$w(Cu)=14.66\%$，标准偏差为 0.07%，测定5次。当矿石到达冶炼厂时，冶炼厂做出的分析结果是：$w(Cu)=14.58\%$、14.61%、14.69% 和 14.64%。问冶炼厂可以接受矿石吗？

解：用方程(3-14)比较两个平均值。为计算普尔标准偏差，先计算 s_2：

$$\bar{x}_2=\frac{14.58\%+14.61\%+14.69\%+14.64\%}{4}=14.63\%$$

$$s_2=\sqrt{\frac{(0.05\%)^2+(0.02\%)^2+(0.06\%)^2+(0.01\%)^2}{4-1}}=0.04\%$$

查表3-2，用置信水平是 95%、自由度为 $7(n_1+n_2-2=5+4-2)$ 时的 t 值

$$\pm ts\sqrt{\frac{n_1+n_2}{n_1n_2}}=(2.36)(0.04\%)\sqrt{\frac{5+4}{(5)(4)}}=0.06\%$$

$$\bar{x}_1-\bar{x}_2=14.66\%-14.63\%=0.03\%$$

因为 $(\bar{x}_1-\bar{x}_2)<\pm ts_p\sqrt{(n_1+n_2)/n_1n_2}$，在两种实验数据的平均值之间没有显著性差异，提炼厂应该接受矿石。

比较两组数据的精密度：F 检验 有时需要比较一组数据的标准偏差 s_1 与另一组数据的标准偏差 s_2 是否有显著性差异。该检验法是通过比较两组数据的方差的比值进行。

$$F_c=\frac{s_{大}^2}{s_{小}^2} \quad (3-15)$$

计算时，规定 $s_{小}^2$ 为分母，$s_{大}^2$ 为分子。如果计算值 $F_c>F_t$，则认为它们存在显著性差异。表3-3列出了 F_t 值。

续表

Number of observations in denominator	Number of observations in numerator						
	3	4	5	6	7	10	∞
5	6.94	6.59	6.39	6.26	6.16	6.00	5.63
6	5.74	5.41	5.19	5.05	4.95	4.78	4.36
7	5.14	4.76	4.53	4.39	4.28	4.10	3.67
10	4.26	3.86	3.63	3.48	3.37	3.18	2.71
∞	2.99	2.60	2.37	2.21	2.09	1.88	1.00

Example 7 The director of a hospital clinical laboratory was trying to decide whether or not to keep a young, recently hired technician. The director decided to see if the new technician's work was of the same quality as that of the other staff. She asked both a senior technician and the new technician to analyze the same sample, using the same procedure, reagents, and instruments. They obtained the following results:

Senior technician	New technician
1.38%	1.28%
1.33%	1.36%
1.34%	1.35%
1.35%	1.40%
1.30%	1.31%

Use the F test to determine if there is a significant difference in the precision of the data.

Solution The standard deviation is a measure of precision. Computing the standard deviation for use in Equation (3-6) gives

$$s_{\text{senior}} = \sqrt{\frac{(0.04\%)^2 + (0.01\%)^2 + (0.00\%)^2 + (0.01\%)^2 + (0.04\%)^2}{5-1}} = 0.029\%$$

And

$$s_{\text{new}} = \sqrt{\frac{(0.06\%)^2 + (0.02\%)^2 + (0.01\%)^2 + (0.06\%)^2 + (0.03\%)^2}{5-1}} = 0.046\%$$

Then

$$F_c = \frac{s_{\text{new}}^2}{s_{\text{senior}}^2} = \frac{(0.046)^2}{(0.029)^2} = 2.5$$

The tabular value, F_t, from Table 3-3 is 6.39. Since F_c does not exceed this value, the hypothesis of no difference is accepted.

Rejecting Data

Occasionally in a set of results we find that one value is much larger or smaller than the others, and we must decide whether to retain or reject the value. Let's follow the case of

表 3-3　置信水平 95% 时的 F_t 值

分母($s_小$)的测定次数	分子($s_大$)的测定次数						
	3	4	5	6	7	10	∞
3	19.00	19.16	19.25	19.30	19.33	19.38	19.50
4	9.55	9.28	9.12	9.01	8.94	8.81	8.53
5	6.94	6.59	6.39	6.26	6.16	6.00	5.63
6	5.74	5.41	5.19	5.05	4.95	4.78	4.36
7	5.14	4.76	4.53	4.39	4.28	4.10	3.67
10	4.26	3.86	3.63	3.48	3.37	3.18	2.71
∞	2.99	2.60	2.37	2.21	2.09	1.88	1.00

例 7　医院检验科的主任在考虑最近聘用的一个年轻化验师的去留问题,主任决定考察一下新化验师和其他化验师工作质量是否一样。她要求新老化验师都用相同的步骤、相同的试剂和相同的仪器分析同样的样品,他们所得到的实验数据如下:

资深化验师	新化验师
1.38%	1.28%
1.33%	1.36%
1.34%	1.35%
1.35%	1.40%
1.30%	1.31%

用 F 检验法检验他们的实验数据是否有显著性差异。

解: 标准偏差是衡量精密度的,用公式(3-6)处理标准偏差

$$s_{老}=\sqrt{\frac{(0.04\%)^2+(0.01\%)^2+(0.00\%)^2+(0.01\%)^2+(0.04\%)^2}{5-1}}=0.029\%$$

和

$$s_{新}=\sqrt{\frac{(0.06\%)^2+(0.02\%)^2+(0.01\%)^2+(0.06\%)^2+(0.03\%)^2}{5-1}}=0.046\%$$

那么

$$F_c=\frac{s_{新}^2}{s_{老}^2}=\frac{(0.046)^2}{(0.029)^2}=2.5$$

由表 3-3 查得 F_t 是 6.39,$F_t > F_c$,故两套数据没有显著性差异。

quantitative analysis student X to see an example of such a situation and how to deal with it. X obtained the following results for $w(\text{Cl})$ in analyzing a laboratory unknown.

Trial	$w(\text{Cl})$
1	12.69%
2	12.58%
3	13.02%
4	12.63%

If he uses all four values, the mean is

$$\bar{x} = \frac{50.92\%}{4} = 12.73\%$$

where if trial number 3 is rejected, the mean is

$$\bar{x} = \frac{37.9\%}{3} = 12.63\%$$

X cannot recall any unique event that occurred during trial 3 that might be responsible for the high value. Although his intuition is to reject the value, X knows that intuition is seldom a good basis for judgment. In preparing for the experiment he remembers reading about the Q test for discordant values. In this test, Q_c is calculated from the equation

$$Q_c = \frac{|\text{questionable value} - \text{nearest numerical value}|}{\text{range}}$$

If Q_c is found to be less than Q_t from a table of statistical constants, the hypothesis of no significant difference is accepted and the value is retained. In X case,

$$Q_c = \frac{|13.02 - 12.69|}{13.02 - 12.58} = \frac{0.33}{0.44} = 0.75$$

The value for Q_t at the 90% confidence level, taken from Table 3-4, is 0.76. Since Q_c is less than Q_t, X is probably better off retaining the value.

Table 3-4 Values of Q_t for rejecting data

Number of observation	Confidence levels		
	90%	96%	99%
3	0.94	0.98	0.99
4	0.76	0.85	0.93
5	0.64	0.73	0.82
6	0.56	0.64	0.74
7	0.51	0.59	0.68
8	0.47	0.54	0.63
9	0.44	0.51	0.60
>9	0.41	0.48	0.57

X's story could go on. Suppose that his grade on this laboratory experiment determines whether he passes or fails the course. With so much at stake, X wisely decides to repeat the

可疑值的取舍(Q 检验法)

有时,一组数据中有一个特别大或特别小的值,我们得决定是保留还是弃去这个值。下面我们来定量分析某学生的实验数据,看看如何解决这个问题。该学生分析未知含量的 Cl 的结果如下:

实 验 编 号	w(Cl)
1	12.69%
2	12.58%
3	13.02%
4	12.63%

如果四个数据都采用,则平均值为:

$$\bar{x}=\frac{50.92\%}{4}=12.73\%$$

如果弃去编号 3,采用其他三个数据,则平均值为:

$$\bar{x}=\frac{37.9\%}{3}=12.63\%$$

该学生想不起来在做 3 号实验时发生了什么特别的事件,使得 3 号实验数据结果偏高,虽然他感觉要去掉 3 号数据,但他也知道仅靠直觉来做事是不行的。想到在准备实验时,他曾看过关于 Q 检验弃去可疑值的有关内容,在这个检验中,可以按下式计算 Q_c:

$$Q_c=\frac{|可疑值-与可疑值最接近的值|}{极差}$$

极差是指最大值与最小值之差,Q_t 为查表 3-4 得到的统计常数,如果 $Q_c < Q_t$,则可疑值就不必弃去,数据应予保留。某同学的 Q_c 计算如下:

$$Q_c=\frac{|13.02-12.69|}{13.02-12.58}=\frac{0.33}{0.44}=0.75$$

查表 3-4,在置信水平 90% 时的 Q_t 是 0.76,$Q_c < Q_t$,故 3 号数据不必弃去。

表 3-4 用于可疑值取舍的 Q_t 值

测定次数	置 信 水 平		
	90%	96%	99%
3	0.94	0.98	0.99
4	0.76	0.85	0.93
5	0.64	0.73	0.82
6	0.56	0.64	0.74
7	0.51	0.59	0.68
8	0.47	0.54	0.63
9	0.44	0.51	0.60
>9	0.41	0.48	0.57

analysis three more times and obtains the following results:

Trial	$w(Cl)$
5	12.81%
6	13.04%
7	12.37%

Indicating that was merely fortuitous that three of the first four results were so closely grouped in value. Using all seven results, he gets a mean of

$$\bar{x} = \frac{89.14\%}{7} = 12.73\%$$

which is the same mean obtained from the first four results.

If time permits, the acquisition of more results is the surest way to a more informed decision. In X's case, it might have turned out that his additional three measurements were 12.61%, 12.66% and 12.60%, which would have clearly established 13.02% as a value that should have been rejected.

3.8 Propagation of Indeterminate Errors

The final result of a determination is always computed from two or more measurements, each of which has an error associated with it. The way in which the individual errors accumulate depends on the type of arithmetic operation performed. It also depends on whether the error is determinate or indeterminate in nature. Since determinate errors are of known direction and magnitude, each individual measurement can be corrected before any arithmetic is performed. The accumulation of indeterminate errors is based on the fact that the individual variances or uncertainties are additive.

Addition and Subtraction

Suppose that a result, R, is to be obtained from the algebraic equation

$$R = A + B - C$$

Where A, B, and C are experimentally measured quantities. If each of these quantities has associated with it an indeterminate error, the equation we really care about is

$$(R \pm r) = (A \pm a) + (B \pm b) + (C \pm c)$$

Where the lower case letters to the error or uncertainty of the measured or calculated value. Statistical theory tells us that the squares of the uncertainties are additive; thus

$$r^2 = a^2 + b^2 + c^2 \quad \text{or} \quad r = \sqrt{a^2 + b^2 + c^2} \tag{3-16}$$

Recalling that indeterminate errors are randomly positive or negative, we must, in any estimate of the overall error, account for the possibility that all of the individual errors were of the same sign (in the same direction). As a result, the uncertainty of C is additive even though C is subtracted from A+B.

For example, Calculate the error in the molecular weight of FeS from the following relative atomic weights: $M(Fe) = 55.847 \pm 0.004$, $M(S) = 32.064 \pm 0.003$

$$r = \sqrt{(\pm 0.004)^2 + (\pm 0.003)^2} = 0.005$$
$$M_W = 87.911 \pm 0.005$$

继续某同学的故事,假如这个实验是该同学的考察性实验,为了保险起见,他多做了三个数据,得到如下结果:

实验编号	$w(Cl)$
5	12.81%
6	13.04%
7	12.37%

数据表明,前4个数据中的3个数据非常接近只是一种偶然,7个数据的平均值为:

$$\bar{x}=\frac{89.14\%}{7}=12.73\%$$

与前4个数据的平均值相同。

如果时间允许,重复更多次的实验不失为更可靠的方法。假如该同学增加的三次实验结果分别为12.61%、12.66%、12.60%,则13.02%就应该弃去了。

3.8 不可定误差(偶然误差)的传递

实验最终的测定结果总是由两个或更多个测定值计算处理而得到,由于每一个测定值都有误差,其误差势必传递到实验最终结果中去。单个测量误差积累的方式不仅取决于计算的类型,也取决于误差在本质上是系统误差还是偶然误差,因为系统误差的正负和大小是可测定的,在计算之前,单个的测定值可以得到校正。不可定误差是单个变化累积起来的,或者说不可定值具有加和性。

加减法

假如一个结果R是从代数方程$R=A+B-C$得到,这里A、B和C是实验测定值,如果每一个测定值都有一个偶然误差(不可定误差),分别为$\pm a$、$\pm b$和$\pm c$,R的误差(不确定值)用$\pm r$表示,则我们真正关心的方程是:

$$(R\pm r)=(A\pm a)+(B\pm b)+(C\pm c)$$

统计学理论告诉我们,分析结果的误差的平方是各测量步骤误差的平方和。故:

$$r^2=a^2+b^2+c^2 \quad \text{或} \quad r=\sqrt{a^2+b^2+c^2} \tag{3-16}$$

偶然误差的正或负值是随机的,在任何总误差的估计中,我们必须考虑单个误差是相同的(方向相同)情况出现的可能性,因此,尽管C误差有可能是被A+B的误差抵消(假如C与A+B误差一正一负),C的不确定值还是要与A、B的不确定值相叠加。

例如,由下列相对原子质量计算FeS摩尔质量的误差,$M(Fe)=55.847\pm0.004$,$M(S)=32.064\pm0.003$:

$$r=\sqrt{(\pm0.004)^2+(\pm0.003)^2}=0.005$$
$$M_W=87.911\pm0.005$$

Multiplication and Division

The error in the result of a multiplication and/or division is calculated in a similar way except that it is the squares of the relative uncertainties that are additive, Hence

$$\left(\frac{r}{R}\right)^2 = \left(\frac{a}{A}\right)^2 + \left(\frac{b}{B}\right)^2 + \left(\frac{c}{C}\right)^2$$

or

$$\frac{r}{R} = \sqrt{\left(\frac{a}{A}\right)^2 + \left(\frac{b}{B}\right)^2 + \left(\frac{c}{C}\right)^2} \qquad (3-17)$$

The absolute error in R can be found by multiplying both sides of Equation by R:

$$r = R\sqrt{\left(\frac{a}{A}\right)^2 + \left(\frac{b}{B}\right)^2 + \left(\frac{c}{C}\right)^2}$$

Example 8 In a titration, the percentage composition analyte is computed from the following equation:

$$\text{analyte \%} = \frac{(cV)_{titrant} \times M_{analyte}}{W_{sample}} \times 100\%$$

Calculate the relative error and the absolute error in the % analyte from the following data.

$$V_{titrant} = (38.04 \pm 0.02) \text{ mL}$$
$$c_{titrant} = (0.1137 \pm 0.0003) \text{ mol/mL}$$
$$M_{analyte} = (74.116 \pm 0.005) \text{ mg/mmol}$$
$$W_{sample} = (800.0 \pm 0.2) \text{ mg}$$

Solution The individual relative errors are

$$\frac{\pm 0.02}{38.04} = \pm 5.2 \times 10^{-4}$$

$$\frac{\pm 0.0003}{0.1137} = \pm 2.6 \times 10^{-3}$$

$$\frac{\pm 0.005}{74.116} = \pm 6.7 \times 10^{-5}$$

$$\frac{\pm 0.2}{800.0} = \pm 2.5 \times 10^{-4}$$

$$\frac{r}{R} = \sqrt{(\pm 5.2 \times 10^{-4})^2 + (\pm 2.6 \times 10^{-3})^2 + (\pm 6.7 \times 10^{-5})^2 + (\pm 2.5 \times 10^{-4})^2}$$

$$= 2.7 \times 10^{-3}$$

The absolute error is given by $r = \frac{r}{R} \times R$

Where

$$R = \text{analyte \%} = \frac{(38.04)(0.1137)(74.116)}{800.0} \times 100\% = 40.07\%$$

Thus

$$r = (2.7 \times 10^{-3})(40.07) = 0.11$$

乘除法

分析结果的相对误差的平方是各测量步骤相对标准误差的平方和。

$$\left(\frac{r}{R}\right)^2 = \left(\frac{a}{A}\right)^2 + \left(\frac{b}{B}\right)^2 + \left(\frac{c}{C}\right)^2$$

或

$$\frac{r}{R} = \sqrt{\left(\frac{a}{A}\right)^2 + \left(\frac{b}{B}\right)^2 + \left(\frac{c}{C}\right)^2} \tag{3-17}$$

R 的绝对误差可以在上述方程的两边同乘以 R 而得到：

$$r = R\sqrt{\left(\frac{a}{A}\right)^2 + \left(\frac{b}{B}\right)^2 + \left(\frac{c}{C}\right)^2}$$

例 8 在滴定中，样品的百分含量可用下式计算：

$$被测物\% = \frac{(cV)_{滴定剂} \times M_{被测物}}{W_{样品}} \times 100\%$$

用下列数据计算样品的相对误差和绝对误差。

$$V_{滴定剂} = (38.04 \pm 0.02) \text{ mL}$$
$$c_{滴定剂} = (0.1137 \pm 0.0003) \text{ mol/mL}$$
$$M_{被测物} = (74.116 \pm 0.005) \text{ mg/mmol}$$
$$W_{样品} = (800.0 \pm 0.2) \text{ mg}$$

解：单个测定的相对误差是：

$$\frac{\pm 0.02}{38.04} = \pm 5.2 \times 10^{-4}$$

$$\frac{\pm 0.0003}{0.1137} = \pm 2.6 \times 10^{-3}$$

$$\frac{\pm 0.005}{74.116} = \pm 6.7 \times 10^{-5}$$

$$\frac{\pm 0.2}{800.0} = \pm 2.5 \times 10^{-4}$$

$$\frac{r}{R} = \sqrt{(\pm 5.2 \times 10^{-4})^2 + (\pm 2.6 \times 10^{-3})^2 + (\pm 6.7 \times 10^{-5})^2 + (\pm 2.5 \times 10^{-4})^2}$$
$$= 2.7 \times 10^{-3}$$

绝对误差是 $r = \dfrac{r}{R} \times R$

此处

$$R = 被测物\% = \frac{(38.04)(0.1137)(74.116)}{800.0} \times 100\% = 40.07\%$$

于是

$$r = (2.7 \times 10^{-3})(40.07) = 0.11$$

Part 4 Experiments

Experiment 1 Checking and Cleaning Equipment

Purpose
1. To learn the general logistics and safety rules in chemlab.
2. To recognize and check the common equipment in chemlab.
3. To learn how to clean the glassware.
4. To know the general sequence to carry on an experiment.

Apparatus and Chemicals
Apparatus: beaker; Erlenmeyer flask; glass rod; dropper; washing bottle; volumetric flask; volumetric pipet; buret; graduated cylinder; thermometer; hotplate; *etc.*

Chemicals: detergent solution; chromic acid cleaning solution; basic alcohol solution.

Procedure
1. Recognizing and Checking the Common Equipment

Recognize every piece of the equipment in your cabinet, and know their names, uses and scales. You should read the marks of volumetric glassware to find out their stated volume and their precisions. Record the names, uses, scales and precisions of the glassware on your notebook. Then check carefully all the labware one by one according to the list of equipment. Take a close look in good light. Look at all parts, especially glass wall, mouth and tip, of the glassware to make sure there aren't any cracks or flaws apparent, and replace the broken one with a new one. Checking equipment before an experiment is essential to a successful lab work.

2. Cleaning Glassware

Take out the equipment from your cabinet. Erase the dust and clean up the cabinet. Then clean every piece of the equipment and put them in order according to their uses, so that it is convenient to fetch them later.

Choose an appropriate method to clean the glassware based on what contaminants, the degree of contamination and the need of the experiment. You can choose an assortment of brushes to clean the glassware based on the sort and scale. Note that:

(1) For the glassware lightly contaminated by water soluble material, rinse with copious volumes of tap water, then rinse 3 – 4 times with distilled water.

(2) For the glassware heavy contaminated, firstly rinse out the contaminants as much as possible with the bulk of tap water, then, add cleaning agent, use the brush to clean the glassware thoroughly, or soak the glassware in the cleaning agent. After cleaning with detergent, rinse the glassware with tap water to get rid of all of the excess soap. Finally, rinse it with distilled water.

Criterion of cleanliness: The surface of the glass is uniform wetted and the water should form

第四部分 实验部分

实验1 仪器的认领和洗涤

实验目的

1. 学习实验室规则和实验室安全守则,了解各项要求的目的和意义。
2. 认识和检查化学实验常用的仪器,明确它们的名称、规格、用途等。
3. 学习一般玻璃仪器的洗涤方法,了解实验用水纯度要求。
4. 了解实验的一般过程和要求,学习如何准备和开始实验。

仪器和试剂

仪器:烧杯;锥形瓶;玻璃棒;滴管;洗瓶;容量瓶;移液管;滴定管;量筒;温度计;电炉等。
试剂:洗涤剂;铬酸洗液;碱性乙醇洗液。

实验步骤

1. 辨认和检查仪器

辨认实验柜中的每一件仪器,知道其名称、用途和规格。对于容量仪器,要查看其刻度,并弄清楚其体积和精密度。将玻璃仪器的名称、用途、规格以及精密度记录在实验记录本上。然后,对照仪器清单,仔细一一检查仪器。要将仪器拿得近一些并对着光线检查。要检查到玻璃仪器的每一个部位,特别是玻璃器壁、口部以及尖嘴,看看有无裂纹和破损,将破损的仪器挑出来,并补充上新的仪器。实验前检查仪器是实验室工作的第一基本功,也是我们做好实验的前提。

2. 洗涤仪器

仪器从实验柜中拿出后,将实验柜里的灰尘仔细抹净,再将仪器一一洗净,根据用途分类放好,仪器的摆放要有条理,以便于日后取用。

根据污垢种类、污染程度以及实验对仪器清洁度的要求选择合适的洗涤方法。刷洗时可选用大小、形状合适的毛刷,毛刷可按所洗涤仪器的类型、规格(口径)大小来选择。注意:

(1) 对于一般轻微污染的仪器,自来水冲洗,蒸馏水荡洗。

(2) 对于污染严重的仪器,先尽可能用自来水将大部分污垢冲走,再加洗涤剂用毛刷进行彻底的清洗,或有针对性地选取合适的洗涤溶液对仪器进行浸泡。用洗涤剂洗过之后,再用自来水冲净,最后蒸馏水荡洗。

a smooth sheath as it runs off of the sides. If there are any spots where a "bubble" forms, you might want to re-clean the glassware.

In addition of tap water and distilled water, detergent solution, detergent powder, chromic acid cleaning solution and basic alcohol are usually used as the cleaning agents in chemlab.

	Principle of decontamination	Contaminant
Detergent solution or detergent powder	Surfactant	Common contaminant
Chromic acid	Oxidant	Contaminant that can be oxidized or dissolved in some acids
Basic alcohol	Base & organic solvent	Contaminant that can dissolved in alkaline solvent

Notes

1. To know requirements for this course and sequence to carry on an experiment.

Preparation of an experiment You must know well the purpose of an experiment beforehand, and inspect whether you gain your aims at last. The pre-lab helps you clear about the experimental objective, principle and procedure. Our experimental purposes include two categories: one deals with the experiment and the other relates to the experimental techniques and methods, for example, to learn a method and to validate a theory, or to learn a technique and a way of thinking. You should do your best to realize the purposes, to comprehend the principles and be clear-headed before you enter the lab. Write the pre-lab after reading the lab materials. The pre-lab should be written on a specially prepared notebook. The pre-lab could be regarded as a kind of design of experiment, not a copy of the textbook. A comprehensive pre-lab is based on your understanding of the lab material, and is concise and methodical to allow the reader understand the experiment at a glance. You could leave some blank to record the observation and measurement in a pre-lab.

Execution of an experiment You should take the time before the lab lecture to check the apparatus and reagents used in this experiment when you enter the lab. List the glassware and the equipment on the lab beach. After the lab lecture, you should carry out the experiment on your own. In this period, you must follow the safety rules at all times. The operations need patience and carefulness, and orderliness. Record all the observation and the measurements directly in your notebook and think about if the results are expected. After the experiment finished, clean up your area and make sure you have cleaned up your equipment and chemicals before you can leave. The students who are on the duty should clean the lab room at last.

Writing of a lab report The goal of lab reports is to document your findings and communicate their significance. The evaluation of any experimental work is based primarily on the contents of a written report. A good lab report does more than present data; it demonstrates the writer's comprehension of the concepts behind the data. Merely recording the expected and observed results is not sufficient; you should also identify how and why differences occurred, explain how they affected your experiment, and show your understanding of the principles that the experiment was designed to

洗净标准:玻璃表面被水均匀润湿,将洗净后的仪器倒立时,水能在玻璃表面形成一层均匀的水膜沿器壁流下。如器壁有附着水珠或有油斑,则应予重洗。

在实验室里,除必备的自来水和蒸馏水外,常用的洗涤液有洗涤剂(洗衣粉)、去污粉、铬酸洗液以及碱性乙醇溶液。

	去污原理	污 染 物
洗涤剂溶液或洗衣粉	表面活性剂	一般性污染以及一些有机污物
铬酸酸性洗液	强氧化性	能被氧化或被酸溶解的污物
氢氧化钠的乙醇溶液	碱性+有机溶剂	能被碱性有机溶剂溶解的污物

注意事项

1. 了解本课程的基本要求及实验的基本过程。

实验准备 每个实验都有其实验目的,实验前要先弄清楚,实验后再检查一下是不是达到了目的。实验目的通常有两方面:一是就这个实验本身提出的具体目的;另外一方面是就实验一般技术和方法提出的要求,比如,学习某种实验方法、印证某种理论,或通过这个实验掌握一定的实验技术、思考方法等等。在进入实验室之前,要力求做到目的明确、理论透彻、思路清楚。在预习的基础上,写好预习报告。预习报告应写在专门准备好的实验记录本上。预习报告相当于设计的实验方案,不要照抄实验教材,要根据自己的理解来写。预习报告要条理清楚、简洁,使实验的内容一目了然,同时,要留一些空白以备填入实验现象和结果。

进行实验 进入实验室后,应利用课前的时间将实验仪器、药品准备好,有关的仪器应整齐地摆放在实验台上。做实验时,应注意安全,独立完成。要有耐心,实验操作时要小心谨慎、有条不紊。要及时将全部实验现象和测得的数据直接记录在记录本上,并思考这些结果与预计的是否一致。实验完毕后,要清理仪器和药品、打扫实验台,做好个人实验清洁之后方可离开实验室。值日生要打扫实验室,做好值日。

写实验报告 写实验报告的目的是为了报告你的实验发现、与他人交流这些发现的意义。人们评价一个实验工作的优劣首先要根据实验报告的内容。好的实验报告不仅仅要报告实验数据,而且要报告作者对这些实验数据背后隐藏的规律的理解。仅仅记录预计的和实际的实验结果是不够的,实验报告还应该写出并解释实际结果与预计结果有什么不同,说明影响实验结果的因素是如何影响实验结果的,以及讨论你对实验设计原理的理解。写实验报告要有条有理。

2. 阅读材料:本教材第一部分的相关内容,包括本课程教学目标、实验的基本要求、实验诚信原则、实验室安全规则、常用实验仪器、仪器的洗涤。反复阅读第一部分的相关内容,仔细理解各项的要求和意义。

3. 洗涤玻璃仪器时,外壁也需要洗净,否则不容易观察内壁是否洗净。一般情况下,已洗净的仪器不能用布或纸巾擦仪器的内壁。用蒸馏水润洗仪器时要遵循"少量多次"原则。

4. 铬酸洗液可以反复使用,直至颜色变绿[$Cr(VI) \rightarrow Cr(III)$]而失效,但在重复使用时要避免被水稀释。铬酸洗液具有强氧化性,其中的浓硫酸具有强烈的腐蚀性,$Cr(VI)$有毒,

examine. You need to organize your ideas carefully and express them coherently.

2. Reading materials: Learning Objectives of this course, General guidelines, Honor Principle, General Safety Rules, General Equipments, Cleaning Laboratory Glassware. Read the related material repeatedly to understand the aim and importance of these chemistry lab requirements.

3. The outside wall of glassware should be also cleaned otherwise it is difficult to judge if the inside wall is clean. Do not wipe the cleaning inside wall with a towel or a piece of tissue paper. Rinsing the glassware with distilled water should be "several times and a little amount per time".

4. Chromic acid cleaning solution can be used repeatedly, but it should be to avoid the solution diluted, until the color changes from orange to green [Cr(VI)→Cr(III)]. Chromic acid cleaning solution is a strong oxidant and a strong corrosive. Be careful to avoid spilling it. Chromium (VI) is highly toxic (mutagenic, carcinogenic). Do not dispose chromic acid down the sink.

Preparation of chromic acid cleaning solution: Weigh 6.5 g sodium dichromate dehydrate, put it into a 250 mL beaker and dissolve it with 20 mL of water. Measure 80 mL of technical grade 98% sulfuric acid with a dry graduated cylinder. Pour the concentrated sulfuric acid into the beaker when carefully stir the solution gently. As long as the stirring is gentle and continuous, the solution will become quite warm. Allow to cool before storing in a glass-stopper reagent bottle.

Questions

1. What cleaning method is suited to the following contaminated glassware?
(1) The glassware adhered MnO_2 on the wall;
(2) The tube coated with a silver mirror film;
(3) The flask coated with grease.

2. What is the difference of the precisions of a 100 mL flask, a 100 mL graduated cylinder and a 100 mL volumetric flask? How do you select the apparatus to prepare a solution of NaOH of 100 mL with $0.1 \text{ mol} \cdot \text{L}^{-1}$?

具有致突变、致癌作用,因此在使用铬酸洗液时,要注意安全,避免洒落,洗涤废液不能随意倒入下水道中。

铬酸洗液的配制:称取 6.5 g 重铬酸钾($K_2Cr_2O_7$)放在一个 250 mL 烧杯中,加入20 mL 水溶解。用干燥量筒量取 80 mL 98% 工业硫酸(浓 H_2SO_4)缓缓倾入烧杯中,并不断用玻璃棒搅拌。继续缓慢搅拌,溶液会很热,待冷至室温,把它转移到具塞玻璃瓶中保存。

思 考 题

1. 下列污物污染玻璃仪器时,用什么洗涤方法?
(1) 附着在瓶壁上的二氧化锰;
(2) 附着在试管上的银镜(金属银);
(3) 在烧瓶中的油污。

2. 100 mL 烧瓶、100 mL 量筒、100 mL 容量瓶的精密度有什么不同?如果需要配制 0.1 mol·L^{-1} NaOH 溶液 100 mL,如何选用这些仪器?

Experiment 2 Weighing Exercise and Preparing Solutions

Purpose

1. To learn the use of an electronic analytical balance.

2. To grasp the weighing methods: direct weighing; weighing by difference; weighing by taring.

3. To grasp the methods of preparing solutions, and to know the precision of measuring apparatus and the significant digits.

4. To learn how to use the lab notebook correctly.

Principle

1. A balance is based on the principle of the lever. Analytical balances are more accurate and precise instruments used to measure masses than table balances. The common analytical balance measures masses to within 0.0001 g. The electronic analytical balances have convenient weighing performance, ease-of-use with touch technology, and comprehensive connectivity options. Today the electronic analytical balances are essentials in many labs.

2. The desired precision of concentration of a chemical depends on the need of the experiment. For example, the requirements are different for the precision to the solution of sulfuric acid used in adjusting pH and used in quantitative determination, so the methods to prepare the solutions should satisfy the precision. If the less accurate solution is prepared, weighing the material by a table balance with the precision 0.05 g and measuring the water by a graduated cylinder will meet the requirement and improve the experiment's efficiency. However, preparing solutions accurately involves weighing a precise amount of dry material or measuring a precise amount of liquid, an analytical balance and a volumetric flask may be used.

Apparatus and Chemicals

Apparatus: electronic analytical balance; table balance; weighing bottle; beaker, 50 mL or 100 mL, 250 mL; volumetric flask, 100 mL.

Chemicals: $K_2Cr_2O_7$ (FW 294.2 g/mol), solid.

Procedure

1. Weighing Exercise

Check the electronic analytical balance to be sure the balance pan is cleaned and check the bubble to be sure the balance is level before turning the balance on.

Direct weighing Place a small dry beaker on the pan of the balance, and read its weight directly.

Weighing by taring Weigh the weighing container (weighing paper, or weighing bottle, small beaker and so on) alone on the balance, then add sample with a clean spatula. Weigh the sample plus weighing container. Subtract the weights to obtain the weight of the sample. The weighing is simplied by using the key of "tare" when the electronic balance is applied. You can read directly the weight of the sample in the container when you previously tare the container. The method of weighing by taring is only valid for samples that are stable in the air on standing. Weigh accurately three samples of $K_2Cr_2O_7$ of 0.2 g. The weight must be in the range of $(0.2 \pm 0.2 \times 10\%)$ g.

Weighing by difference The sample in the weighing bottle is weighed and then a portion is

实验 2　称量练习和溶液的配制

实验目的

1. 学会电子分析天平的使用方法；
2. 学会直接称量法、减重称量法和固定称量法；
3. 掌握配制溶液的方法，了解测量仪器的精密度、有效数字的意义；
4. 学习如何作正确规范的实验记录。

实验原理

1. 天平是基于杠杆原理的仪器。分析天平的质量称量结果要比托盘天平准确和精密。普通分析天平可以精确称至 0.0001 g。电子分析天平的操作简便，当被称物放在秤盘上后，几乎立即就能用数字显示出质量数值，它还可以与打印机、计算机、记录仪等联用。如今，电子分析天平是许多实验室必备的实验仪器。

2. 实验中有时对不同溶液的浓度精密度有不同要求，例如，调节酸度用的硫酸溶液与定量分析中用的标准溶液在浓度精密度要求上是不同的，在配制不同浓度精密度要求的溶液时要按不同方法进行配制。配制一般浓度精密度要求不高的溶液时，用托盘天平称量样品、用量筒量取水的体积，就可以满足实验的要求，而且这样操作简便，能提高实验的效率。而配制浓度精密度要求高的溶液时，要求实验时所称取的质量和体积皆精确，这就要用到分析天平、容量瓶等仪器。

仪器和试剂

仪器：电子分析天平；托盘天平；称量瓶；烧杯，50 mL 或 100 mL，250 mL；容量瓶，100 mL。

试剂：$K_2Cr_2O_7$（摩尔质量为 294.2 g/mol），固体。

实验步骤

1. 分析天平称量练习

称量前应先检查天平是否水平，秤盘是否清洁。

直接称量　直接将需要称量的物品，如一个干燥小烧杯，放到秤盘上读数即可。

固定称量　先将空的称量容器（如称量纸、称量瓶或小烧杯等）放在天平秤盘上称量，再用干净的药匙把药品加入到称量容器中，称得样品加容器的总质量，而样品的质量等于总质量减去空的容器的质量。使用电子天平时，可用"除皮键"简化称量过程。电子分析天平有除皮功能，因此用电子天平称量时，可以先将称量容器除皮，则加在容器中的样品质量可以直接从天平显示器读出，而不需要另外计算。固定称量法适用于称量在空气中稳定的样品。准确称取 0.2 g $K_2Cr_2O_7$ 试样三份，要求称样量在 (0.2±0.2×10%) g 范围内。

减重称量　将样品装在称量瓶中，先称样品加称量瓶的总质量；之后，将样品从称量瓶中移出一部分（如用敲击的办法），移出的部分要定量转移到容器中；再将称量瓶和剩余样品

removed (e. g., by tapping) and quantitatively transferred to a vessel. The weighing bottle and sample are reweighed and from the difference in weight, the weight of sample is calculated. Similarly, the weighing procedure is simplied by using the key of "tare" when the electronic balance is applied. The method of weighing by difference is especially appropriate for samples that absorb water or CO_2 from the air on standing. Weigh accurately three samples of $K_2Cr_2O_7$ of 0.2 g by difference. The weight must be in the range of $(0.2 \pm 0.2 \times 10\%)$ g.

2. Preparing Solutions

Preparing Solution A: 100 mL $K_2Cr_2O_7$ solution with the concentration 0.1 mol·L^{-1} Weigh 2.9 g of $K_2Cr_2O_7$ by a table balance and place it into a 250 mL beaker, then measure 100 mL of distilled water by a graduated cylinder and add the water to the beaker, stir the mixture until the solid of $K_2Cr_2O_7$ is dissolved.

Preparing Solution B: 100 mL $K_2Cr_2O_7$ solution with the concentration 0.1000 mol·L^{-1} Weigh accurately 2.942 g of $K_2Cr_2O_7$. Pour it into a small beaker, add about 20 mL of water, stir the mixture until the potassium dichromate is dissolved completely, then transfer the solution completely (quantitatively) to a 100 mL volumetric flask, finally add water to bring the volume up to the mark, then stop the flask with the stopper; turn it up and down to mix the solution.

Data records

a. Weight of the small beaker $m_{beaker} =$

b. Weighing the samples of $K_2Cr_2O_7$ by the taring method (not using the key of "tare")

Trial number	I	II	III
Wt. of the container m_1/g			
Wt. of ($K_2Cr_2O_7$ & container) m_2/g			
Wt. of the sample of $K_2Cr_2O_7$/g			

c. Weighing the samples of $K_2Cr_2O_7$ by the difference method (not using the key of "tare")

Trial number	I	II	III
Wt. of (bottle & sample) before removing m_1/g			
Wt. of (bottle & sample) after removing m_2/g			
Wt. of the sample removed (m_1-m_2)/g			

d. Concentration of the solutions prepared

Trial number	A	B
Wt. of $K_2Cr_2O_7$/g		
Vol. of the solution/mL		
Conc. of the solution/mol·L^{-1}		

称重,则被移出样品的质量等于这两个质量之差。同样,使用电子天平时可用"除皮键"简化称量过程。使用电子分析天平的除皮功能,在从称量瓶中倾倒样品之前,可以先将"称量瓶+样品"除皮,则被倒出来的样品质量可以直接从天平显示器读出。减重称量法特别适合于在空气中不稳定的样品,如易吸潮、吸CO_2的样品。用减重法准确称取 0.2 g $K_2Cr_2O_7$ 试样三份,要求称样量在$(0.2\pm0.2\times10\%)$ g 范围内。

2. 溶液的配制

配制 100 mL 浓度为 0.1 mol·L^{-1} 的 $K_2Cr_2O_7$ 溶液 A　用托盘天平称取 $K_2Cr_2O_7$ 固体 2.9 g,置于 250 mL 烧杯中,用量筒量取 100 mL 蒸馏水加入,搅拌至溶解。

配制 100 mL 浓度为 0.1000 mol·L^{-1} 的 $K_2Cr_2O_7$ 溶液 B　用分析天平准确称取 $K_2Cr_2O_7$ 固体 2.942 g,置于小烧杯中,加约 20 mL 蒸馏水搅拌至溶解,将溶液完全的、定量的转移至 100 mL 容量瓶中,最后将容量瓶定容至刻度线,塞紧瓶塞,上下颠倒摇动使溶液混合均匀。

实验数据记录

a. 小烧杯的质量　$m_{烧杯}=$

b. 固定称量法称取 0.2 g $K_2Cr_2O_7$ 试样(此表格适用于未使用"除皮键")

序　号	I	II	III
器皿质量　m_1/g			
($K_2Cr_2O_7$+器皿)质量　m_2/g			
$K_2Cr_2O_7$ 的质量/g			

c. 减重法称取 0.2 g $K_2Cr_2O_7$ 试样(此表格适用于未使用"除皮键")

序　号	I	II	III
(称量瓶+试样)的质量(倾出前)　m_1/g			
(称量瓶+试样)的质量(倾出后)　m_2/g			
倾出试样的质量　(m_1-m_2)/g			

d. 所配制溶液的浓度

序　号	A	B
溶质 $K_2Cr_2O_7$ 的质量/g		
溶液的体积/mL		
溶液浓度/mol·L^{-1}		

Notes

1. Reading materials Part 2 and Part 3 in the textbook: Digital balance and weighing, significant digit, graduated cylinder, volumetric flask, notebook.

2. Do not move the balance after zeroing. Do not bump or place objects on the bench.

3. Do not place any chemical directly onto the balance pan. Mass powders on paper or dishes. Handle objects with tongs, tweezers, gloves, or paper to prevent fingerprints.

4. Let hot objects cool before massing.

5. Always close the chamber door when reading.

6. Mass hygroscopic materials rapidly since they will absorb water during massing or by difference method.

7. When making repetitive massing always is used the same procedure.

8. DO NOT go off and leave spilled chemicals on or around the balance! Report any spill to the instructor so that she may clean it up in a proper manner, re-calibrating if necessary.

9. When massing or preparing solutions the proper apparatus are used, e. g. , an analytical balance or a table balance, a volumetric flask or a graduated cylinder, according to the precision of the experiment.

Questions

1. Why must the chamber always be clean dry? Why must the weighing container be kept clean and the out wall of the container clean and dry in weighing?

2. Why should the chamber door be close when reading?

3. What is the effect when a little of sample is spilled outside of the weighing container? What is the effect using a spoon to take the sample out from the weighing bottle when weighing by difference?

4. How do you use significant figures in a weighing calculation?

注意事项

1. 阅读材料:本教材第二和第三部分的相关内容,包括天平及称量、有效数字、量筒、容量瓶、实验记录。
2. 当天平调水平之后,不要挪动天平,也不要用重物撞击天平台。
3. 称量时,不得将需称量的药品、试剂直接放在天平秤盘上。用于称量的器皿一定要清洁、干燥。要用称量纸或器皿称量粉末样品,不要用手直接接触被称量物品,要使用镊子或戴手套或用纸,防止手上污物沾到被称量物上。
4. 热的样品放冷后才能在天平上进行称量。
5. 读取称量数据时一定要关上天平门。
6. 称取易吸水的样品时要快速,尽量采用减重法称量。
7. 平行称量时,要采用相同的称量过程。
8. 称量时不要将药品洒在桌面,更不要将药品洒到天平里或天平周围!如果不慎将药品洒在天平里,要报告老师处理。
9. 称量和配制溶液时,要根据实验的要求选择不同精密度的仪器(分析天平或台秤、量筒或容量瓶)。

思 考 题

1. 为什么在称量时,天平箱内必须保持清洁、干燥?称量用的器皿为什么也必须保持清洁、干燥?
2. 为什么读数时天平箱的门必须是关闭的?
3. 减重称量时,如果样品洒落在小烧杯外面时,对称量结果有什么影响?用药匙将称量瓶中的样品取到小烧杯里进行减重称量,对称量结果有什么影响?
4. 在称量记录和计算中,如何正确运用有效数字?

Experiment 3 Determination of Crystal Water in Barium Chloride

Purpose
1. To practice the use of analytical balance.
2. To grasp the method and principle of determination of water by volatilizatical gravimetry.
3. To know the meaning of constant weight.

Principle
The evaporating pressure of $BaCl_2 \cdot 2H_2O$ is 0.17 kPa at 20 ℃ and 1.57 kPa at 35 ℃. So the crystal water in $BaCl_2 \cdot 2H_2O$ generally is stable unless the air is very dry. But $BaCl_2 \cdot 2H_2O$ will lost its crystal water at 113 ℃ to form anhydrous $BaCl_2$, which is nonvolatilizable and nondecomposable. The experiment can be carried out at a higher temperature than 113 ℃.

$$\text{Theoretical content of crystal water }(\%) = \frac{2M_{H_2O}}{M_{BaCl_2 \cdot 2H_2O}} \times 100\%$$

$$= \frac{2 \times 18.05}{244.27} \times 100\% = 14.78\%$$

$$\text{Practical content of crystal water }(\%) = \frac{\text{weight lost (g)}}{\text{sample weight (g)}} \times 100\%$$

Apparatus and Chemicals
Apparatus: analytical balance; weighing bottles; dry oven; crucible clamp; desiccator.
Chemicals: $BaCl_2 \cdot 2H_2O$ (A.R).

Procedure
1. Drying of the weighing bottles to constant weight

After cleaning three lower weighing bottles with 3 cm diameter, put them into the dry oven at 115 ℃, then transfer the bottles to the desiccators by crucible and cool to room temperature. Weigh the bottles accurately, and record the readings. Reheat and reweigh them until successive weighings agree to within 0.3 mg.

2. Determination of the water of crystallization of the sample

Grind the sample of $BaCl_2 \cdot 2H_2O$ into powder. Weigh accurately about 1 g of the $BaCl_2 \cdot 2H_2O$ powder and add it to the bottle with constant weight. Spread the powder to be a thin layer lower than 5 mm. Then heat the sample in the oven at 115 ℃ for 1 hour. Allow it to cool for a few minutes and then place it in the desiccator. When it has reached room temperature weigh the bottle accurately, and then put it back into the oven. Reheat and reweigh it until the sample reaches constant weigh.

实验 3　氯化钡结晶水的测定

实验目的

1. 通过本实验进一步巩固分析天平的使用。
2. 掌握挥发重量法测定水分的原理和方法。
3. 掌握恒重的意义。

实验原理

$BaCl_2 \cdot 2H_2O$ 中结晶水的蒸气压，20℃时为 0.17 kPa(1.3 mmHg)，35 ℃时为 1.57 kPa (11.8 mmHg)。所以氯化钡除在特别干燥气候中外，一般情况下所含两分子结晶水十分稳定。$BaCl_2 \cdot 2H_2O$ 于 113℃失去结晶水，无水氯化钡不挥发，也不易变质。故干燥温度可高于 113 ℃。

$BaCl_2 \cdot 2H_2O$ 结晶水含量的计算：

$$理论结晶水含量(\%) = \frac{2M_{H_2O}}{M_{BaCl_2 \cdot 2H_2O}} \times 100\%$$

$$= \frac{2 \times 18.05}{244.27} \times 100\% = 14.78\%$$

$$实际结晶水含量(\%) = \frac{失重}{W_{样品(g)}} \times 100\%$$

仪器和试剂

仪器：分析天平；称量瓶（扁平型）；电热干燥箱；坩埚钳；干燥器。
试剂：$BaCl_2 \cdot 2H_2O$ 样品。

实验步骤

取直径为 3 cm 的扁型称量瓶 3 个，洗净，于电热干燥箱中，在 115℃干燥后，置于干燥器中放冷至室温（30 min）后，称重。重复上述条件，再烘，放冷，称重。至连续两次干燥后的重量差小于 0.3 mg 为达到恒重要求。

将 $BaCl_2 \cdot 2H_2O$ 样品在研钵中研成粗粉，分别精密称取 3 份试样，每份约 1 g，置已恒重的称量瓶中，使样品平铺于瓶底（厚度不超过 5 mm），称量瓶盖斜放于瓶口，置称量瓶于电热干燥箱中 115 ℃干燥 1 h，移至干燥器中，盖好称量瓶盖，放置 30 min，冷至室温，称其重量。再重复上述操作，直至恒重。

取平行操作 3 份的数据，求出结晶水的百分含量，计算平均值及相对标准偏差。

Data and Results

Table 1 Determination of the crystal water in BaCl$_2$ · 2H$_2$O

No.		1	2	3
Wt. of weighing bottle being dried/g	First			
	Second			
	Third			
Wt. of sample & bottle before dried/g				
Wt. of sample & bottle being dried/g	First			
	Second			
	Third			
Const. Wt. of sample/g				
Wt. of cryst. water/g				
Cryst. water/%				
Averg. cryst. water/%				
RSD/%				

This table should be simpled when you tare the weighing bottle with using the key "tare" on the electron balance.

Notes

1. Know about the parallel principle in drying a sample to constant weight about cool time, temperature, and so on.

2. Weigh sample quickly to avoid absorbting water during weighing.

3. Before using the desiccator, change the old desiccant with fresh desiccant. Use the desiccator properly. To open and close the desiccator, slide the lid sideways with a steady pressure. Never jerk or lift the lid upwards. Carry the desiccator with one hand on the lid, to prevent it from sliding off.

4. In order to evaporating utterly, the sample should be put a thin layer in the bottom of the bottle.

5. Before closing the desiccator, leave the desiccator lid ajar for about 10 minutes, so that a partial vacuum will not form in the desiccator after the air inside has cooled.

Questions

1. Why should the sample be grinded into powder? Is it "the finer, the better"?

2. What is the meaning of constant weight and how to heat a sample to constant weight?

3. Why must the empty bottle be constant weight before it is loaded?

4. If we add 1 g BaCl$_2$ to the desiccator and use the following drying agents to absorb the water in sample, which one can make the sample full dry?

(1) 2 g NaOH;

(2) 3 g CaCl$_2$ + 5 g CaCl$_2$ · H$_2$O;

(3) 10 g CaBr$_2$ · 6H$_2$O.

实验数据与结果

表 1　$BaCl_2 \cdot 2H_2O$ 中结晶水含量测定

编　　号		1	2	3
空称量瓶恒重/g	第一次干燥			
	第二次干燥			
	第三次干燥			
（干燥前样品＋瓶）重/g				
（样品＋瓶）恒重/g	第一次干燥			
	第二次干燥			
	第三次干燥			
（干燥恒重后样品）重/g				
结晶水重量/g				
结晶水百分含量/%				
结晶水平均百分含量/%				
相对标准偏差/%（RSD/%）				

注：电子分析天平有除皮"tare"功能，因此用电子天平测定时，可以对恒重后的称量瓶进行除皮，直接读出干燥前后以及干燥过程中的"样品"重，而不需要读"样品＋瓶"重，则以上表格可以简化。

注 意 事 项

1. 对于要求恒重的称量，应注意平行原则，即同批次样品的实验条件，如扁型称量瓶（或加样品后）在烘箱中干燥温度、在干燥器中冷却时间等，应保持一致。

2. 在称扁型称量瓶与样品时，要盖好扁型称量瓶盖子，称量速度要快，以免称量过程中样品吸湿。

3. 正确使用干燥器，要注意干燥器中干燥剂是否失效。干燥器磨口处涂的凡士林应薄而均匀，打开盖子应采用推开方法。一般不需要将干燥器盖子完全打开，只打开到可以放入和取出器皿为度。搬动干燥器应用双手拿干燥器两侧底和盖的边缘，不可采用抱或托干燥器等不正确的操作，以免干燥器的盖子滑落打破。

4. 样品要均匀地铺在扁型称量瓶底部，以便样品中水分得到挥发。

5. 从烘箱中取出的扁型称量瓶置干燥器中冷却时，先不要把干燥器的盖子盖严，10分钟之后再盖上，以防干燥器内的空气冷却后形成部分负压，使干燥器难于打开。

思 考 题

1. 粗颗粒的样品为什么要研碎？是否研得愈细愈好？
2. 什么叫恒重？如何才能达到恒重？
3. 空称量瓶为何要干燥至恒重？
4. 若将 1 g $BaCl_2 \cdot 2H_2O$ 样品放在小容积的干燥器中，分别用以下物质中的一种作为干燥剂，保持 20 ℃放置，最后能否达到样品全部风化成无水物？

(1) 2 g 粒状 NaOH；

(2) 3 g $CaCl_2$ ＋5 g $BaCl_2 \cdot H_2O$；

(3) 10 g $CaBr_2 \cdot 6H_2O$。

Experiment 4 Preparation of Ferrous Ammonium Sulfate Hexahydrate

Purpose

1. To learn how to prepare double salts.
2. To master the operation of water bath and filter by suction.
3. To practice washing solid by decanting.
4. To understand the analytical method of impurity limitation.

Principle

The ferrous ion is converted into ferric ion easily when it is exposed to the air, but the ferrous ion in ferrous ammonium sulfate hexahydrate (FAS) cannot be easily oxidized.

Ferrous sulfate ($FeSO_4$), which can be obtained by reacting iron powder with diluted sulfuric acid (H_2SO_4), reacts with ammonium sulfate [$(NH_4)_2SO_4$] in equimolar ratio in aqueous solution. Ferrous ammonium sulfate hexahydrate [$FeSO_4 \cdot (NH_4)_2SO_4 \cdot 6H_2O$] with less solubility crystallizes from the solution as pale blue monoclinic crystal.

$$Fe + H_2SO_4 = FeSO_4 + H_2 \uparrow$$
$$FeSO_4 + (NH_4)_2SO_4 + 6H_2O = FeSO_4 \cdot (NH_4)_2SO_4 \cdot 6H_2O$$

Apparatus and Chemicals

Apparatus: platform balance; thermostatic water bath; evaporating dish; erlenmeyer flask; filter flask; buchner funnel; graduated cylinder, 10 mL; comparison tube.

Chemicals: Iron powder; ammonium sulfate [$(NH_4)_2SO_4$];
H_2SO_4, 3 mol·L^{-1}; HCl, 3 mol·L^{-1};
Na_2CO_3, 10%; KSCN, 0.1 mol·L^{-1}.

Procedure

1. Preparation of $FeSO_4 \cdot (NH_4)_2SO_4 \cdot 6H_2O$

(1) 2 g iron powder and 20 mL 10% Na_2CO_3 are put into an Erlenmeyer flask which is heated on a hotplate for about 10 minutes. Remove the solution by decanting; wash the iron powder in the same way with distilled water.

(2) Add 20 mL 3 mol·L^{-1} H_2SO_4 into the Erlenmeyer flask, which is heated over the water bath at 60 – 70 ℃ until the reaction is completed. Filter by suction while hot; wash the filter residue with 5 mL warm water. The filtrate is put into a clean evaporating dish. On the basis of the remains of the iron powder, calculate the ferrous sulfate's yield.

(3) For each gram ferrous sulfate, 0.75 g solid ammonium sulfate is added into the evaporating dish. Stir the mixture to dissolve the solid. Then the evaporating dish is heated over the boiling water bath until a layer of tiny crystals can be observed. Cool the concentrated solution to R.T. and filter by suction. Dry the pale blue crystal and weigh it. Calculate the percentage yield of the product.

2. Purity examination of the product

(1) Dissolve 2 g ferrous ammonium sulfate hexahydrate with 30 mL of oxygen-free distilled water in a 50 mL comparison tube. Add 3 mol·L^{-1} HCl 4 mL and 0.1 mol·L^{-1} KSCN 2 mL. Fill it to the mark with oxygen-free distilled water and mix the solution.

(2) Compare the color with that of the series of standard samples to determine the purity

实验 4　硫酸亚铁铵的制备

实验目的

1. 了解复盐的制备方法。
2. 掌握水浴加热和减压过滤等操作。
3. 练习倾泻法洗涤固体的方法。
4. 了解产品杂质限度的分析方法。

实验原理

一般亚铁盐在空气中都易被氧化,但形成复盐硫酸亚铁铵(FAS)后却比较稳定,不易被氧化。

铁屑易与稀硫酸反应,生成硫酸亚铁:

$$Fe + H_2SO_4 = FeSO_4 + H_2 \uparrow$$

硫酸亚铁与等物质的量的硫酸铵在水溶液中相互作用,便生成溶解度较小、浅蓝色的硫酸亚铁铵 $FeSO_4 \cdot (NH_4)_2SO_4 \cdot 6H_2O$:

$$FeSO_4 + (NH_4)_2SO_4 + 6H_2O = FeSO_4 \cdot (NH_4)_2SO_4 \cdot 6H_2O$$

仪器和试剂

仪器:天平;恒温水浴;蒸发皿;锥形瓶;抽滤瓶;布氏漏斗;量筒,10 mL;比色管。

试剂:固体　碎铁屑,硫酸铵

　　　酸　　H_2SO_4,3 mol·L^{-1};HCl,3 mol·L^{-1}

　　　盐　　Na_2CO_3,10%;KSCN,0.1 mol·L^{-1}

实验步骤

1. 制备步骤

铁屑的净化(去油污)　在台秤上称取 2 g 铁屑于锥形瓶中,然后加入 20 mL 10% Na_2CO_3 溶液,在电炉上微微加热约 10 min,用倾泻法去碱液,用水把铁屑冲洗干净。

硫酸亚铁的制备　往盛有铁屑的锥形瓶中加入 20 mL 3 mol·L^{-1} H_2SO_4,在水浴上加热,使铁屑和硫酸反应至不再有气泡冒出为止。趁热抽滤,用 5 mL 热水洗涤残渣。滤液转至蒸发皿中,将锥形瓶中的以及滤纸上的未反应的铁屑用滤纸吸干后称重。从反应的铁屑的量求算出生成的 $FeSO_4$ 的产量。

硫酸亚铁铵的制备　根据上面计算出来的 $FeSO_4$ 的产量,按照 $FeSO_4$ 与 $(NH_4)_2SO_4$ 质量比为 1:0.75 称取固体硫酸铵,加到硫酸亚铁溶液中,水浴上蒸发浓缩至表面出现晶体膜为止,自然冷却后,便得到硫酸亚铁铵晶体。过滤除去母液,把晶体留在蒸发皿中晾干。称重,计算产率。

2. 产品纯度鉴定

grade of the product.

Preparation of the standard samples

Add a solution containing ferric ion (the content of Fe^{3+} in the various solutions shown below) into comparison tubes respectively. The following procedure is the same as that of the product.

Grade I : 0.10 mg
Grade II : 0.20 mg
Grade III : 0.40 mg

Notes

1. The iron powder containing a trace quantity of arsenic As may give out poisonous gas AsH_3, so the reaction for ferrous sulfate should be carried out in the fume cupboard.

2. After ammonium sulfate is added to the evaporating dish, the mixture must be stirred thoroughly until the ammonium sulfate dissolves.

3. The time for evaporation should not be too long and the concentrate should be kept under room temperature for a while to produce the crystal of $FeSO_4 \cdot (NH_4)_2SO_4 \cdot 6H_2O$.

4. Calculation the theoretical yield of $FeSO_4 \cdot 7H_2O$ and FAS

$$\text{Percentage yield} = \frac{\text{actual yield}}{\text{theoretical yield}} \times 100\%$$

Questions

1. Between the iron powder and the sulfuric acid in the secondary step, which one should be present in considerable excess?

2. Why should we filter the solution while it is still hot when the reaction for ferrous sulfate is completed? Why do we need warm water to wash the filter residue?

3. Why should the pH of the solution be kept at 2 or 3 during evaporation?

4. How to get rid of the oil from the iron powder?

5. How to determine whether the reaction to produce ferrous sulfate is completed?

6. The solution of ferrous sulfate is easy to be oxidized. What measures can be taken to prevent the oxidation during the procedure?

7. How to calculate the yield of the product?

8. How to get oxygen-free distilled water? Why should the water to dissolve the ferrous ammonium sulfate hexahydrate be oxygen-free distilled water?

9. List sources of errors that would contribute toward lowering percent yield of FAS.

Fe^{2+} 的限量检查 称 2 g 产品于 50 mL 比色管中,先加入约 30 mL 不含氧的蒸馏水使之溶解。再加 4 mL 3 mol·L^{-1} HCl 和 2 mL 0.1 mol·L^{-1} KSCN,继续加不含氧的蒸馏水至 50 mL 刻度。摇匀,所呈现的红色与标准试样比较,检验产品级别。

标准试样的制备

取含有下列数量 Fe^{3+} 的溶液若干毫升:
Ⅰ级试剂:0.10 mg
Ⅱ级试剂:0.20 mg
Ⅲ级试剂:0.40 mg
放入 50 mL 比色管中,然后与产品同样处理,得到标准溶液系列。

注意事项

1. 由于铁屑含有杂质砷,本实验在合成过程中有剧毒气体 AsH_3 放出,它能刺激和麻痹神经系统。故实验需在通风橱中进行。
2. 在 $FeSO_4$ 溶液中加入固体 $(NH_4)_2SO_4$ 后,必须充分搅动至 $(NH_4)_2SO_4$ 完全溶解后,才能进行蒸发浓缩。
3. 加热浓缩时间不宜过长。浓缩到一定体积后,需在室温放置一段时间,以待结晶析出、长大。
4. 计算 $FeSO_4·7H_2O$ 和 FAS 的理论产量。

$$产率(\%) = \frac{实际产量}{理论产量} \times 100\%$$

思 考 题

1. 本实验反应中是铁过量,还是 H_2SO_4 过量?为什么要这样选择?
2. 反应结束后,为什么要趁热过滤?为什么需用热水洗涤残渣?
3. 为什么本实验在蒸发浓缩时,溶液应控制在酸性(pH=2~3)?
4. 如何除去废铁屑表面的油污?
5. 合成硫酸亚铁时,怎样判断反应已进行完全?
6. $FeSO_4·7H_2O$ 溶液在空气中很容易被氧化,在制备硫酸亚铁铵的过程中,怎样防止 Fe^{2+} 氧化成 Fe^{3+}?
7. 怎样计算硫酸亚铁铵的产率?是根据铁的用量还是硫酸铵的用量?
8. 如何制备不含氧的蒸馏水?为什么配制硫酸亚铁铵试液时要用不含氧的蒸馏水?
9. 说明影响硫酸亚铁铵产率不高的因素有哪些。

Experiment 5 Preparation of Zinc Gluconate

Purpose
1. To learn how to prepare Zinc gluconate.
2. To learn the operation of heating over the water bath and filter by suction.
3. To learn the crystallization method with ethanol as a solvent.

Principle
Zinc gluconate is obtained when calcium gluconate reacts with zinc sulfate by equal molar:

$$Ca(C_6H_{11}O_7)_2 + ZnSO_4 = Zn(C_6H_{11}O_7)_2 + CaSO_4 \downarrow$$

This is a double decomposition reaction, in which the new complex of zinc gluconate is formed when calcium ion Ca^{2+} in complexation with gluconate is substituted by zinc ion Zn^{2+}. The reaction goes further toward completion since calcium sulfate $CaSO_4$ precipitates from the equilibrium system to reduce the concentration of Ca^{2+} and SO_4^{2-}. After $CaSO_4$ solid is removed by filtering, the filtration is condensed. No solid of zinc gluconate shows because of its large solubility in water. Anhydrous alcohol is added to reduce the solubility of $Zn(C_6H_{11}O_7)_2$ and to force the solid to form.

Apparatus and Chemicals
Apparatus: platform balance; thermostatic water bath; filtration assembly; measuring cylinder; beaker; evaporating dish; hotplate.

Chemicals: calcium gluconate; heptahydrate zinc sulfate; 95% ethanol.

Procedure

1. Preparation of Zinc Gluconate

Measure 80 mL water to a beaker, heat it to 80 – 90 ℃, then add 13.4 g of $ZnSO_4 \cdot 7H_2O$ and dissolve it completely. Put the beaker into a water bath boiler and keep the temperature at 90 ℃, add 20 g of calcium gluconate gradually and then stir the mixture constantly. After 20 minutes, filter it by vacuum filter (double filter paper). The filtrate is transferred to an evaporating dish (filter cake discarded). It is concentrated and became the dope (the volume of approximately 20 mL). Then the filtrate is cooled to room temperature. 20 mL 95% ethanol is added (to reduce the solubility of zinc gluconate) and it is stirred continually. A lot of colloidal zinc gluconate is separated out. Pour the liquid out from the mixture. Add another 20 mL 95% ethanol into the solid and stir it again. After stirring completely, the precipitate slowly changes into crystal. Then by filtering, the crude product is obtained.

2. Recrystallization of crude Zinc Gluconate

The crude product is dissolved by adding 20 mL water at 90 ℃, filtered warmly, then the filtrate is cooled to room temperature. Add 20 mL 95% ethanol, stir strongly, and the crystal is separated out. After filtered and baked at 50 ℃, pure product is obtained.

Notes

1. The reaction must be undertaken in a water bath at the constant temperature of 90 ℃. Too high a temperature may lead to the decomposition of zinc gluconate while too low a temperature may lower the speed of the reaction.

2. When crystallized with ethanol as solvent, amount of zinc gluconate may appear at the

实验 5　葡萄糖酸锌的制备

实验目的

1. 学习葡萄糖酸锌(治疗人体缺锌药物)的制备方法。
2. 学会水浴加热和减压过滤等操作。
3. 学会用更换溶剂进行结晶的方法。

实验原理

葡萄糖酸钙与等物质的量的硫酸锌反应如下：

$$Ca(C_6H_{11}O_7)_2 + ZnSO_4 = Zn(C_6H_{11}O_7)_2 + CaSO_4 \downarrow$$

这是一个复分解反应,锌离子 Zn^{2+} 替代葡萄糖酸钙 $Ca(C_6H_{11}O_7)_2$ 配合物中的钙离子 Ca^{2+} 生成葡萄糖酸锌 $Zn(C_6H_{11}O_7)_2$ 配合物。由于反应体系中有硫酸钙沉淀的生成,使反应进行得较为完全。过滤除去硫酸钙沉淀,滤液蒸发浓缩时不能直接产生葡萄糖酸锌固体,原因是葡萄糖酸锌在水中的溶解度太大,要加入无水乙醇使其溶解度降低,促使葡萄糖酸锌固体生成。

仪器和试剂

仪器:台秤;恒温水浴;抽滤装置;量筒;烧杯;蒸发皿;电炉。
试剂:葡萄糖酸钙,$Ca(C_6H_{11}O_7)_2$;$ZnSO_4 \cdot 7H_2O$;95% 乙醇。

实验步骤

1. 葡萄糖酸锌的生成

量取 80 mL 蒸馏水置烧杯中,加热至 80~90 ℃,加入 13.4 g $ZnSO_4 \cdot 7H_2O$ 使之完全溶解,将烧杯放在 90 ℃ 的恒温水浴中,再逐渐加入葡萄糖酸钙 20 g,并不断搅拌。在 90 ℃ 水浴上静置保温 20 min。抽滤(用两层滤纸),滤液移至蒸发皿中(滤渣为 $CaSO_4$,弃去),将滤液在沸水浴上浓缩至黏稠状(体积约为 20 mL,如浓缩液有沉淀系 $CaSO_4$,需过滤掉)。滤液冷至室温,加 20 mL 95% 乙醇(降低葡萄糖酸锌的溶解度),并不断搅拌,此时有大量的胶状葡萄糖酸锌析出,充分搅拌后,用倾泻法除去乙醇液。于胶状沉淀中再加 20 mL 95% 乙醇,充分搅拌后,沉淀慢慢转变成晶体状,抽滤至干,即得粗品(母液、乙醇需回收)。

2. 重结晶

粗品加水 20 mL,加热(90 ℃)至溶解,趁热抽滤,滤液冷至室温,加 20 mL 95% 乙醇,充分搅拌,结晶析出后,抽滤至干,即得精品,在 50 ℃ 下烘干。

注意事项

1. 反应需在 90 ℃ 恒温水浴中进行。这是因为温度过高,葡萄糖酸锌会分解;温度过低,则反应速率降低。

beginning. Chopstick, easy to stir, often replaces glass rod.

3. The filtrate should be concentrated at a boiled water bath.

Questions

1. Refer to other materials and understand the important role of trace zinc playing in human body.

2. Design the flow chart of preparation of zinc gluconate.

3. Why must we keep the temperature at 90 ℃ when zinc sulfate reacts with calcium gluconate?

4. By how many means can zinc gluconate be crystallized?

2. 用酒精为溶剂进行结晶时,开始有大量胶状葡萄糖酸锌析出,不易搅拌,可用竹棒代替玻璃棒进行搅拌。

3. 滤液需在沸水浴中浓缩。

思 考 题

1. 查阅有关资料,了解微量元素锌在人体中有怎样的重要作用。
2. 设计葡萄糖酸锌制备的流程图。
3. 为什么葡萄糖酸钙和硫酸锌的反应需保持在 90 ℃ 的恒温水浴中?
4. 葡萄糖酸锌可以用哪几种方法进行结晶?

Experiment 6 Preparation of Medicinal Sodium Chloride and Examination of Impurities Limitation

Purpose

1. To master the principle and method of how to prepare medicinal sodium chloride.

2. To learn how to check whether the ions precipitate completely.

3. To master vacuum filter (filter with suction), evaporating a solution and crystallization and recrystallization.

4. To learn how to identify and examine medicines.

Principle

1. Sodium chloride is soluble in aqueous solution, so the impurities in sodium chloride can be removed with the processes showed below:

(1) The insoluble impurities are removed by filtration.

(2) Some soluble impurities can be removed by precipitation basing on their chemical properties. For example, sulfate can be separated as $BaSO_4$ by addition solution of $BaCl_2$; Ca^{2+}, Mg^{2+}, Fe^{3+}, Ba^{2+}, etc. can be removed as insoluble precipitates by addition of Na_2CO_3.

(3) Some low content soluble impurities, such as K^+, I^-, Br^-, having different solubility from sodium chloride, can be removed by recrystallization. They will be retained in the mother liquid and moved away.

2. The limit tests of barium, sulfate, potassium, calcium and magnesium are carried out in comparison tubes by addition of corresponding precipitate agents. Under same conditions, any opalescence produced in the test solutions should be not pronounced than that of the reference solutions.

3. Heavy metals (comprised the ions of Pb, Bi, Cu, Hg, Sb, Sn, Co, Zn and other metals) can be colored by sulfide ion under the specified test conditions. The test of heavy metallic impurities is carried out by comparing the color of solutions with the corresponding reference solutions under same conditions.

$$Pb^{2+} + S^{2-} \longrightarrow PbS \downarrow$$

Apparatus and Chemicals

Apparatus: hotplate; evaporating dish; buchner funnel; filter flask; platinum wire; beakers, 250 mL; platform balance; color-comparison tubes, 25 mL, 50 mL.

Chemicals: bromothymol blue; raw sodium chloride; concentrated hydrochloric acid, 2 mol·L^{-1}; diluted hydrochloric acid, 0.02 mol·L^{-1}; sulfuric acid, 6 mol·L^{-1}; acetic acid, 0.1 mol·L^{-1}; sodium hydroxide solution, 2 mol·L^{-1}, 0.02 mol·L^{-1}; sodium carbonate, saturated solution; $BaCl_2$, 25%; KI, 10%; Ca^{2+} TS; $AgNO_3$, 0.25 mol·L^{-1}; KBr, 10%; Mg^{2+} TS; zinc uranylacetate solution; $(NH_4)_2C_2O_4$ TS; Na_2HPO_4 TS; NH_4Cl TS; ammonia

实验 6　药用氯化钠的制备、性质及杂质限度检查

实验目的

1. 掌握药用氯化钠的制备原理和方法。
2. 学会检查沉淀是否完全的操作。
3. 正确掌握减压抽滤、蒸发浓缩、结晶和重结晶技术。
4. 了解药品的鉴定、检出方法。

实验原理

1. 食盐是能溶于水的固态物质。对于其中所含杂质的去除方法基本如下：
（1）机械杂质（如泥沙）可采取过滤法除去。
（2）一些可溶于水的杂质根据其性质借助于化学方法除去。如加入 $BaCl_2$ 溶液可使 SO_4^{2-} 生成 $BaSO_4$ 沉淀，加 Na_2CO_3 溶液可使 Ca^{2+}、Mg^{2+}、Fe^{3+}、Ba^{2+} 等离子生成难溶碳酸盐沉淀，先后过滤除去。
（3）少量可溶性杂质如 K^+、I^-、Br^- 等离子，可根据其盐与 NaCl 在不同温度下溶解度的不同，在重结晶时，使其残留在母液中弃去。

2. 钡盐、硫酸盐、钾盐、钙镁盐的限度检查，是根据沉淀反应原理，样品管和标准管在相同条件下进行比浊试验，样品管不得比标准管更深。

3. 重金属系指 Pb、Bi、Cu、Hg、Sb、Sn、Co、Zn 等金属离子，它们在一定条件下能与 H_2S 或 Na_2S 作用而沉淀。中国药典规定是在弱酸性条件下进行，用稀醋酸调节。实验证明，在 pH＝3 时，HgS 沉淀最完全。重金属的检查，是在相同条件下进行比色试验。反应式为：

$$Pb^{2+} + S^{2-} \rightarrow PbS \downarrow$$

仪器和试剂

仪器：电炉；蒸发皿；布氏漏斗；抽滤瓶；铂丝棒；烧杯，250 mL，两只；天平；奈氏比色管（25 mL、50 mL）。

试剂：溴百里酚蓝指示剂。
　　　固体：固体 NaCl 粗盐；KI-淀粉试纸。
　　　酸：HCl（浓、稀，2、0.02 mol·L^{-1}）；H_2SO_4（6 mol·L^{-1}）；HAc（0.1 mol·L^{-1}）。
　　　碱：NaOH（2、0.02 mol·L^{-1}）；饱和 Na_2CO_3 溶液；氨试液。
　　　盐：25% $BaCl_2$ 溶液；KI 10%，钙盐溶液，$AgNO_3$（0.25 mol·L^{-1}），KBr 10%，镁盐溶液，醋酸铀酰锌溶液，$(NH_4)_2C_2O_4$ 试液，Na_2HPO_4 试液，NH_4Cl 试液。

TS; KI-starch test paper.

Procedure

1. Purification of NaCl

Raw NaCl 50 g is purified and the procedure is illustrated as the following flow chart.

2. Identification test

An aqueous solution (1 in 10) of the product is prepared for the identification test.

Sodium

Fame test   Prepare a platinum wire by burning it on a non-luminous flame after moistening it with concentrated hydrochloric acid until the flame is colorless. Moisten the test solution on the platinum wire; it imparts an intense yellow color to a non-luminous flame.

Precipitation   Add one or two drops of solution in a tube, acidified with three drops of 3 mol·L^{-1} acetic acid, and add 10 drops of zinc uranylacetate TS, if necessary, rub the inside wall of the test tube with a glass rod. A yellow precipitate is formed.

$$Na^+ + Zn^{2+} + 3UO_2^{2+} + 8Ac^- + HAc + 9H_2O \rightarrow NaAc \cdot Zn(Ac)_2 \cdot 3UO_2(Ac)_2 \cdot 9H_2O \downarrow + H^+$$

Chloride

1~2 drops of the tested solution yield a white, curdy precipitate with 2 drops of 0.25 mol·L^{-1} silver nitrate TS. On addition of 6 mol·L^{-1} ammonia TS, the precipitate dissolve. White precipitate is formed again by acidifing the solution with 6 mol·L^{-1} nitric acid.

$$Cl^- + Ag^+ = AgCl \downarrow$$
$$AgCl \downarrow + 2NH_3 = Ag(NH_3)_2^+ + Cl^-$$
$$Ag(NH_3)_2^+ + Cl^- = AgCl \downarrow + 2NH_3$$

3. Limit test of impurity

The product should be tested as described below.

(1) Appearance of solution

Dissolve 5 g in 25 mL of distilled water. The solution should be clear and colorless.

(2) Acidity or alkalinity

Dissolve 5 g in carbon dioxide-free water, and dilute with the same solvent to 50 mL. Add 2 drops of bromothymol blue TS; Not more than 0.2 mL of 0.02 mol·L^{-1} hydrochloric acid (HCl) or 0.1 mL of 0.02 mol·L^{-1} sodium hydroxide (NaOH) is required to change the color of the solution.

The content of free acid or alkali, if there is any, in medicinal NaCl should be not more than the limit. And the pH ranges of bromothymol blue is 6.6 - 7.6, yellow to blue.

(3) Iodide (I^-) and bromide (Br^-)

Dissolve 1 g in 3 mL of distilled water and add 1 mL of chloroform. Cautiously introduce, dropwise, with constant agitation, dilute chlorine (Cl_2) TS (1 in 2); the chloroform does not acquire a violet, yellow, or orange color.

实验步骤

1. NaCl 的提纯

粗食盐 50 g,主要成分:NaCl。杂质成分:有机物、泥沙以及 Ca^{2+}、Mg^{2+}、Fe^{3+}、SO_4^{2-}、Br^-、I^- 等。将食盐小火炒至有机物炭化,移烧杯中,按照"食盐精制的简要流程图"对 NaCl 进行提纯。

2. 鉴别反应

制备氯化钠溶液(1∶10)供下列鉴定用。

钠盐

(1) 焰色反应 取铂丝,用浓盐酸湿润后在无色火焰中灼烧,至火焰不显色为止(表示铂丝已洁净)。后蘸取氯化钠溶液,置无色火焰中燃烧,火焰出现持久鲜黄色。

(2) 沉淀反应 取氯化钠溶液 1~2 滴,用 3 $mol·L^{-1}$ HAc 3 滴使酸化,加醋酸铀酰锌试液 10 滴,用玻璃棒摩擦管壁,即渐渐析出醋酸铀酰锌钠黄绿色沉淀。反应如下:

$$Na^+ + Zn^{2+} + 3UO_2^{2+} + 8Ac^- + HAc + 9H_2O \rightarrow NaAc·Zn(Ac)_2·3UO_2(Ac)_2·9H_2O\downarrow + H^+$$
(醋酸铀酰锌钠)

氯化物

与银离子生成氯化银沉淀。

取氯化钠溶液 1~2 滴,加 0.25 $mol·L^{-1}$ 硝酸银试液 2 滴,即产生白色凝乳状沉淀。滴加 6 $mol·L^{-1}$ 氨试液,沉淀溶解,再加 6 $mol·L^{-1}$ 硝酸至显酸性,又有白色沉淀生成。

$$Cl^- + Ag^+ = AgCl\downarrow$$
$$AgCl\downarrow + 2NH_3 = Ag(NH_3)_2^+ + Cl^-$$
$$Ag(NH_3)_2^+ + Cl^- = AgCl\downarrow + 2NH_3$$

3. 杂质限度检查

成品氯化钠须进行以下各项质量检查试验。

(1) 溶液的澄清度

取本品 5 g,加水至 25 mL,应溶解成无色澄明的溶液。

(2) 酸碱度

取本品 5 g,加新鲜蒸馏水 50 mL 溶解后,加溴百里酚蓝指示液 2 滴。如溶液呈黄色,滴加 0.02 $mol·L^{-1}$ NaOH 溶液使其变蓝,所消耗的 0.02 $mol·L^{-1}$ NaOH 溶液不得超过 0.1 mL(约 2 滴)。如果显蓝色或绿色,滴加 0.02 $mol·L^{-1}$ HCl 溶液使其变黄色,所消耗的 0.02 $mol·L^{-1}$ HCl 溶液不得超过 0.2 mL。

药用氯化钠可能夹杂少量酸或碱,所以药典把它限制在很小范围内。氯化钠为强酸强碱盐,在水溶液中应呈中性。溴百里酚蓝指示液的变色范围是 pH 6.6~7.6,由黄色到蓝色。

Control test: Add 1 mL of 10% iodide TS and bromide TS in two tubes respectively, the following step is similar to that of the test solution. The color of chloroform in one tube (containing the iodide TS) is violet while the other tube is yellow or orange.

$$2Br^- + Cl_2 \rightarrow Br_2 + 2Cl^-$$
$$2I^- + Cl_2 \rightarrow I_2 + 2Cl^-$$

(4) Barium (Ba^{2+})

Dissolve 4 g in 20 mL of distilled water, filter if necessary, and divide the solution into two equal portions. To one portion add 2 mL of diluted sulfuric acid, and to the other add 2 mL of water; the solution should be equally clear after standing for 2 hours.

(5) Calcium (Ca^{2+}) and magnesium (Mg^{2+})

Dissolve 4 g in 20 mL of distilled water, add 2 mL of ammonia TS and divide the mixture into two equal portions. Treat one portion with 1 mL of ammonium oxalate TS and the other portion with 1 mL of dibasic sodium phosphate TS and a few drops of ammonium chloride TS; no opalescence is produced within 5 minutes.

Control test:

Calcium Pipet 1 mL of calcium TS to a tube, alkalied with ammonia TS, add ammonium oxalate TS; white crystal is precipitated.

$$Ca^{2+} + C_2O_4^{2-} \rightarrow CaC_2O_4 \downarrow$$

Magnesium Pipet 1 mL of magnesium TS to a tube, add a few drops of ammonia TS and ammonium chloride, drop in dibasic sodium phosphate (Na_2HPO_4) TS; white precipitate is separated from the solution.

$$Mg^{2+} + HPO_4^{2-} + NH_4^+ + OH^- \rightarrow MgNH_4PO_4 \downarrow + H_2O$$

(6) Sulfate (SO_4^{2-})

Reference preparation Into a 50 mL color-comparison tube pipet 1 mL of standard potassium sulfate solution, and dilute with water to about 25 mL. Add 1 mL of 1 mol·L^{-1} hydrochloric acid and 3 mL of barium chloride TS, dilute with water to volume and mix.

Test preparation Into a 50 mL color-comparison tube place 5 g of the product, and dissolve the solid with about 25 mL of distilled water, add 1 mL of 0.1 mol·L^{-1} HCl, filter if necessary. Add 3 mL of barium chloride TS, dilute with water to volume and mix.

Procedure Allow the two tubes to stand for 10 minutes, and view downward. Any opalescence produced in the latter tube is not more pronounced than that of the standard tube.

Standard potassium sulfate Dissolve 0.1810 g of potassium chloride with distilled water in 1000 mL volumetric flask, dilute with water to volume and mix. This solution contains the equivalent of 0.1 mg of sulfate per mL.

(7) Iron

Dissolve 5 g with 35 mL of distilled water in a 50 mL color-comparison tube, add 4 mL of dilute hydrochloric acid and approximate 50 mg of ammonium persulfate, add 3 mL of 30% ammonium thiocyanate (NH_4SCN) solution and sufficient water to produce 50 mL, mix well. Any color produced is not more intense than that of a reference solution using 1.5 mL of standard

（3）碘化物与溴化物

取本品 1 g，加蒸馏水 3 mL 溶解后，加氯仿 1 mL，并注意滴加用等量蒸馏水稀释的氯水试液，随滴随振荡，氯仿层不得显紫堇色、黄色或橙色。

对照试验：分别取碘化物和溴化物溶液各 1 mL，分置于 2 只试管内，各加氯仿 1 mL，并注意滴加用等量蒸馏水稀释的氯水试液，随滴随振荡，氯水试液氧化 I^- 释出 I_2，使氯仿层显紫红色。氯水试液氧化 Br^- 释出 Br_2，使氯仿层显黄色或橙黄色。

$$2Br^- + Cl_2 \rightarrow Br_2 + 2Cl^-$$
$$2I^- + Cl_2 \rightarrow I_2 + 2Cl^-$$

（4）钡盐

取本品 4 g，用蒸馏水 20 mL 溶解，过滤，滤液分为两等份，一份中加稀硫酸 2 mL，另一份中加蒸馏水 2 mL，静置 2 h，两溶液应同样透明。

（5）钙盐与镁盐

取本品 4 g，加水 20 mL 溶解后，加氨试液 2 mL 摇匀，分成两等份。一份加草酸铵试液 1 mL，另一份加磷酸氢二钠试液 1 mL 和几滴 NH_4Cl 溶液，5 分钟内均不得发生浑浊。

对照试验

钙盐：取钙盐溶液 1 mL，滴加氨水至显微碱性，加草酸铵试液 1 mL，溶液有白色结晶析出。反应式为：

$$Ca^{2+} + C_2O_4^{2-} \rightarrow CaC_2O_4 \downarrow （白）$$

镁盐：取镁盐溶液 1 mL，加几滴氨水和 NH_4Cl，加 Na_2HPO_4 1 mL，有白色结晶析出。反应式为：

$$Mg^{2+} + HPO_4^{2-} + NH_4^+ + OH^- \rightarrow MgNH_4PO_4 \downarrow （白） + H_2O$$

（6）硫酸盐

本品含硫酸盐，依下法检查，如发生浑浊，与标准硫酸钾溶液 1 mL 制成的对照标准液比较，不得更浓（0.002%）。

取 50 mL 奈氏比色管两只，甲管中加标准硫酸钾溶液 1 mL，加蒸馏水稀释至约 25 mL 后，加 1 mol·L^{-1} HCl 1 mL，加 25% $BaCl_2$ 溶液 3 mL，再加适量水使成 50 mL，摇匀，放置 10 分钟。

取本品 5 g 置于乙管中，加水溶解至约 25 mL，溶液应透明，如不透明可过滤，于滤液中加 1 mol·L^{-1} HCl 1 mL，加 25% $BaCl_2$ 溶液 3 mL，用蒸馏水稀释使成 50 mL，摇匀，放置 10 分钟。

甲乙两管放置 10 分钟后，置于比色架上，在光线明亮处双眼由上而下透视，比较两管的浑浊度，乙管发生的浑浊度不得高于甲管。

标准硫酸钾溶液的制备　精密称取在 105 ℃ 干燥至恒重的硫酸钾 0.1810 g，置烧杯中溶解，然后转入 1000 mL 的容量瓶中，多次洗涤烧杯，洗液一并倒入容量瓶中，加蒸馏水稀释至刻度，摇匀即得。每 1 mL 相当于 0.1 mg 的 SO_4^{2-}；每 1 mL 也相当于 0.0811 mg 的钾。故可用于鉴定 SO_4^{2-} 和钾盐。

iron solution, that is the content of iron in the product should be less than 0.0003%.

$$Fe^{3+} + 3SCN^- \rightarrow Fe(SCN)_3$$

Standard iron solution Dissolve 863.0 mg of ferric ammonium sulfate in water, add 2 mL of dilute hydrochloric acid, and dilute with water to 1000 mL. Pipet 10 mL of this solution into a 100 mL volumetric flask, add 0.5 mL of dilute hydrochloric acid, dilute with water to volume, and mix. This solution contains the equivalent of 0.01 mg of iron per mL.

(8) Potassium (K^+)

Dissolve 5 g of product with distilled water in a color-comparison tube, and adjust with 2 drops of acetic acid to a pH between 5 and 6. Add 2 mL of 0.1 mol·L^{-1} sodium tetraphenylboron solution and dilute with water to 50 mL, mix. Any opalescence produced is not more pronounced than that of a reference solution using 0.5 mL of standard potassium sulfate solution (0.01%).

$$K^+ + B(C_6H_5)_4^- \rightarrow KB(C_6H_5)_4 \downarrow$$

Standard potassium sulfate solution Dissolve 2.228 g of potassium sulfate, previously dried at 105 ℃ to constant weight, with water in 1000 mL volumetric flask, dilute to volume, and mix. This solution contains the equivalent of 1 mg of potassium per mL.

Sodium tetraphenylboron solution Triturate 1.5 g of sodium tetraphenylboron with 10 mL of water, then add 40 mL of water, triturate again and filter.

(9) Heavy metals

Reference preparation Into a 50 mL color-comparison tube pipet 1 mL of standard lead solution (10 μg of Pb), add 2 mL of dilute acetic acid and dilute with water to 25 mL.

Test preparation Dissolve 5 g of the product with 20 mL of distilled water in another 50 mL color-comparison tube, add 2 mL of dilute acetic acid.

Procedure To each of the two tubes add 10 mL of hydrogen sulfide TS, dilute to 50 mL and mix. The tubes are allowed to stand for 10 minutes in dark place. If the color of the solution from the standard preparation is not darker than that of the solution from the standard preparation, it pronounces that the content of heavy metals is not more than the limits.

(7) 铁盐

取本品 5 g,置于 50 mL 奈氏比色管中,加蒸馏水 35 mL 溶解,加稀 HCl 4 mL,过硫酸铵 50 mg,再加 30% 硫氰酸铵溶液 3 mL,适量蒸馏水稀释成 50 mL,摇匀。如显色,立即与标准溶液 1.5 mL 用同法处理后制得的标准管颜色比较,不得更深,即样品中铁含量不得超过 0.0003%。反应式为:

$$Fe^{3+} + 3SCN^- \rightarrow Fe(SCN)_3 (血红色)$$

标准铁盐溶液的制备 精密称取未风化的硫酸铁铵 0.8630 g,溶解后转入 1000 mL 容量瓶中,加盐酸 2 mL,用蒸馏水稀释至刻度,摇匀。精密量取 10 mL,置于 100 mL 容量瓶中,加稀 HCl 0.5 mL,用蒸馏水稀释至刻度,摇匀即得(每 1 mL 相当于 0.01 mg 的 Fe)。

(8) 钾盐

取本品 5 g,置奈氏比色管中,加蒸馏水 20 mL 溶解,加 3 mol·L^{-1} 醋酸 2 滴(使 pH 为 5~6),加 0.1 mol·L^{-1} 四苯硼钠溶液 2 mL,加水至刻度,使成 50 mL,如显浑浊,与标准硫酸钾 0.5 mL 制成的标准溶液比较,不得更浓(0.01%)。反应式为:

$$K^+ + B(C_6H_5)_4^- \rightarrow KB(C_6H_5)_4 \downarrow (白)$$

标准硫酸钾溶液的制备 精密称取 105 ℃ 干燥至恒重的硫酸钾 2.228 g,溶解后转入 1000 mL 容量瓶中,加适量蒸馏水溶解,加水至刻度,摇匀即得(每 1 mL 相当于 1 mg 的 K)。

四苯硼钠溶液的制备 取四苯硼钠[NaB(C$_6$H$_5$)$_4$] 1.5 g,置研钵中,加水 10 mL 研磨后,再加水 40 mL 研习,用质密的滤纸过滤即得。

(9) 重金属

取 50 mL 比色管两只,于第一管中加标准铅溶液(含 Pb 0.01 mg/mL)1 mL,加稀醋酸 2 mL,加水适量使成 25 mL。于第二管中加样品 5 g,加水 20 mL 溶解后,加稀醋酸 2 mL。再于两管中分别加硫化氢试液 10 mL,摇匀,在暗处放置 10 min,同置白纸上,自上面透视,第二管中显出的颜色与第一管比较,不得更深(含重金属不得超过 2 ppm)。

药典 NaCl 的重金属检查项目规定:"取样品 5 g,加蒸馏水 20 mL 溶解后,加稀醋酸 2 mL 与水适量使成 25 mL,依法检查,含重金属不得超过 2 ppm。"但并没有直接给出标准铅溶液的取用量,须按下述次序自行计算:

先根据供试品的取用量和重金属的限量,求得重金属的毫克数。

已知条件:NaCl 的用量为 5 g,其中含重金属的限量为 2 ppm。

则:允许重金属 Pb 的毫克数

$$5 \times 1000 \times 0.000002 = 0.01 \text{ mg(Pb)}$$

再计算标准铅溶液的取用量。

因每 1 mL 标准铅溶液含 Pb 0.01 mg,需 0.01 mg Pb 作标准进行对照,应取标准铅溶液的体积为:0.01 mg/(0.01/1 mL) = 1 mL。

铅贮备液的制备 精密称取在 105 ℃ 干燥至恒重的硝酸铅 0.1598 g,加硝酸 5 mL 与水 50 mL,溶解后,按规定配制成 1000 mL,摇匀,即得(每 1 mL 相当于 81 mg 的 Pb)。

标准铅溶液的制备 精密量取铅贮备液 10 mL,置 100 mL 容量瓶中,加水稀释至刻度,摇匀,即得(每 1 mL 相当于 0.01 mg 的 Pb)。标准铅溶液应新鲜配制。配制与贮存用的玻璃容器均不得含有铅。

Flow Chart of Preparation of Medicinal NaCl

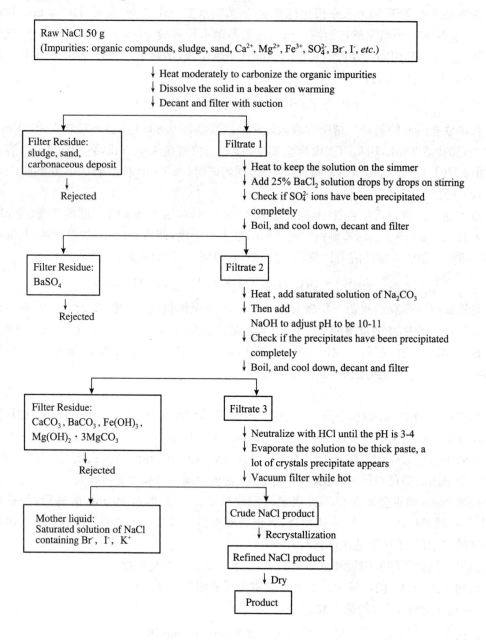

Notes

1. Water used to dissolve the raw material should be proportioned to the material. Too much water will introduce difficult to the following evaporation.

2. When we remove the impurities by addition precipitate agents, the time of boiling the solution should not be too long. Otherwise, some of NaCl will separate from the hot solution. If some NaCl crystal showing, some few distilled water should be added to the solution.

3. During the evaporation, the crystal membrane of NaCl on the surface of the condensed solution should be ruptured with a parallel lying glass rod, otherwise, the crystal will splatter

药用氯化钠制备流程图

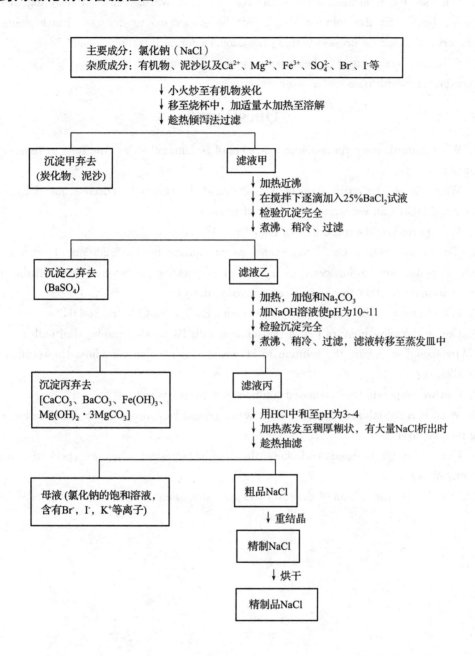

注意事项

1. 将粗食盐加水(自来水)到全部溶解(其量根据食盐溶解度计算)为限,用水量不宜过多,以免给以后蒸发浓缩带来困难。

everywhere.

4. Some soluble impurities, such as Br^-, I^- and K^+, will be removed away together with the mother liquid. So the solution should not be evaporated to dryness. Furthermore, the resulting crystal should be pressed with a glass stopper during filtration.

5. Use the comparison tubes correctly. And compare the sample tubes with the corresponding standard tubes under same conditions.

Questions

1. What impurities are contained in raw solid of sodium chloride? And how to remove these impurities?

2. Why, during the purification of NaCl, should the agents be added sequentially: $BaCl_2$, Na_2CO_3, and HCl? Can we change the order of agents?

3. How to remove the impurities of K^+, Br^-, I^- and other ions?

4. To remove SO_4^{2-}, Ca^{2+}, Mg^{2+} *etc.* as precipitate by the addition of corresponding precipitate agents, what influences does heating or not heating the solution have on the result? How to determine whether these ions are removed entirely?

5. How to remove the excess precipitate agents: $BaCl_2$, Na_2CO_3 and NaOH?

6. During the adjustment of pH of the solution with HCl, what can we deal with the excess HCl? Why should we adjust the solution to be weakly acidic? Can we adjust the solution to be weakly alkaline?

7. Can we evaporate the condensed solution to dryness? Why?

8. What is recrystallization? How much water should be appropriate to dissolve the product during recrystallization?

9. Can tap water be used to dissolve the resulting product when we check the impurity limitation? Why?

10. Summarize the results of the tests and give conclusion on how to get a higher yield.

2. 在加沉淀剂过程中,溶液煮沸时间不宜过长,以免水分蒸发而使 NaCl 晶体析出。若发现液面有晶体析出时,可适当补充些蒸馏水。

3. 加热浓缩时,当大量 NaCl 晶体析出时,要不断搅拌(玻璃棒尽量平放在溶液中)以破坏表层薄膜,防止 NaCl 晶体外溅。

4. 浓缩时不可蒸发至干,要保留少量水分,以使 K^+、Br^-、I^- 等离子随母液去掉,并在抽滤时用玻璃瓶盖尽量将晶体压干。

5. 正确使用奈氏比色管,注意平行条件,用水稀释至刻度后再摇匀。

思 考 题

1. 海盐除含有 NaCl 外,还含有哪些杂质?如何用化学方法除去?

2. 食盐精制过程中,加试剂的次序,为什么必须先加 $BaCl_2$,再加 Na_2CO_3,最后加盐酸,是否可以改变加入的次序?

3. 粗食盐中所含 K^+、Br^-、I^- 等离子是怎样除去的?

4. 在加入沉淀剂使 SO_4^{2-}、Ca^{2+}、Mg^{2+} 等离子转入沉淀以除去它们的时候,加热与不加热对分离操作各有何影响?如何检查这些离子已沉淀完全?

5. 如何除去过量的沉淀剂 $BaCl_2$、Na_2CO_3 和 NaOH?

6. 在调 pH 过程中,若加入的盐酸过量怎么办?为何要调成弱酸性(碱性行吗)?

7. 在浓缩过程中,能否把溶液蒸干?为什么?

8. 何谓重结晶?根据所得粗品 NaCl 晶体的量,应加入多少水使之溶解为宜?

9. 在检查产品纯度时,能否用自来水溶解食盐,为什么?

10. 根据实验,总结本实验中提高 NaCl 精品的产率和质量的关键。

Experiment 7 Synthesis of Potassium Aluminum Sulfate (Alum)

Purpose

1. To learn how to convert the aluminum metal from a beverage can into the chemical compound potassium aluminum sulfate, $KAl(SO_4)_2 \cdot 12H_2O$, commonly referred to as *alum*.

2. To learn the identification of alum.

Principle

One of the interesting properties of aluminum is that it is amphoteric, meaning it will dissolve in both strong, aqueous acids and strong, aqueous bases. In both cases, the formation of hydrogen gas is observed:

$$2Al(s) + 6H^+(aq) \rightarrow 2Al^{3+}(aq) + 3H_2(g)$$
$$2Al(s) + 6H_2O(l) + 2OH^-(aq) \rightarrow 2Al(OH)_4^-(aq) + 3H_2(g)$$

1. The first reaction to be carried out is that of aluminum and potassium hydroxide, KOH:

$$2Al(s) + 2KOH(aq) + 6H_2O(l) \rightarrow 2Al(OH)_4^-(aq) + 2K^+(aq) + 3H_2(g)$$

Adding sulfuric acid, H_2SO_4 to the resulting solution will cause the compound $Al(OH)_3$ to precipitate; however, $Al(OH)_3$ is also amphoteric and will re-dissolve when more acid is added:

$$2K^+(aq) + 2Al(OH)_4^-(aq) + H_2SO_4(aq) \rightarrow 2K^+(aq) + 2Al(OH)_3(s) + 2H_2O(l) + SO_4^{2-}(aq)$$
$$2Al(OH)_3(s) + 3H_2SO_4(aq) \rightarrow 2Al^{3+}(aq) + 3SO_4^{2-}(aq) + 6H_2O(l)$$

Crystals of the double salt $KAl(SO_4)_2 \cdot 12H_2O(s)$, or alum, will form upon cooling this final solution since the solubility of alum in water decreases as the temperature is lowered:

$$K^+(aq) + Al^{3+}(aq) + 2SO_4^{2-}(aq) + 12H_2O(l) \rightarrow KAl(SO_4)_2 \cdot 12H_2O(s)$$

2. In order to confirm that your synthesis of alum resulted in the desired product, we need to perform a qualitative analysis of the compound. In this case, chemical reactions will be performed with an alum sample that will confirm the presence of K^+, Al^{3+} and SO_4^{2-}.

In the first test, alum will be reacted with barium chloride in an aqueous solution, which will result in the following chemical reaction:

$$SO_4^{2-}(aq) + Ba^{2+}(aq) \rightarrow BaSO_4(s)$$

The precipitation of solid material, $BaSO_4$, from a solution of alum upon reaction with barium chloride is a positive test for the presence of sulfate ion, SO_4^{2-}.

The second test is performed to confirm the presence of potassium and is a flame test. Potassium is volatilized at the very high temperature of a flame (about 1000 ℃) at which point it imparts a bluish-purple color to the flame. After a few seconds in the flame, the sulfur dioxide will be driven off of your alum sample, and the solid material remaining will consist of aluminum oxides.

The third test confirms the presence of aluminum ion and involves its reaction with potassium hydroxide. A wispy, gelatinous precipitate of $Al(OH)_3$ will form upon addition of a small amount of KOH to the aqueous alum solution. Further addition of KOH will cause the precipitate to re-dissolve.

3. It is possible to calculate this percent yield in the following manner

(1) Calculate the amount of substance of reactant, in this case, aluminum.

$$\text{amount of substance of Al} = \frac{\text{mass of Al (g)}}{\text{molar mass of Al (g/mol)}}$$

(2) From the stoichiometry of the reaction, determine the expected amount of substance of

实验 7　硫酸铝钾(明矾)的合成

实验目的

1. 学会用金属铝合成化合物硫酸铝钾 $KAl(SO_4)_2 \cdot 12H_2O$(明矾)。
2. 学会明矾的定性分析。

实验原理

铝的两性是其有趣的性质之一：铝既能溶于强酸也能溶于强碱,两种反应都能产生氢气。

$$2Al(s)+6H^+(aq) \rightarrow 2Al^{3+}(aq)+3H_2(g)$$
$$2Al(s)+6H_2O(l)+2OH^-(aq) \rightarrow 2Al(OH)_4^-(aq)+3H_2(g)$$

1. 硫酸铝钾的制备：本实验采用饮料罐金属铝合成硫酸铝钾。第一步反应是铝与氢氧化钾 KOH 反应：

$$2Al(s)+2KOH(aq)+6H_2O(l) \rightarrow 2Al(OH)_4^-(aq)+2K^+(aq)+3H_2(g)$$

在上述反应后的溶液中加入硫酸,会产生 $Al(OH)_3$ 沉淀。

$$2K^+(aq)+2Al(OH)_4^-(aq)+H_2SO_4(aq) \rightarrow 2K^+(aq)+2Al(OH)_3(s)+2H_2O(l)+SO_4^{2-}(aq)$$

然而,$Al(OH)_3$ 是两性的,加较多的酸后将会再次溶解。

$$2Al(OH)_3(s)+3H_2SO_4(aq) \rightarrow 2Al^{3+}(aq)+3SO_4^{2-}(aq)+6H_2O(l)$$

因为硫酸铝钾在水中的溶解度随着温度的降低而降低,冷却溶液,复盐 $KAl(SO_4)_2 \cdot 12H_2O$ 的晶体就会生成。

$$K^+(aq)+Al^{3+}(aq)+2SO_4^{2-}(aq)+12H_2O(l) \rightarrow KAl(SO_4)_2 \cdot 12H_2O(s)$$

2. 硫酸铝钾的定性分析：为了确定合成的产品为硫酸铝钾(明矾),需要进行化合物的定性分析,通过试样的化学反应,能够确定 K^+、Al^{3+} 和 SO_4^{2-} 的存在。

(1) 与 $BaCl_2$ 反应产生沉淀,鉴定 SO_4^{2-} 存在。硫酸铝钾将和 $BaCl_2$ 在水溶液中进行下列反应：

$$SO_4^{2-}(aq)+Ba^{2+}(aq) \rightarrow BaSO_4(s)$$

如果反应体系中有 $BaSO_4$ 沉淀产生,反应呈阳性,则表示有 SO_4^{2-}。

(2) 焰色反应,确证钾的存在。在高温火焰(大约 1000 ℃)下,钾被挥发,在此温度下产生蓝紫色火焰。几秒钟后,硫酸铝钾试样在火焰中将产生二氧化硫,剩下的产物由氧化铝组成。

product. A review of the above reactions reveals that there is a 1∶1 relationship between the aluminum containing reactant and the aluminum containing product in all cases. Therefore, one mole of aluminum metal should produce one mole of alum.

$$\text{amount of substance of alum} = \text{amount of substance of Al} \times \frac{1 \text{ mole alum}}{1 \text{ mole Al}}$$

(3) From the expected amount of substance of product, calculate the expected mass of product or the theoretical yield.

Theoretical yield = [amount of substance of $KAl(SO_4)_2$] × [molar mass of $KAl(SO_4)_2$, g/mol]

(4) Calculate the percent yield by dividing your actual yield by the theoretical yield and multiplying by 100%.

$$\text{Percent yield} = \frac{\text{mass of } KAl(SO_4)_2 \text{ (g)}}{\text{theoretical yield (g)}} \times 100\%$$

Apparatus and Chemicals

Apparatus: beaker, 250 mL×2; filter flask; Buchner funnel; graduated cylinder, 100 mL; wire loop; Bunsen burner; watchglass.

Chemicals: beverage can or aluminum strip or aluminum sheet; H_2SO_4, 9 mol·L^{-1}; KOH, 1.4 mol·L^{-1}; $BaCl_2$, 0.5 mol·L^{-1}; HCl, concentrated; ethanol; litmus paper.

Procedure

1. Synthesis of alum

Obtain a piece of scrap aluminum from a beverage can and cut it into thin strips. Accurately weigh out about 0.5 g of these strips and place them in a 250 mL beaker. Carefully add 25 mL of 1.4 mol·L^{-1} potassium hydroxide, KOH solution. If any of this solution should spill, clean it up immediately.

Proceed with the reaction in a fume hood. Set the beaker on a hot plate and turn the dial to medium high. Heat the mixture for about 30 minutes in order to dissolve as much of the aluminum as possible. Stir frequently in order to keep the metal pieces from floating to the top of the froth. If necessary, add water to keep the volume constant. Do not allow bubbling to become so vigorous that the froth spills over the side of the beaker. The hydrogen gas (flammable!) generated by the reaction will be carried away by the suction of the fume hood. The reaction may splatter so heating should be slow and care should be exercised at all times.

After most of the metal has dissolved, the fizzing around each piece should stop. There may still be some undissolved solids present such as paint and varnish particles. Turn the hot plate off and gravity filter the hot solution through filter paper in a funnel. Rest the funnel in a funnel holder. Collect the filtrate in a 250 mL beaker and cool to room temperature.

Acidify the solution slowly by carefully adding 10 mL of 9 mol·L^{-1} H_2SO_4 in 2−3 mL portions with continuous stirring. Caution! Considerable heat is generated by this reaction. A precipitate of aluminum hydroxide, $Al(OH)_3$, will form during the initial addition of acid; however, this solid will re-dissolve upon further acidification.

If any white lumps still remain after addition of all the sulfuric acid, gently heat the mixture, with stirring, until the solution becomes clear. Remove the heat and let the solution cool for a few minutes. Make an ice bath by filling a large beaker 1/2−3/4 full with ice and adding cold water. Place the cooled beaker in the bath for about 20 minutes. Crystals of alum should form in the beaker. If no crystals grow, stir the solution and try to induce crystal formation by scratching the

(3) 铝与 KOH 的反应,确定铝的存在。将少量 KOH 加到硫酸铝钾溶液中,会产生束状 $Al(OH)_3$ 胶体沉淀,继续加 KOH,沉淀将再次溶解。

3. 产率的计算

(1) 计算反应物铝的物质的量:

$$铝的物质的量 = \frac{铝的质量(g)}{铝的摩尔质量(g/mol)}$$

(2) 根据反应的计量关系,确定产品的物质的量。上述反应中,在反应物中的铝和产物中的铝的关系是 1:1,因此,1 摩尔的金属铝能产生 1 摩尔的硫酸铝钾。

$$硫酸铝钾的物质的量 = 铝的物质的量 \times \frac{1\ 摩尔明矾}{1\ 摩尔铝}$$

(3) 从预期的产品的物质的量,计算所期望的产品的质量或理论产量。

$$理论产量 = (硫酸铝钾的物质的量) \times (硫酸铝钾的摩尔质量,g/mol)$$

(4) 计算产率:

$$产率 = \frac{硫酸铝钾的质量(g)}{理论产量(g)} \times 100\%$$

仪器和试剂

仪器:烧杯,250 mL×2;抽滤瓶;布氏漏斗;量筒,100 mL;铂丝圈;酒精喷灯;表面皿。
试剂:饮料罐或铝片或铝条;
$\quad\quad H_2SO_4$,9 mol·L^{-1};
$\quad\quad$KOH,1.4 mol·L^{-1};
$\quad\quad BaCl_2$,0.5 mol·L^{-1};
$\quad\quad$浓 HCl;酒精;石蕊试纸。

实验过程

1. 硫酸铝钾(明矾)的合成

将铝制饮料罐剪成细条。准确称取 0.5 g 置于 250 mL 的烧杯中,小心加入 1.4 mol·L^{-1} KOH 溶液 25 mL。如果有溶液溢出,要立即擦净。

把烧杯放到电热板上,将旋钮扭至中挡,加热 30 分钟,应使尽可能多的铝溶解。为防止金属片随泡沫浮起,要频繁进行搅动溶液。必要时,加水保持溶液体积。要防止泡沫泛滥而溢出烧杯。反应应在通风橱中进行,通风橱能抽走反应所产生的易燃氢气。加热应小心而缓慢,以防止反应溶液的溅失。

当大部分的金属溶解后,就不再产生气体,但可能会存在一些未溶解的固体(如漆包着的小颗粒)。此时,关上电热板旋钮,常压过滤热溶液。将漏斗放在漏斗架上,用 250 mL 烧杯盛接滤液,冷却至室温。

分几次小心加入 9 mol·L^{-1} H_2SO_4,边搅拌边加入,每次 2~3 mL,共加入 10 mL,使溶液酸化,注意:反应将产生大量的热!开始加酸酸化时会产生 $Al(OH)_3$ 沉淀,但是,继续

inside of the beaker with a glass rod.

Pour your suspension of crystals onto the filter paper. Allow the liquid to drain completely, then wash out the beaker with 10 mL of 1 : 1 ethanol - water solution in order to completely transfer any crystals still left in it. Pour this mixture (in which the alum is not very soluble) over the crystals in the funnel and again let the liquid drain. Finally, wash the crystals with 10 mL of ethanol. Leave the vacuum on for about 5 minutes in order to draw air through the crystals to dry them. Carefully scrape the crystals off the filter paper and onto a watchglass.

Label a clean, *dry* vial with the word "Alum", your name and the date, and leave a space to enter the total mass produced. Weigh the vial and cap to the nearest 0.001 g and record the weight in your notebook. This is the tare weight of the container. Transfer your dried alum crystals to the vial, weigh again to the nearest 0.001 g and record this weight in your notebook. Enter the total mass of alum produced on the label of the vial.

2. Chemical tests for qualitative analysis of alum

(1) Transfer enough crystals of alum to just cover the pointed end of your spatula to a small test tube, and dissolve them in 6 drops of distilled water (use a medicine dropper). Remove a drop of this solution from the test tube using a glass stirring rod, and transfer it to a piece of blue litmus paper. Record the color of the litmus paper. Add one or two drops of $0.5 \text{ mol} \cdot \text{L}^{-1}$ $BaCl_2$ solution to the alum solution in the test tube, and record your observations.

(2) Place a small amount of your solid alum product on a watchglass and carry it to the section of the laboratory which has been set up for the flame tests. Your instructor will provide you with a wire loop. Clean the loop by dipping it in a solution of concentrated HCl and placing it in the Bunsen burner flame until it glows red hot. Now, touch the loop, while still hot, to the sample of alum on the watchglass such that some of the crystals stick to the loop. Hold the crystals in the flame for several seconds (until the solid glows), and record your observations (especially the color of the flame).

(3) Place 1 drop of $1.4 \text{ mol} \cdot \text{L}^{-1}$ KOH solution in a small test tube and add 9 drops of water. This dilution will result in a solution which has a concentration of $0.14 \text{ mol} \cdot \text{L}^{-1}$ KOH. Transfer a spatula tip full of alum crystals to a small, clean test tube, and dissolve in 5 drops of water. While stirring with a small glass rod, add 1 to 4 drops of $0.14 \text{ mol} \cdot \text{L}^{-1}$ KOH solution and observe what happens. Now add 1 to 4 drops of the more concentrated $1.4 \text{ mol} \cdot \text{L}^{-1}$ KOH solution, and observe what happens.

Clean up your work table, fume hood and sink. Do not leave scraps of aluminum can in the sink. Hand in the labeled vial that contains your product to your instructor before leaving.

Notes

1. KOH is highly caustic; take extreme care not to splatter and avoid contact with skin and eyes.

2. Ethanol is flammable. Before lighting a Bunsen burner, check to be sure that there is no ethanol in the vicinity.

3. Concentrated HCl is extremely corrosive and should be handled with care. If any HCl spills on your skin, wash the area immediately with running water.

酸化沉淀将会再溶解。

如果全部硫酸都加完后仍有少量沉淀,温和加热、搅拌,直到溶液澄清。关掉电源,使溶液冷却几分钟,在一个大烧杯中加入二分之一到四分之三的冰块和冷水做成冰浴,把前述冷却下来的烧杯放到冰浴中 20 分钟,应有晶体生成。如果没有生成晶体,搅拌溶液并用玻璃棒摩擦烧杯壁促进晶体生成。

将含有晶体的悬浊液倒入漏斗中,用 10 mL 1∶1 的乙醇—水溶液洗涤烧杯(明矾在乙醇—水的混合物中难溶),以便将烧杯中余下的晶体完全转移到漏斗中,洗涤烧杯,将洗涤液倒入漏斗并过滤,最后,用 10 mL 乙醇洗涤晶体,抽滤,为尽量抽干液体,抽滤 5 分钟,再小心地将晶体刮到表面皿上。

在一个洁净干燥的小瓶上写上"明矾"、姓名和实验日期,留出写产品产量的位置。称量小瓶和瓶盖,准确至 0.001 g,在记录本上记下数据,这是容器的皮重,将干燥的明矾转移到小瓶里,称重,准确至 0.001 g,在记录本上记下数据,在标签上写上明矾的质量。

2. 明矾的定性分析

(1) 用小刮刀取一定量的明矾置于小试管中,用 6 滴蒸馏水将其溶解。用玻璃棒从试管中取一滴试液滴在蓝色石蕊试纸上,记录石蕊试纸的颜色。向装有明矾溶液的试管中加 1~2 滴 0.5 mol·L^{-1} $BaCl_2$ 溶液,记录所观察的现象。

(2) 取少量固体明矾于表面皿上做焰色反应,实验室将提供铂丝圈。清洁铂丝圈的方法是:将铂丝蘸浓盐酸然后在酒精喷灯上烧至通红为止。用热的铂丝在表面皿上蘸明矾试样,然后放在喷灯的火焰上直到固体发光,记录实验现象,特别要注意火焰的颜色。

(3) 取一滴 1.4 mol·L^{-1} KOH 溶液,加 9 滴蒸馏水,得到浓度为 0.14 mol·L^{-1} 的 KOH 溶液。取一刮刀明矾固体于一洁净的小试管中,加水 5 滴,在玻棒的搅拌下加入 1~4 滴 0.14 mol·L^{-1} KOH 溶液,观察实验现象。然后再加入 1~4 滴更浓的 1.4 mol·L^{-1} KOH 溶液,观察实验现象。

收拾实验台、通风橱和水槽,不要把饮料罐废屑丢入水槽。离开实验室之前将装有明矾产品、贴有标签的瓶子交给老师。

注意事项

1. KOH 具腐蚀性,实验过程要特别小心 KOH 溶液飞溅,避免与眼睛和皮肤接触。
2. 酒精易燃,在点燃酒精喷灯之前,要进行检查,以确保灯附近没有酒精。
3. 浓盐酸有很强的腐蚀性,使用时应该小心,若不小心弄到皮肤上,应立即用水冲洗。

思 考 题

1. 讨论定性分析第一个实验的实验现象:石蕊试纸实验说明什么问题?从 $BaCl_2$ 实验中可以得到什么结论?

2. 从焰色反应中能看到什么?能确定试样中存在钾离子吗?将焰色反应残留的固体溶解并加入 $BaCl_2$ 溶液,将会发生什么现象?请解释。

Questions

1. Discuss your observations from the first chemical test. What did the test with litmus paper tell you? What conclusion can you draw from the test with $BaCl_2$?

2. What did you see in the flame test, and does this confirm the presence of potassium ion in your sample? What do you think would happen if you could dissolve the solid that remains after the flame test and added $BaCl_2$ solution to it? Explain.

3. What did you see in the reaction of alum with KOH, and did it confirm the presence of aluminum? What did you see when more KOH was added? Write a chemical equation for the reaction of the aluminum hydroxide compound, $Al(OH)_3$, with KOH.

4. List sources of error that would contribute toward lowering the percent yield of alum.

5. List sources of error that would contribute toward producing a percent yield of greater than 100%.

6. Calculate the theoretical yield of alum and the percent yield that you obtained from the preparation.

3. 明矾和 KOH 反应有什么现象？该现象能证明铝存在吗？在明矾溶液中加入过量的 KOH 后又会发生什么现象？写出 Al(OH)$_3$ 与 KOH 反应的化学方程式。

4. 指出可能导致明矾产率不高的原因。

5. 指出可能导致明矾产率超过 100% 的原因。

6. 按照实验过程计算明矾的理论产量和实际收率。

Experiment 8 Qualitative Analysis of Cations (1)

Purpose

1. To know the principles and methods in qualitatively analysis of cations.
2. To learn the general operation techniques in qualitatively analysis.
3. To learn the identifications of Ag^+, Fe^{3+} and Co^{2+}.

Principle

1. Separation and idenfication of Ag^+

The first step in the separation of these three cations involves the precipitation of the silver ion as the insoluble silver chloride salt. This precipitation is accomplished simply through the addition of hydrochloric acid.

$$Ag^+(aq) + HCl(aq) \rightarrow AgCl(s) \downarrow + H^+(aq) \qquad (1)$$

Since neither iron (Ⅲ) nor cobalt (Ⅱ) forms an insoluble chloride salt, the silver can then be separated from the other two cations by centrifuging the test tube and decanting the supernatant.

A test is next performed to confirm the presence of silver in the precipitate that remains in the test tube. First, ammonia is added, which results in dissolution of the silver (Ⅰ) compound.

$$AgCl(s) + 2NH_3(aq) \rightarrow Ag(NH_3)_2^+(aq) + Cl^-(aq) \qquad (2)$$

Adding nitric acid will shift the equilibrium to the left as the acid reacts with the ammonia, NH_3, causing the silver ion to re-precipitate as silver chloride.

2. Separation and idenfication of Fe^{3+}

The supernatant that was decanted earlier contains both Fe^{3+} and Co^{2+}. Addition of ammonia to this solution causes the iron (Ⅲ) ion to form an insoluble hydroxide and precipitate out of solution. Centrifugation and decantation leaves $Fe(OH)_3$ solid which can be dissolved in hot hydrochloric acid.

$$Fe^{3+}(aq) + 3NH_3(aq) + 3H_2O(l) \rightarrow Fe(OH)_3(s) \downarrow + 3NH_4^+(aq) \qquad (3)$$

There are two confirmation tests for iron (Ⅲ). In the first, addition of potassium thiocyanate solution produces a blood red-color if Fe^{3+} is present.

$$Fe^{3+}(aq) + SCN^-(aq) \rightarrow Fe(SCN)^{2+}(aq, \text{blood red}) \qquad (4)$$

In the second test, ammonia is first added until the solution is only weakly acidic, and this is followed by the addition of potassium ferrocyanide, $K_4[Fe(CN)_6]$, solution. The dark blue solid that appears when iron (Ⅲ) is present is a dye known as Prussian blue. A suspension of this compound is used in some blue inks.

$$4Fe^{3+}(aq) + 3K_4[Fe(CN)_6](aq) \rightarrow Fe_4[Fe(CN)_6]_3(s, \text{blue}) \downarrow + 12K^+ \qquad (5)$$

3. Separation and idenfication of Co^{2+}

The presence of the cobalt ion, which now remains in the decanted supernatant solution, can be confirmed in a simple test. The basic solution is made strongly acidic with the addition of hydrochloric acid. Addition of a solution of potassium nitrite, KNO_2, will produce an insoluble yellow compound, potassium hexanitritocobaltate (Ⅲ).

$$Co^{2+}(aq) + 7NO_2^-(aq) + 3K^+(aq) + 2H^+(aq) \rightarrow NO(g) + H_2O(l) + K_3[Co(NO_2)_6](s, \text{yellow}) \qquad (6)$$

The reaction in Eq. 6 produces nitric oxide, NO, which is a colorless gas. Nitric oxide reacts with oxygen in the air to produce nitrogen dioxide, NO_2, which may be visible as a red-brown gas.

$$2NO(g, \text{colorless}) + O_2(g) \rightarrow 2NO_2(g, \text{red-brown}) \qquad (7)$$

实验 8　阳离子定性分析(1)

实验目的

1. 了解阳离子定性分析的基本原理和方法。
2. 学会定性分析的基本操作。
3. 学会 Ag^+，Fe^{3+} 和 Co^{2+} 离子的鉴定。

实验原理

1. Ag^+ 的分离与鉴定

分离 Ag^+，Fe^{3+} 和 Co^{2+} 三种阳离子，第一步是将银离子以不溶于水的氯化物形式沉淀下来。这种沉淀通过在溶液中加入 HCl 而得到。

$$Ag^+(aq) + HCl(aq) \rightarrow AgCl(s)\downarrow + H^+(aq) \tag{1}$$

因为 Fe^{3+} 和 Co^{2+} 都不能形成不溶的氯化物（$FeCl_3$、$CoCl_2$ 皆易溶），所以离心沉降后，转移出离心试管中的上清溶液，便可将 Ag^+ 与另外两种离子分离。

接下来是确定留在试管中的沉淀里有无银离子存在。首先，加入氨水，沉淀溶解。

$$AgCl(s) + 2NH_3(aq) \rightarrow Ag(NH_3)_2^+(aq) + Cl^-(aq) \tag{2}$$

再加入 HNO_3，由于 HNO_3 与氨的反应，平衡将向左移动，Ag^+ 将以 AgCl 沉淀的形式再次析出。

2. Fe^{3+} 的分离与鉴定

在上述经离心分离后的含有 Fe^{3+} 和 Co^{2+} 的离心液中加入氨水，产生 $Fe(OH)_3$ 沉淀。

$$Fe^{3+}(aq) + 3NH_3(aq) + 3H_2O(l) \rightarrow Fe(OH)_3(s)\downarrow + 3NH_4^+(aq) \tag{3}$$

离心沉降后，将上清液转移出去，得到 $Fe(OH)_3$ 固体。将 $Fe(OH)_3$ 固体溶于热 HCl 溶液。

Fe^{3+} 的鉴定方法有两种。第一种方法是，向待测试的溶液中加入 KSCN 溶液，如果试液产生血红色，表示有 Fe^{3+}。

$$Fe^{3+}(aq) + SCN^-(aq) \rightarrow Fe(SCN)^{2+}(aq,血红色) \tag{4}$$

第二种方法，先向待测试的溶液中加入 NH_3，将溶液调成弱酸性，然后加入 $K_4[Fe(CN)_6]$ 溶液，如果产生深蓝色沉淀，则表示有 Fe^{3+}。这种深蓝色的沉淀是一种染料，称为普鲁士蓝。普鲁士蓝的悬浊液可用做蓝墨水。

$$4Fe^{3+}(aq) + 3K_4[Fe(CN)_6](aq) \rightarrow Fe_4[Fe(CN)_6]_3(s,蓝)\downarrow + 12K^+ \tag{5}$$

3. Co^{2+} 的鉴定

分离了 $Fe(OH)_3$ 的离心液中只剩下 Co^{2+}，Co^{2+} 可用简单的实验来鉴定。加入 HCl 使碱性溶液呈强酸性，再加入 KNO_2，生成一种不溶的黄色化合物六硝基合钴(Ⅲ)酸钾。

$$Co^{2+}(aq) + 7NO_2^-(aq) + 3K^+(aq) + 2H^+(aq) \rightarrow NO(g) + H_2O(l) + K_3[Co(NO_2)_6](s,黄)\downarrow \tag{6}$$

反应 6 中产生无色的一氧化氮气体(NO)，NO 与空气中的氧气反应可生成棕红色的二氧化

Apparatus and Chemicals

Apparatus: test tubes; test tube racks; beaker, 500 mL(waste), 250 mL; droppers; watchglass; hotplate; centrifuge; water bath.

Chemicals: 0.1 mol·L^{-1} silver nitrate, $AgNO_3$; 0.1 mol·L^{-1} iron(Ⅲ) nitrate, $Fe(NO_3)_3$ (aq); 0.1 mol·L^{-1} cobalt(Ⅱ) nitrate, $Co(NO_3)_2$(aq); 6 mol·L^{-1} potassium nitrite, KNO_2; HCl(aq), 6 mol·L^{-1}; HNO_3(aq), 6 mol·L^{-1}; H_2SO_4(aq), 6 mol·L^{-1}; HAc, 6 mol·L^{-1}; NH_3(aq), 6 mol·L^{-1}; KSCN(aq), 0.1 mol·L^{-1}; $K_4Fe(CN)_6$(aq), 0.1 mol·L^{-1}; Litmus or pH Paper.

Procedure

Make a table in your notebook *before* coming to lab. Enter all your observations in this table. You will obtain 2 mL of a solution that contains Ag^+, Fe^{3+} and Co^{2+} in a test tube.

1. Separating and confirming the presence of Ag^+ ion

Transfer 15 drops of the solution in the test tube to a smaller centrifugal test tube. Add 6 drops of 6 mol·L^{-1} HCl solution to the 15 drops of the cation solution to bring about precipitation. Be sure to stir (with a clean stirring rod) the contents of the test tube. Centrifuge the test tube for 2 minutes. Add one drop of 6 mol·L^{-1} HCl to the test tube and carefully observe the supernatant. If more precipitate is formed upon this addition, add another drop or two of HCl and centrifuge again. Re-test the solution for complete precipitation.

When it has been determined that precipitation of silver chloride is complete, decant the supernatant into a clean test tube, label it and set it aside. Wash the precipitate that remains in the test tube by adding 10–15 drops of water to it. Stir the precipitate well and centrifuge the test tube. Decant and discard the supernatant. Repeat this procedure to wash the precipitate one more time.

Add 8 drops of 6 mol·L^{-1} NH_3 solution to the washed precipitate and stir well. Record your observations. Add 6 mol·L^{-1} HNO_3 solution with mixing until the solution is acidic when tested with litmus paper. Record your observations.

2. Separating and confirming the presence of Fe^{3+} ion

Add a sufficient amount of 6 mol·L^{-1} NH_3 solution to the supernatant that was set aside earlier until the solution is basic when tested with litmus paper. Stir well and be sure the solution is *strongly* basic! Centrifuge the test tube and test for complete precipitation. When it has been determined that the precipitation of iron (Ⅲ) hydroxide is complete, decant the supernatant into a clean test tube, label it and set it aside. Wash the remaining precipitate with water.

Add 10 drops of 6 mol·L^{-1} HCl to the washed precipitate. Heat the mixture in a boiling water bath to dissolve the precipitate, if necessary. When the precipitate has dissolved, test with litmus paper to be sure it is strongly acidic.

Separate the solution into two approximately equal portions. To one portion, add 3 drops of 0.1 mol·L^{-1} KSCN solution and record your observations. To the other portion, add 6 mol·L^{-1} NH_3 until the solution, which was strongly acidic, now tests only weakly acidic. If the solution becomes basic, add 6 mol·L^{-1} acetic acid solution dropwise until it is weakly acidic. Add 3 drops of 0.1 mol·L^{-1} $K_4[Fe(CN)_6]$ solution and record your observations.

3. Confirming the presence of Co^{2+} ion

To the supernatant that was set aside, add 6 mol·L^{-1} HCl until the solution tests acidic. In

氮(NO_2)。

$$2NO(g,无色)+O_2(g)\rightarrow 2NO_2(g,棕红) \tag{7}$$

仪器和试剂

仪器:试管;试管架;烧杯,500 mL(做废液缸),250 mL;滴管;表面皿;电炉;离心机;水浴锅。

试剂:硝酸银,$AgNO_3$(aq),0.1 mol·L^{-1};硝酸铁,$Fe(NO_3)_3$(aq),0.1 mol·L^{-1};硝酸钴,$Co(NO_3)_2$(aq),0.1 mol·L^{-1};亚硝酸钾,KNO_2,6 mol·L^{-1};HCl(aq),6 mol·L^{-1};HNO_3(aq),6 mol·L^{-1};H_2SO_4(aq),6 mol·L^{-1};HAc,6 mol·L^{-1};NH_3(aq),6 mol·L^{-1};KSCN(aq),0.1 mol·L^{-1};$K_4Fe(CN)_6$(aq),0.1 mol·L^{-1};石蕊试纸或 pH 试纸。

实验步骤

进实验室前在记录本上制作一个表格,该表格用于填写实验时所观察到的现象。每人将领到 2 mL 溶液,溶液中含有 Ag^+、Fe^{3+} 和 Co^{2+}。

1. Ag^+ 的分离和鉴定

取 15 滴溶液加到离心试管中,加 6 滴 6 mol·L^{-1} HCl,有沉淀产生,用一干净的玻棒搅拌沉淀,离心沉降 2 分钟,向试管中的上层清液加入一滴 6 mol·L^{-1} HCl,仔细观察,如果有沉淀,再加入 1~2 滴 HCl 并离心沉降,再次进行上述检查,直到沉淀完全。

当确定 AgCl 沉淀完全时,将上层清液转移到一干净的试管中,贴上标签放到一旁。向原离心试管的沉淀加入 10~15 滴蒸馏水,充分搅拌沉淀,离心分离,弃去上层清液。重复上述操作一次或多次,使洗涤沉淀干净。

向洗涤过的沉淀加入 8 滴 NH_3 溶液并充分搅拌,记录所观察的现象,再加入 6 mol·L^{-1} HNO_3 溶液至混合液中使溶液呈酸性(用石蕊试纸检查),记录实验现象。

2. Fe^{3+} 的分离和鉴定

向步骤 1 中得到的、放置一旁的离心液中,加入足够量的 6 mol·L^{-1} NH_3,直到溶液用石蕊试纸检查呈碱性。充分搅拌确保溶液呈强碱性!离心分离,并检查沉淀是否完全。当 Fe^{3+} 以 $Fe(OH)_3$ 的形式沉淀完全时,转移上层清液至另一干净试管中,贴上标签放到一旁,余下的沉淀用蒸馏水进行洗涤。

在经过洗涤的 $Fe(OH)_3$ 沉淀中加入 10 滴 6 mol·L^{-1} HCl,必要时可在沸水浴上加热促进其溶解。当沉淀溶解后,用石蕊试纸检查,确保溶液为强酸性。

将溶液大致分为两部分,其中一部分加入 3 滴 0.1 mol·L^{-1} KSCN 溶液,记录所观察到的现象;另一部分加入 6 mol·L^{-1} NH_3 使溶液由强酸性变为弱酸性,如果 NH_3 溶液加过量使溶液变成碱性了,则逐滴加入 6 mol·L^{-1} HAc 将溶液再调成弱酸性。加入 3 滴 0.1 mol·L^{-1} $K_4[Fe(CN)_6]$ 溶液,记录所观察到的现象。

3. Co^{2+} 的鉴定

向步骤 2 中放置一旁的离心液中,加入 6 mol·L^{-1} HCl 直到溶液呈酸性。在通风橱中(因实验产生有毒气体!)加入 6 滴 6 mol·L^{-1} KNO_2(氧化剂,有刺激性),充分混合,记录所

a fume hood (*the gas generated is toxic*!), add 6 drops of 6 mol·L^{-1} KNO$_2$ solution (oxidizer, irritant), mix well and record your observations. When the evolution of gas has subsided (this may require several minutes), centrifuge the test tube and discard the supernatant so that the precipitate and its color can be clearly observed. The appearance of a yellow precipitate confirms the presence of Co^{2+}.

Obtain from your instructor an unknown solution that contains one or more of the cations Ag$^+$, Fe^{3+} and Co^{2+}. Repeat the above procedure using 15 drops of this solution. Record all your observations, then fill in the data sheet and hand it in to your TA before leaving the laboratory.

Questions

1. After the silver ion is precipitated as silver chloride, the precipitate is washed with water. Why is this washing performed, and what *exactly* is being washed away? Similarly, the iron (Ⅲ) hydroxide precipitate is washed with water. What is being washed away in this case?

2. Eq. 7 represents a redox reaction. Which species is oxidized (write the element and the oxidation numbers before and after oxidation)? Which species is reduced (write the element and the oxidation numbers before and after reduction)?

3. What will happen if ammonia is added to a mixture of Ag$^+$, Fe^{3+} and Co^{2+}? Describe the subsequent steps that could be performed to separate the three cations.

Techniques for Qualitative Analysis

The process of finding out *what* compounds are contained in a sample is called *qualitative analysis*. For example, you will use *qualitative* analysis techniques in order to determine what metallic cations are contained in aqueous solutions that are provided for you. These techniques will allow you to both separate the cations in the mixture as well as to identify them. The various salts of the cations that you will be studying have varying solubility in water. The differences in solubility of these salts can be exploited in such a way as to allow for separation of the cations. As an example, let's examine the cations calcium, Ca^{2+}, and sodium, Na$^+$. Calcium chloride is readily soluble in water, while calcium hydroxide is highly insoluble in water. By contrast, sodium chloride and sodium hydroxide are both readily soluble in water. A mixture of calcium chloride and sodium chloride can be treated with an aqueous solution that contains hydroxide ions. The result of such treatment will be the formation of a solid precipitate which is composed of insoluble calcium hydroxide. The sodium hydroxide, on the other hand, remains dissolved in the water. The solid material that contains the calcium ions can be separated from the aqueous solution containing the sodium ions using a technique such as filtration. In summary, the addition of an appropriate chemical reagent to an aqueous mixture of cations can selectively cause one or more of the cations to form a solid precipitate while one or more of the cations will remain dissolved in the water, thus allowing for separation.

1. Avoiding Contamination As you follow the procedures for separating and identifying the cations in a mixture, it will be extremely important to avoid contamination of the samples and chemicals at all times. The presence of a contaminant that contains a cation will lead to a false positive result; therefore, it is critical that all glassware be thoroughly cleaned before it is used and should remain clean throughout the procedure. Laying a glass stirring rod on a dirty lab bench, for instance, could lead to contamination. Touching the tip of a medicine dropper to the side of a test tube could cause contaminants to be picked up and transferred to another solution. In

观察到的现象。当产生的气体逐渐减少(大约需要几分钟),离心沉降并弃去上层溶液以便清楚地观察沉淀的颜色。如有黄色沉淀,表示 Co^{2+} 存在。

从指导教师处领取一份未知液,未知液含有 Ag^+、Fe^{3+} 和 Co^{2+} 中的一种或多种。取 15 滴溶液重复上述过程,将实验现象写在记录纸上,离开实验室前交给老师。

思 考 题

1. 当银以氯化物的形式沉淀后,沉淀要用水洗涤,为什么? 洗去的是什么? 同样,$Fe(OH)_3$ 也要用水洗,这种情况下洗去的是什么?

2. 反应 7 表示一个氧化还原反应,在反应中,哪种物质被氧化(写出氧化前后的元素符号和氧化数)? 哪种物质被还原(写出还原前后的元素符号和氧化数)?

3. 在 Ag^+、Fe^{3+} 和 Co^{2+} 的混合溶液中加入氨水会发生什么反应? 写出分离这三种阳离子的步骤。

定性分析操作技术

发现样品中有什么化合物的过程叫做定性分析。以阳离子定性分析为例,检测溶液中的金属阳离子用到定性分析技术,这些技术包括阳离子的分离和鉴定。不同阳离子的盐在水中有不同的溶解性,它们溶解性的差别可用于阳离子分离。例如,分离 Na^+ 和 Ca^{2+},由于氯化钙易溶于水,而氢氧化钙难溶于水,与之形成对照的,是氯化钠和氢氧化钠都易溶于水,在含有氯化钠和氯化钙的水溶液中加入 OH^- 后,将有难溶于水的氢氧化钙沉淀生成,而氢氧化钠溶于水,却还留在溶液中,采用合适的分离技术就可以将含有钙离子的固体物质和含有钠离子的水溶液分离。总之,在阳离子的混合溶液中加入某种适当的化学试剂,选择性地生成一种或多种固体沉淀,而另一种或多种的阳离子仍溶于水中而得到分离。

1. 避免污染

在进行混合阳离子的分离和鉴定的任何操作步骤中,都要特别注意避免试样和化学试剂的污染。含有被分析阳离子的污染物会导致假阳性的错误结果,因此,实验前要将所使用的仪器做彻底的清洗,整个实验过程中要保持所使用仪器的清洁,这些是非常关键的。例如,将实验用的玻璃棒放在脏的实验台面上,就会带来污染;当使用滴管吸取和转移溶液时将滴管的尖嘴伸入试管内,接触到试管的内壁时也会引起污染;此外,实验过程的所有用水都必须是实验室中的蒸馏水,来自水龙头的自来水含有多种离子,会产生错误的现象,使人得出错误的结论。

2. 仪器和药品

经过分离,用于鉴定阳离子实验所需要的样品量很少,因此,实验时需用到最小号的试管(10 mm×75 mm)、细小的玻棒和滴管。不需要用量筒测量所需溶液的体积,通过数滴管中滴出的液体滴数就可以方便地判断溶液的体积。一般来说,1 毫升溶液大约为 20 滴,因此实验步骤常表述为"加入多少滴液体"。如前所述,非常重要的是实验时绝不要污染滴管,比如,不让滴管尖嘴接触其他溶液液面或试管内壁。

用于定性分析的化学试剂是水溶液,装在滴瓶中,每个滴瓶都贴上了标签,不要将滴瓶

addition, all water used for these procedures must be distilled water. Tap water from the regular faucets contains a wide variety of ions and could result in misleading observations.

2. Equipment and Chemicals The tests that you will perform to identify the cations after they have been separated require only very small amounts of sample. Therefore, you will work with the smallest size test tubes available (10 mm×75 mm), thin glass stirring rods and medicine droppers. Rather than measuring volumes of liquids with a graduated cylinder, it will be easier to simply count the number of drops of liquid dispensed from a medicine dropper. In general, it is safe to assume that 20 drops from a medicine dropper is equivalent to 1 mL of solution. You will usually be told exactly how many drops to add in any given step of the procedure. As mentioned previously, it is very important that the medicine droppers never become contaminated, for instance, by allowing the tip to touch either the surface of a solution or the side of a test tube.

The chemical reagents for the qualitative analysis procedures are aqueous solutions that are contained in dropper bottles. The caps of these bottles are labeled. Do not put the wrong dropper into a bottle, or the entire contents of the bottle will become contaminated! When you have finished using the bottle, screw the cap back on *tightly*. If the cap is not on tight, the next person to grab it by the dropper will knock the bottle over. In addition, some of the water will evaporate if the cap is not on tight, and this will alter the concentration of the solution.

3. Mixing Solutions After a reagent has been added to a solution in a test tube, it is very important that the solutions are stirred together thoroughly. Reactions cannot proceed to completion unless the reactants come into intimate contact with each other. Use a *thin* stirring rod to mix the liquids so that the contents of the test tube do not overflow. Of course, the stirring rod must be rinsed well with distilled water before being used.

In many cases during the separation of cations, an acid or base solution is added to a sample until the pH of the solution becomes either acidic or basic. In these instances, litmus paper is used to monitor the pH. A drop of the solution being tested is removed from the test tube using a glass stirring rod. The rod is touched to the litmus paper, and the color of the paper indicates whether the pH is basic (blue) or acidic (red). The litmus paper must never be placed directly into the solution! It is possible to use litmus paper to observe if a solution is weakly acidic or basic as opposed to strongly acidic or basic. A strong acid or base will turn litmus paper a more intense color than a weak acid or base will.

4. Forming, Separating and Washing Precipitates When an appropriate chemical reagent is added to a solution containing a mixture of cations, a precipitate will form which is composed of the insoluble salt(s) of one or more of the cations. This solid precipitate can be separated from the aqueous solution, which contains the soluble salt(s) of the remaining cations, by filtration; however, a more convenient technique for the separation of a solid from a liquid is centrifugation. A test tube containing the mixture is placed in a centrifuge, which spins the mixture at a very high speed. The resulting centrifugal force packs the solid material tightly into the bottom of the test tube. The liquid supernatant can then be easily poured away, or *decanted*, from the solid.

Before the liquid supernatant is decanted from the solid material, it is a good idea to check for complete precipitation. After centrifuging the mixture, add one or two more drops of the reagent which caused precipitation to occur. Watch the supernatant carefully to see if more precipitate forms. If no more precipitate forms, then precipitation is complete and the supernatant may be poured off. If more solid does appear in the supernatant, add another drop or two of the reagent

的滴管放错到其他试剂瓶中,否则瓶中的液体将会被污染!取完溶液后,要把试剂瓶的瓶盖盖紧,如果瓶盖没有盖紧的话,下一个使用的人可能会失手弄翻它,此外,瓶盖没盖紧会使瓶中的水蒸发,使溶液浓度发生改变。

3. 溶液的混合

向试管中的溶液中加入另一种试剂后,要彻底搅拌溶液,这个步骤非常重要,如果反应物之间不能很好地接触,就不能反应完全。混合溶液的时候要使用细的玻棒,以免试管中液体溢出(粗玻璃棒体积大)。当然,玻棒在使用之前必须用蒸馏水冲洗。

阳离子的分离过程中,常常要加入酸或碱调 pH 值,使溶液变为酸性或碱性,在这种情形下,要用石蕊试纸检查溶液的 pH 值。检查 pH 值时,用玻棒蘸一滴溶液滴到石蕊试纸上,试纸的颜色将指示溶液是碱性(蓝色)或是酸性(红色),石蕊试纸在强酸或强碱中比在弱酸或弱碱中显示出更深的颜色,石蕊试纸不能直接放到溶液中,这样可能会使本来是弱酸或弱碱的溶液得出强酸或强碱的结论。

4. 沉淀的生成、分离和洗涤

向含有阳离子混合物的溶液中加入某种适当的化学试剂会生成沉淀,沉淀由一种或多种阳离子的不溶盐组成。这种固体沉淀通过过滤就可以与其他阳离子的可溶盐水溶液实现分离,然而,离心分离是更为方便的沉淀与溶液分离技术。将含有固、液混合物的试管放入离心机,当混合物高速旋转时,其所产生的离心力使固体物沉积在试管的底部,而液体则很容易被倒掉,达到与固体分离的目的。

在将液体与固体分离前,有必要检查沉淀是否完全。当离心分离后,向上清液加入 1~2 滴沉淀剂,观察上层离心液中是否有沉淀产生,如果没有沉淀产生,说明沉淀完全,上层清液可以转移,如果上层离心液中有沉淀出现,则仍需要 1~2 滴沉淀剂,然后彻底搅拌试管中的沉淀,再离心沉降 2 分钟。沉淀是否完全的检查过程应该一直持续到上层清液中加入沉淀剂后没有固体产生为止。

当液体从固体分离出来后,试管底部的固体中会残留少量离心液,为了将残留的溶液和可能的一些杂质完全除去,可以用蒸馏水洗涤沉淀,在试管中加入 10~15 滴水,用玻棒充分搅拌,离心沉降 2 分钟,然后将洗涤液倒掉。在进行下一个步骤开始之前,固体通常要洗涤 2 次。

5. 用标签做标记

在定性分析实验过程中,将会用到许多装有各种不同内容物的试管,因此,用清楚、恰当的标签标记试管显得尤为重要。如果试管没有标签,有用的溶液和固体很容易不小心丢掉。另外,当你的试管和其他同学试管放在同一台离心机中时,应该在自己的试管上标记上自己的名字。

6. 实验记录

必须一如既往地在实验记录本中完整地、如实地记录定性分析实验过程的实验现象。将实验现象整理成表格是很有用的。颜色变化、是否产生浑浊、气体的生成都是必须仔细记录的相关现象。当两个溶液混合发生反应的时候,混合前的两种溶液的颜色以及混合后发生的变化都要记录下来。记住,"清澈"和"无色"是有区别的,说液体"清澈"表示没有固体物存在,而"无色"则是缺乏色彩的最佳描述词,例如,如果两种无色溶液混合且生成白色的沉淀,实验现象的正确描述是"无色和浑浊的"混合物。实验记录需要用钢笔填写。

solution, thoroughly stir the contents of the test tube, and place it back in the centrifuge for two minutes. The process of checking for complete precipitation should continue until no solid appears in the supernatant upon addition of the reagent.

After the liquid has been decanted from the solid, a small amount of this solution will remain behind in the test tube with the solid. In order to completely remove this liquid and any contaminants that it contains, the precipitate is washed with water. Ten to fifteen drops of water are added to the test tube, and the mixture of solid and liquid is thoroughly stirred with a glass rod. The test tube is place in a centrifuge for two minutes, and then the water is decanted away from the solid and may be disposed of down the drain. Solids are usually washed two times before proceeding to the next step in the procedure.

5. Labeling During the qualitative analysis procedures, you will be manipulating a large number of test tubes with a variety of contents. It is vital that everything be clearly and properly labeled. Solutions and solids can be inadvertently discarded if test tubes do not have accurate labels. Also, before you place one of your test tubes in the centrifuge with other students' test tubes, you must label it with your name.

6. Laboratory Notebook As always, you must record very complete and descriptive observations in your lab notebook throughout the qualitative analysis procedures. You might find it useful to organize your observations in a table. Color changes, the presence or absence of cloudiness and the formation of bubbles are all relevant observations that must be carefully noted. You should record the color of solutions *before* they are added to each other, as well as what happens after they are mixed together. Remember that there is a difference between the terms *clear* and *colorless*. To say that a liquid is "clear" indicates that no solid material is present, while the absence of color is best described by the word "colorless". For instance, if two non-colored solutions are mixed together and a white precipitate forms, the correct observation is a "colorless and cloudy" mixture. You should hand in the carbon copies of your notebook.

7. Hazardous Chemical Byproducts Some of the chemical reagents and cations that you will work with during the qualitative analysis procedures are toxic. These materials present a hazard to the environment and must be collected in the Laboratory Byproducts jars. Those chemicals that are not hazardous may be safely poured down the drain. Read carefully the instructions provided for each experiment, and listen to your instructor's pre-lab talk to learn which materials may be disposed of in the sink and which must be collected in a Byproducts jar. Ask your instructor if you have any questions regarding the proper disposal of a chemical.

7. 危险的废弃物

在定性分析过程中,我们所用到的一些化学试剂和阳离子是有毒的,这些物质对环境有危害,必须用废液缸收集。那些没有危害的物质可以倒入水槽中。仔细阅读每个实验的说明,实验前仔细聆听指导教师的介绍,学会并了解哪些物质可以直接倒入水槽,哪些必须倒入废液缸。处置化学品的有关问题要咨询指导老师。

Experiment 9 Qualitative Analysis of Cations (2)

Purpose
1. To analyze a mixture of several substances involving a systematic separation of cations.
2. To learn techniques of separations and confirmation tests to identify an unknown cation.
3. To understand the principle and the procedure behind the separation for various cations.

Principle

Qualitative analysis is an analytical procedure in which the question "what is present?" is answered. In a systematic qualitative analysis scheme, generally, each substance present is separated from the other substances. Then a confirmatory test is used to prove that the isolated substance is the expected one. Many reagents are used in qualitative analysis, but only a few are involved in nearly every group procedure. The four most commonly used reagents are 6 mol·L^{-1} HCl, 6 mol·L^{-1} HNO_3, 6 mol·L^{-1} NaOH, 6 mol·L^{-1} NH_3. Understanding the uses of the reagents is helpful when planning an analysis.

You will first analyze a solution known to contain specific ions before proceeding to the analysis of an unknown solution in this experiment. The "known" solution contains six cations of Ag^+, Al^{3+}, Cu^{2+}, Fe^{3+}, Mn^{2+} and Zn^{2+}. And the second solution contains any combination of six different cations. The "unknown" solution is analyzed to determine which ions are present and which are absent. This experiment is carried out on a semi-micro scale. Very small quantities of reagents are used.

Cleanliness and a great deal of care are necessary to obtain good results. The diagram of the procedure is presented as shown in the next page. The purpose of such a diagram is to help you understand what is happening in a given reaction. Read the directions carefully. Don't just follow directions "cook book" style, but make an effort to understand the chemical principles behind the procedures. While going through the steps of the analysis, keep a copy of the appropriate flow chart available for reference. It will help to give the "total picture" of where each analysis is and where it is heading. It is important to keep good records of your observations.

Apparatus and Chemicals

Apparatus: test tubes; test tube racks; beaker, 500 mL (waste), 250 mL; water bath; hotplate; centrifuge.

Chemicals: Unknown Cation Solution (may contain some or all of the following) 0.1 mol·L^{-1} silver nitrate, $AgNO_3$; 0.1 mol·L^{-1} iron (Ⅲ) nitrate, $Fe(NO_3)_3$ (aq); 0.1 mol·L^{-1} copper (Ⅱ) nitrate, $Cu(NO_3)_2$ (aq); 0.3 mol·L^{-1} $Al(NO_3)_3$; 0.3 mol·L^{-1} zinc nitrate, $Zn(NO_3)_2$ (aq); 0.1 mol·L^{-1} $Mn(NO_3)_2$.

Other Chemicals: 6 mol·L^{-1} HCl(aq); 6 mol·L^{-1} HNO_3(aq); 6 mol·L^{-1} H_2SO_4(aq); 6 mol·L^{-1} NH_3(aq); 6 mol·L^{-1} NaOH (aq); 6% H_2O_2(aq); 0.5 mol·L^{-1} KSCN (aq); 0.1 mol·L^{-1} $K_4Fe(CN)_6$(aq); 0.1 mol·L^{-1} $AgNO_3$ (aq); $NaBiO_3$ (s); 0.1 mol·L^{-1} $BaCl_2$(aq); aluminon reagent; litmus or pH paper.

实验 9　阳离子定性分析(2)

实验目的

1. 系统分离分析几种物质的混合液。
2. 学习未知阳离子的分离和鉴定技术。
3. 掌握几种阳离子分离和鉴定的原理和步骤。

实验原理

定性分析是回答"样品中有什么"的分析过程。在系统的定性分析流程中,一般情况下样品中存在的每种物质要先与其他物质进行分离,然后再对所分离得到的物质进行鉴定,看是不是所希望得到的。定性分析会用到很多种试剂,但是每次分离几乎都要用到的试剂(阳离子分组试剂)只有几种。6 mol·L^{-1} HCl、6 mol·L^{-1} HNO$_3$、6 mol·L^{-1} NaOH、6 mol·L^{-1} NH$_3$ 是最常用的分组试剂,明白这些试剂的使用原理对设计分析方案很有帮助。

本实验先分析一个由已知离子组成的溶液,然后再分析一个由未知离子所组成的溶液。"已知"溶液含有六种阳离子:Ag^+、Al^{3+}、Cu^{2+}、Fe^{3+}、Mn^{2+}、Zn^{2+},"未知"溶液由这六种离子中的某些离子组成,要求分析出"未知"溶液含有哪些离子、不含哪些离子。本实验是半微量实验,实验所用的试剂只需要很少的量。

要想获得好的实验结果,实验时就必须做到整洁和小心仔细。本实验过程的流程图如下页所示,这个流程图可以帮助同学们明白实验过程中会发生哪些变化。请仔细阅读实验步骤,认真理解实验流程背后的化学原理,这样,做实验时就不会是简单的"按方抓药"。在展开实验分析时,要准备一张合适的实验流程图以供实验参考,这会让你对分析步骤有一个整体印象,让你能清楚知道实验进行到了哪一步、下一步的工作又是什么。实验时要将观察到的实验现象做好记录,这很重要。

仪器与试剂

仪器:试管;试管架;烧杯,500 mL(做废液缸),250 mL;水浴锅;电炉;离心机。

试剂:未知阳离子溶液(含有以下离子中的几种或全部)

　　硝酸银,AgNO$_3$(aq),0.1 mol·L^{-1};硝酸铁,Fe(NO$_3$)$_3$(aq),0.1 mol·L^{-1};

　　硝酸铜,Cu(NO$_3$)$_2$(aq),0.1 mol·L^{-1};硝酸铝,Al(NO$_3$)$_3$(aq),0.3 mol·L^{-1};

　　硝酸锌,Zn(NO$_3$)$_2$(aq),0.3 mol·L^{-1};硝酸锰,Mn(NO$_3$)$_2$(aq),0.1 mol·L^{-1}。

其他试剂:

　　HCl(aq),6 mol·L^{-1};HNO$_3$(aq),6 mol·L^{-1};H$_2$SO$_4$(aq),6 mol·L^{-1};

　　NH$_3$(aq),6 mol·L^{-1};NaOH(aq),6 mol·L^{-1};H$_2$O$_2$(aq),6%;

　　KSCN(aq),0.5 mol·L^{-1};K$_4$Fe(CN)$_6$(aq),0.1 mol·L^{-1};AgNO$_3$(aq),0.1 mol·L^{-1};

　　NaBiO$_3$(s);BaCl$_2$(aq),0.1 mol·L^{-1};铝试剂;

　　石蕊试纸或 pH 试纸。

Diagram of the procedure

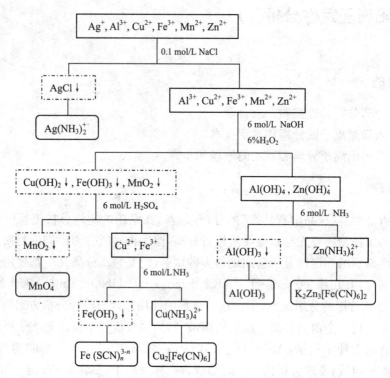

Procedure

Note that the following directions are written for a "known" solution that contains all of the cations. An "unknown" solution will probably not form all of the products described in this procedure. Make note of any differences in the "unknown" solution as it is analyzed. Use 22 drops of unknown unless instructed otherwise.

Your instructor will tell you how to obtain your sample containing the following ions: Ag^+, Fe^{3+}, Cu^{2+}, Mn^{2+}, Zn^{2+} and Al^{3+}. Then check the pH of the solution with test papers.

Part A: Separation of silver from solution

Step 1 Separation and Confirmation of Silver

1. To the solution containing the ions, add 15 drops of $0.1 \text{ mol} \cdot L^{-1}$ sodium chloride, NaCl, solution. Shake well. You should see a white precipitate, silver chloride, form. Centrifuge until the liquid on top of the precipitate (the supernatant) is clear (approximately 1 minute). Silver ions are precipitated with the addition of any soluble chloride to silver chloride AgCl:

$$Ag^+ (aq) + Cl^- (aq) \rightarrow AgCl (s)$$

2. The AgCl is a white precipitate quite insoluble in water and dilute acids. Test to see if the Ag^+ ion is completely precipitated by adding another drop or two of $0.1 \text{ mol} \cdot L^{-1}$ NaCl. If more precipitate forms, add $0.1 \text{ mol} \cdot L^{-1}$ NaCl until no more precipitate occurs, then centrifuge and test again.

3. Carefully draw off the supernatant with a Pasteur pipet and save in a test tube labeled "Step 2". The supernatant contains the remainder of the ions except for silver. It is better to leave a small amount of the supernatant in the tube than to draw any of the precipitate into the pipet. Set the tube with the clear liquid aside.

实验流程图

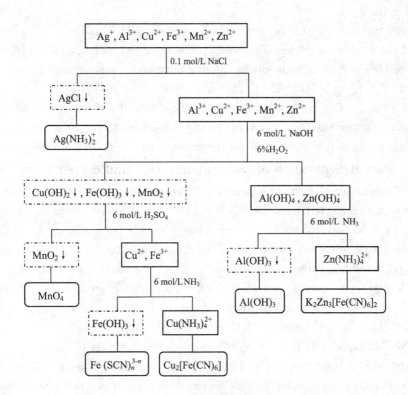

实验步骤

下列实验步骤是针对含有全部六种阳离子的"已知"溶液,"未知"溶液可能不完全包括这六种离子,注意分析"未知"溶液时的实验现象与分析"已知"溶液时有哪些区别。在没有另外说明的情况下,未知溶液需要取 22 滴。

取 Ag^+、Fe^{3+}、Cu^{2+}、Mn^{2+}、Zn^{2+}、Al^{3+} 溶液各 4 滴混合,制成"已知"的分析样品溶液。取完溶液后要用 pH 试纸测定溶液的 pH 值。

第一部分 银离子 Ag^+ 的分离

步骤 1 银离子 Ag^+ 的分离与鉴定

1. 在含有离子的溶液中加入 15 滴 $0.1\ mol \cdot L^{-1}$ 氯化钠溶液,充分振摇试管,有氯化银白色沉淀生成。离心分离,直至沉淀上面的溶液(上清液)是澄清的(大约需要 1 分钟)。加入可溶性氯化物,银离子会以氯化银 AgCl 的形式沉淀下来:

$$Ag^+(aq) + Cl^-(aq) \rightarrow AgCl(s)$$

2. AgCl 是不溶于水和稀酸的白色沉淀。向上清液中再加入 1 到 2 滴 $0.1\ mol \cdot L^{-1}$ NaCl,检验 Ag^+ 是否完全沉淀。如果还有沉淀生成,要继续加入 $0.1\ mol \cdot L^{-1}$ NaCl,直至不再有新的沉淀出现为止。然后,离心沉降,再进行检验。

3. 用吸管将上清液小心吸取出来,并存放在一支干净的试管中,贴上标签"步骤 2",上清液中含有除银离子外的全部其他离子。在用吸管吸取上清液时,最好留少量的上清液在

4. Wash the white precipitate with 20 drops of distilled water by stirring, centrifuging, and drawing off the wash liquid with a pipet. Discard the wash. Be careful not to draw up the precipitate into the pipet.

5. To confirm that the precipitate is AgCl, carefully add 6 mol·L^{-1} aqueous NH$_3$ with shaking until the solid is completely dissolved (about 20 drops). The AgCl can be distinguished from the insoluble chlorides in ammonia by the formation of the silver complex ion:

$$AgCl(s) + 2NH_3(aq) \rightarrow Ag(NH_3)_2^+(aq) + Cl^-(aq)$$

Now add 6 mol·L^{-1} nitric acid, HNO$_3$ (Caution!) until the solution is acidic (about 20 drops). Caution: the test tube may get very warm. Check with litmus to see that the solution is acidic. A white precipitate of AgCl confirms the presence of silver. Discard this precipitate.

Part B: Separation of iron, manganese, and copper from solution

Step 2 Precipitation of Iron, Manganese, and Copper

1. To the solution saved from step 1, labeled "Step 2", add 10 drops of 6% hydrogen peroxide, H$_2$O$_2$ (Caution!). While stirring, add 6 mol·L^{-1} sodium hydroxide, NaOH (Caution!), until the solution is basic (about 10 drops), and then add 3 drops more. Stir the solution and place the test tube in a boiling water bath. A dark precipitate should form of iron hydroxide, Fe(OH)$_3$, manganese dioxide, MnO$_2$, and copper hydroxide, Cu(OH)$_2$. Boil the solution for 2 minutes or more and centrifuge out the solid.

$$Fe^{3+}(aq) + 3OH^-(aq) \rightarrow Fe(OH)_3(s, \text{reddish solid})$$
$$Mn(OH)_2(s) + H_2O_2(aq) \rightarrow MnO_2(s, \text{dark brown}) + 2H_2O$$
$$Cu^{2+}(aq) + 2OH^-(aq) \rightarrow Cu(OH)_2(s, \text{blue, masked by dark precipitate})$$

2. Draw off the supernatant with a Pasteur pipet, and save it in a test tube labeled "Step 6". Save the precipitate for step 3. The solution should contain Al(OH)$_4^-$, aluminum hydroxide ions, and Zn(OH)$_4^{2-}$, zinc hydroxide ions, still in solution. The excess NaOH is needed to form the hydroxide complex ions of Al and Zn and insure that they stay in solution.

$$Al^{3+}(aq) + 4OH^-(aq) \rightarrow Al(OH)_4^-(aq)$$
$$Zn^{2+}(aq) + 4OH^-(aq) \rightarrow Zn(OH)_4^{2-}(aq)$$

Step 3 Separation and Confirmation of Manganese

1. Wash the solid from step 2, labeled "Step 3", with 10 drops of water and 10 drops of NaOH. Centrifuge and use a Pasteur pipet to discard the wash. The washing will insure that all of the aluminum and zinc ions have been removed from the precipitate.

2. To the precipitate, which may contain Fe(OH)$_3$, MnO$_2$, and Cu(OH)$_2$, add 5 drops of water. Add 6 mol·L^{-1} sulfuric acid, H$_2$SO$_4$ (Caution!) until acidic when tested with litmus paper (about 5 drops). Stir the solution for a minute and centrifuge out any undissolved solid, which should be brown to black MnO$_2$. Decant the liquid with a pipet into a test tube. The liquid may contain Fe^{3+} and Cu^{2+}. Save the supernatant for "Step 4" and set aside for future tests. Save the precipitate. Wash the precipitate with 20 drops of 6 mol·L^{-1} H$_2$SO$_4$, centrifuge, and discard the wash.

$$Fe(OH)_3(s) + 3H^+(aq) \rightarrow Fe^{3+}(aq) + 3H_2O$$
$$Cu(OH)_2(s) + 2H^+(aq) \rightarrow Cu^{2+}(aq) + 2H_2O$$

3. Add 20 drops of 6 mol·L^{-1} HNO$_3$ (Caution!) to the precipitate and stir. Slowly dissolve small amounts of sodium bismuthate, NaBiO$_3$, until no more will dissolve. Make sure there is a small excess of solid on the bottom of the test tube. Centrifuge. The solution will be purple in the

试管中以防止沉淀被吸进吸管中。将此盛放上清液的试管放置一旁。

4. 向白色沉淀中加入 20 滴蒸馏水用来洗涤沉淀。搅拌,离心沉降,然后用吸管吸取洗涤液,将洗涤液弃去,吸取洗涤液时要小心,不要将沉淀吸进吸管中。

5. 为了鉴定此沉淀为 AgCl,向沉淀中小心加入 6 mol·L^{-1} NH$_3$ 水(大约 20 滴),振摇试管直至沉淀全部溶解。AgCl 能溶于氨水生成银氨配离子,这样可以将 AgCl 与其他不溶性氯化物区别开来。

$$AgCl(s) + 2NH_3(aq) \rightarrow Ag(NH_3)_2^+(aq) + Cl^-(aq)$$

向银氨配离子溶液中加入 6 mol·L^{-1} 硝酸 HNO$_3$(要小心加入!)直至溶液酸化(大约 20 滴)。小心:试管会变得很热。用石蕊或 pH 试纸检测溶液是否呈酸性。此时有 AgCl 白色沉淀出现,示有银离子存在。将沉淀弃去。

第二部分 铁离子、锰离子和铜离子的分离

步骤 2 沉淀铁离子、锰离子和铜离子

1. 向从步骤 1 中得到的、标有"步骤 2"的溶液中,加入 10 滴 6% 过氧化氢 H$_2$O$_2$(要小心加入!)。边搅拌边加入 6 mol·L^{-1} 氢氧化钠,直至溶液碱化(大约 10 滴),然后再多加 3 滴 NaOH 溶液。搅拌溶液,将试管放入沸水浴中加热;有黑色沉淀生成,沉淀中包括氢氧化铁 Fe(OH)$_3$、二氧化锰 MnO$_2$ 和氢氧化铜 Cu(OH)$_2$。再继续将试管在沸水浴中加热 2 分钟或更长时间,离心分离出固体。

$$Fe^{3+}(aq) + 3OH^-(aq) \rightarrow Fe(OH)_3(s, 红棕色)$$
$$Mn(OH)_2(s) + H_2O_2(aq) \rightarrow MnO_2(s, 深棕色) + 2H_2O$$
$$Cu^{2+}(aq) + 2OH^-(aq) \rightarrow Cu(OH)_2(s, 蓝色, 颜色被深色沉淀掩盖)$$

2. 用吸管将上清液吸取出来,保存上清液并标记"步骤 6",保存沉淀并标记为"步骤 3"。此时,上清液中应含有 Al(OH)$_4^-$ 和 Zn(OH)$_4^{2-}$。加入 NaOH 时需要过量,以确保 Al^{3+} 和 Zn^{2+} 能与 OH$^-$ 生成它们的氢氧根配离子,使其留在溶液里:

$$Al^{3+}(aq) + 4OH^-(aq) \rightarrow Al(OH)_4^-(aq)$$
$$Zn^{2+}(aq) + 4OH^-(aq) \rightarrow Zn(OH)_4^{2-}(aq)$$

步骤 3 锰离子的分离与鉴定

1. 用 10 滴蒸馏水和 10 滴 NaOH 洗涤从步骤 2 中所得的、标记为"步骤 3"的固体,离心分离并用吸管弃去洗涤液。洗涤时要确保沉淀中所含的铝离子和锌离子被全部去除。

2. 沉淀中可能含有 Fe(OH)$_3$、MnO$_2$ 和 Cu(OH)$_2$,向沉淀中加入 5 滴水,再加入 6 mol·L^{-1} 硫酸 H$_2$SO$_4$(要小心)直至溶液酸化(大约 5 滴),用石蕊试纸检测。搅拌溶液 1 分钟,离心使溶液中所含的棕色或黑色的 MnO$_2$ 沉淀分离出来。用吸管取出上清液,上清液中含有 Fe^{3+} 和 Cu^{2+},将其标记为"步骤 4",放置一旁用于后面的测试。沉淀用 20 滴 6 mol·L^{-1} H$_2$SO$_4$ 洗涤,离心分离,将洗涤液弃去。

$$Fe(OH)_3(s) + 3H^+(aq) \rightarrow Fe^{3+}(aq) + 3H_2O$$
$$Cu(OH)_2(s) + 2H^+(aq) \rightarrow Cu^{2+}(aq) + 2H_2O$$

3. 向沉淀中加入 20 滴 6 mol·L^{-1} HNO$_3$(要小心)并搅拌,慢慢加入铋酸钠固体直至铋酸钠固体不再溶解为止,保证试管底部有少量过量的固体。离心分离,上清液若含有 MnO$_4^-$ 会呈现紫红色,示有锰元素存在。将溶液弃去。

presence of MnO_4^-, which confirms the presence of manganese. Discard.

$$2MnO_2(s)+3NaBiO_3(s)+10H^+(aq) \rightarrow 2MnO_4^-(aq)+3Bi^{3+}(aq)+3Na^+(aq)+5H_2O$$

Step 4 Separation and Confirmation of Iron

1. The solution from step 3, labeled "Step 4" should contain Fe^{3+} and Cu^{2+} ions. Add aqueous NH_3 until precipitation is complete. A brown precipitate should form of $Fe(OH)_3$. Centrifuge. Decant the supernatant, which should contain Cu^{2+}. A slight blue tinge indicates that copper is present in solution. Save the liquid for later in a test tube labeled "Step 5".

$$Cu(OH)_2(s)+4NH_3(aq) \rightarrow Cu(NH_3)_4^{2+}(aq)+2OH^-(aq)$$

2. Wash the precipitate with 20 drops of water and 20 drops of aqueous NH_3. Discard the wash. Add 15 drops of H_2O and 15 drops of $6 \text{ mol} \cdot L^{-1}$ H_2SO_4 to dissolve the precipitate. Add a few drops of $0.5 \text{ mol} \cdot L^{-1}$ KSCN solution to the Fe^{3+} ion to produce the intensely red $Fe(SCN)_n^{3-n}$ ion in solution. Discard.

$$Fe^{3+}(aq)+nSCN^-(aq) \rightarrow Fe(SCN)_n^{3-n}(aq, \text{blood red})$$

Step 5 Confirmation of Copper

To the solution from step 4, labeled "step 5", add $6 \text{ mol} \cdot L^{-1}$ HCl until the blue color fades and the solution is acidic when tested with litmus paper. Add 2 or 3 drops of $0.1 \text{ mol} \cdot L^{-1}$ potassium ferrocyanide, $K_4Fe(CN)_6$, which forms a reddish brown to pink precipitate. You may need to centrifuge to see the color clearly. This is a sensitive test for copper, giving a red coloration even in extremely dilute solutions. Discard.

$$2Cu^{2+}(aq)+[Fe(CN)_6]^{4-}(aq) \rightarrow Cu_2[Fe(CN)_6](s, \text{reddish brown})$$

Part C: Separation of aluminum and zinc from solution

Step 6 Separation and Confirmation of Aluminum

1. Return to the solution from step 2, labled "step 6". Since copper ions will interfere with tests for aluminum and zinc, it is important to have removed them thoroughly. If there is a blue tinge to the supernatant, there are still copper ions in solution and you will need to do an additional step. If a blue color is present, it is likely that the solution is deficient in hydroxide ions. Add 3 drops of $6 \text{ mol} \cdot L^{-1}$ NaOH. Boil the solution for 2 minutes or more, then centrifuge. Decant. Discard the solid. If the solution is still not colorless, repeat this procedure.

2. Once the solution containing aluminum and zinc ions is colorless, add about 10 drops of $6 \text{ mol} \cdot L^{-1}$ HNO_3 (Caution!) until acidic when tested with litmus. Place the solution in a boiling water bath for a minute or two. Add $6 \text{ mol} \cdot L^{-1}$ aqueous NH_3 (Caution!), drop by drop, until the solution is basic when tested with litmus. Add 3 more drops of $6 \text{ mol} \cdot L^{-1}$ aqueous NH_3. Stir the mixture for a minute. If aluminum is present, a light, translucent, gelatinous precipitate of $Al(OH)_3$ should be present. It may be hard to see, so look closely. Centrifuge and save the supernatant that contains zinc ions in a tube labeled "Step 7".

$$Al^{3+}(aq)+3NH_3(aq)+3H_2O \rightarrow Al(OH)_3(s)+3NH_4^+(aq)$$
$$Zn^{2+}(aq)+4NH_3(aq) \rightarrow Zn(NH_3)_4^{2+}(aq)$$

3. Dissolve the precipitate with $6 \text{ mol} \cdot L^{-1}$ HNO_3. Add 5 drops of water and 2 drops of aluminon reagent and stir thoroughly. At this point, the solution may be colored because of the aluminon. Add $6 \text{ mol} \cdot L^{-1}$ aqueous NH_3 (Caution!) drop by drop, stirring well, until the solution is basic when tested with litmus. If Al^{3+} is present, the precipitate of $Al(OH)_3$ forms and adsorbs the aluminon from the solution producing a red precipitate of $Al(OH)_3$. This is called a "lake". The supernatant is essentially colorless. The test for aluminum is not the red solution but rather

$$2MnO_2(s)+3NaBiO_3(s)+10H^+(aq) \rightarrow 2MnO_4^-(aq)+3Bi^{3+}(aq)+3Na^+(aq)+5H_2O$$

步骤 4　铁离子的分离与鉴定

1. 步骤 3 所得、标记为"步骤 4"的溶液应含有 Fe^{3+} 和 Cu^{2+}，加入氨水 NH_3 直至沉淀完全，所生成的棕色沉淀为 $Fe(OH)_3$。离心分离，上清液呈现淡蓝色说明铜离子的存在，将含有 Cu^{2+} 的上清液吸取出来，标记为"步骤 5"，并保存。

$$Cu(OH)_2(s)+4NH_3(aq) \rightarrow Cu(NH_3)_4^{2+}(aq)+2OH^-(aq)$$

2. 沉淀用 20 滴蒸馏水、20 滴氨水洗涤，将洗涤液弃去。加入 15 滴 H_2O 和 15 滴 6 mol·L^{-1} H_2SO_4 使沉淀溶解，再加入几滴 0.5 mol·L^{-1} KSCN 溶液，Fe^{3+} 则生成 $Fe(SCN)_n^{3-n}$，溶液呈血红色。将溶液弃去。

$$Fe^{3+}(aq)+nSCN^-(aq) \rightarrow Fe(SCN)_n^{3-n}(aq, 血红色)$$

步骤 5　铜离子的鉴定

向从步骤 4 中得到的、标记为"步骤 5"的溶液中加入 6 mol·L^{-1} HCl 直至蓝色褪去，用石蕊试纸检验溶液呈酸性。再加入 2~3 滴 0.1 mol·L^{-1} 亚铁氰化钾[$K_4Fe(CN)_6$]，应有红棕色或浅粉色沉淀生成。若想清楚地观察沉淀的颜色，可能需要进一步离心。这个反应是铜离子鉴别的灵敏实验，即使铜离子溶液的浓度非常小也能观察到溶液变红色。将沉淀弃去。

$$2Cu^{2+}(aq)+[Fe(CN)_6]^{4-}(aq) \rightarrow Cu_2[Fe(CN)_6](s, 红棕色)$$

第三部分　铝离子和锌离子的分离

步骤 6　铝离子的分离与鉴定

1. 现在转向从步骤 2 所得的、标记为"步骤 6"的溶液。由于铜离子会干扰铝离子和锌离子的鉴定，所以完全除去铜离子很关键。如果此上清液呈现蓝色，说明溶液中还有铜离子存在，则需要另加除去铜离子的操作。溶液为蓝色可能是溶液中的氢氧根离子浓度不够，向溶液中加入 3 滴 6 mol·L^{-1} NaOH，沸水浴 2 分钟以上，然后离心分离，取出上清液，将固体弃去。如果溶液仍有颜色，重复上述过程。

2. 当含有铝离子、锌离子的溶液为无色溶液时，向溶液中加入大约 10 滴 6 mol·L^{-1} HNO_3（小心加入！），直至溶液用石蕊试纸检验呈酸性。将溶液置于沸水浴加热 1~2 分钟。逐滴加入 6 mol·L^{-1} 氨水，直至溶液用石蕊试纸检验呈碱性，再多加 3 滴 6 mol·L^{-1} 氨水，搅拌混合物 1 分钟。如果溶液中含有铝离子，则有明亮的、半透明、胶状沉淀 $Al(OH)_3$ 生成，这个沉淀不太容易观察，需要将试管拿近一些观察。离心分离出上清液，将上清液保存在一根干净的、标记有"步骤 7"的试管中。

$$Al^{3+}(aq)+3NH_3(aq)+3H_2O \rightarrow Al(OH)_3(s)+3NH_4^+(aq)$$
$$Zn^{2+}(aq)+4NH_3(aq) \rightarrow Zn(NH_3)_4^{2+}(aq)$$

3. 用 6 mol·L^{-1} HNO_3 将氢氧化铝沉淀溶解，加入 5 滴蒸馏水、2 滴铝试剂，充分搅拌，此时，溶液可能由于铝试剂的加入而显色。逐滴加入 6 mol·L^{-1} 氨水，充分搅拌，直至用石蕊试纸检测溶液呈碱性。如果有 Al^{3+} 存在，将会生成 $Al(OH)_3$ 沉淀，而 $Al(OH)_3$ 沉淀会从溶液中吸附铝试剂，使 $Al(OH)_3$ 沉淀呈现红色，这种红色是一种"深红色"。此时，其上清液则必定呈现无色。这种检测方法得到的不是红色溶液而是红色沉淀[$Al(OH)_3$ 及其吸附的铝试剂染料]。离心使沉淀聚沉，观察沉淀的颜色是否为红色。弃去沉淀和溶液。

the red precipitate [Al(OH)$_3$ and adsorbed aluminon dye]. Centrifuge to concentrate the precipitate and check to see if the color of the precipitate is red. Discard all.

Step 7   Confirmation of Zinc

Make the solution from step 6 slightly acidic when tested with litmus with 6 mol·L^{-1} HCl (Caution!) added drop by drop. Add 2 or 3 drops of 0.1 mol·L^{-1} K$_4$Fe(CN)$_6$ and stir. If zinc is present, a white to greenish precipitate of K$_2$Zn$_3$[Fe(CN)$_6$]$_2$ forms. Centrifuge it to make the precipitate more compact for examination. Discard all.

* Zinc may also be confirmed by adding a few drops of NaOH to the original sample until the solution tests basic with litmus and testing it with dithizone test paper. A red-violet color to the test paper is produced if Zn^{2+} is present. NaOH gives an orange color to the test paper that must not be mistaken for the red-violet color with Zn^{2+}. A known sample of Zn^{2+} should always be used side by side with the unknown when using the dithizone test paper. Discard the test paper.

Notes

1. Several of the chemicals used are toxic and corrosive. Do not ingest chemicals. Avoid contacting with the chemicals. Wash spills immediately using large amounts of water. Wash hands before leaving the laboratory.

2. Stirring Solutions

Each time a reagent is added to a test tube, the solution needs to be stirred. It is important to mix the solutions at the top and the bottom of the test tube. A stirring rod that is flattened at the bottom can be used as a plunger to effectively mix solutions in narrow test tubes.

3. Separating Solids from Solutions

Centrifuge solutions so that the solid is packed at the bottom of the test tube. Never fill centrifuge tubes to capacity. Keep liquid levels at least 1 cm from the top. Label all centrifuge tubes before inserting to avoid mix-up. Place tubes in a symmetrical fashion, the objective being to keep the rotor balanced. Fill all tubes to the same height. If only one tube needs to be centrifuged, achieve balance by inserting an additional tube (labeled as a "blank") containing the same volume of liquid. Follow manufacturer's directions.

Questions

1. In each question, a test is carried out to determine the presence or absence of several ions. Only those listed may be present. State if the tests indicate each ion is present, absent, or undetermined.

(a) Test for Ag$^+$, Cu^{2+}, Fe^{3+}

6 mol·L^{-1} HCl is added to a solution that may contain the three ions. A white precipitate forms.

Ions present:_____, Ions absent:_____, Ions undetermined:_____.

(b) Test for Cu^{2+}, Ag$^+$, and Zn^{2+}

6 mol·L^{-1} HCl is added to a solution that may contain the three ions. No precipitate forms. The addition of 6 mol·L^{-1} NaOH until the solution is basic results in no formation of precipitate.

Ions present:_____, Ions absent:_____, Ions undetermined:_____.

(c) Test for Cu^{2+}, Fe^{3+}, and Zn^{2+}

6 mol·L^{-1} NaOH is added to a clear solution that may contain the three ions until the

步骤 7　锌离子的鉴定

向步骤 6 产生的溶液中,逐滴滴加 6 mol·L^{-1} HCl,使溶液呈弱酸性,用石蕊试纸检验。加入 2~3 滴 0.1 mol·L^{-1} K$_4$Fe(CN)$_6$,搅拌。如果溶液有锌离子存在,则有白色或带绿色的沉淀 K$_2$Zn$_3$[Fe(CN)$_6$]$_2$ 生成,离心使沉淀聚沉下来以便更好观察沉淀的颜色。将沉淀与溶液弃去。

* 锌离子也可以用双硫腙试纸进行鉴定。向原始溶液中加入几滴 NaOH 溶液,使溶液呈碱性,用石蕊试纸检验;然后,溶液用双硫腙试纸测试,如果试纸显紫红色则说明溶液中有 Zn^{2+} 存在。NaOH 溶液能使双硫腙变为橙色,但并不干扰 Zn^{2+} 与双硫腙产生的紫红色。在用双硫腙试纸检测一个未知溶液中是否含有 Zn^{2+} 时,应当同时对一个已知含有 Zn^{2+} 的样品进行测定,以此作为对照。

注意事项

1. 实验中所用试剂有的是有毒或有腐蚀性的,使用时要小心,不要使化学药品进入口中,应避免与化学药品的直接接触;当化学药品洒出来时,要用大量水冲洗;离开实验室时要将手洗净。

2. 溶液的搅拌

每次将一种试剂加入试管中的时候,试管中底部与上部溶液的混合均匀很关键,因此溶液都需要搅拌。由于试管底部很窄,而玻璃棒下端是平的,玻璃棒搅不到试管底部的混合物时,将玻璃棒在试管中进行上下运动会得到较好的搅拌效果。

3. 固体与溶液的分离

离心分离能使固体沉降在试管的底部。向离心试管中装入液体时不能装满,液面到试管口的距离要不少于 1 cm;离心管要做好标记,以免与其他试管混淆;离心管在离心机中要对称放置,保持离心机工作马达平衡;每根离心管所装溶液的高度要保持一致;如果只有一个试管需要离心,则可以用一根装着同样体积液体的试管记作"空白"来平衡离心机。使用离心机时请阅读离心机制造商提供的说明书,要按照说明书的要求进行操作。

思 考 题

1. 根据实验现象判断下面待测溶液中哪些离子一定存在、哪些离子一定不存在、哪些离子不能确定是否存在。

(a) Ag$^+$、Cu^{2+}、Fe^{3+}

向可能含有以上三种离子的溶液中加入 6 mol·L^{-1} HCl,有白色沉淀生成。

存在的离子:_____,不存在的离子:_____,不能确定的离子:_____。

(b) Cu^{2+}、Ag$^+$、Zn^{2+}

向可能含有以上三种离子的溶液中加入 6 mol·L^{-1} HCl,没有沉淀生成,向溶液中加入 6 mol·L^{-1} NaOH 使溶液呈碱性,也没有沉淀生成。

存在的离子:_____,不存在的离子:_____,不能确定的离子:_____。

(c) Cu^{2+}、Fe^{3+}、Zn^{2+}

向可能含有以上三种离子的澄清溶液中加入 6 mol·L^{-1} NaOH 使溶液呈碱性,有深色沉

solution is basic. A dark precipitate forms. The precipitate totally dissolves in 6 mol · L^{-1} H$_2$SO$_4$. The addition of 6 mol · L^{-1} NH$_3$ to this acidic solution until it is basic results in a clear solution containing a dark precipitate. The resulting dark precipitate completely dissolves in 6 mol · L^{-1} H$_2$SO$_4$.

Ions present: _____ , Ions absent: _____ , Ions undetermined: _____ .

2. Consult a solubility table. Suggest other ions which could be precipitated when chloride ion is added.

3. Write an outline scheme for the separation of Ag$^+$, Cu^{2+}, and Al^{3+} ions.

4. Write the balanced net ionic equations for the reactions that occur when
(a) solutions of AgNO$_3$ and KCl are mixed
(b) solutions of Fe(NO$_3$)$_3$ and KOH are mixed
(c) solutions of NaNO$_3$ and KCl are mixed

5. Give the chemical formula for each of the following:
(a) a blue complex ion
(b) a colorless complex ion
(c) a purple complex ion
(d) a white precipitate

6. Why is H$_2$O$_2$ needed in the separation of manganese?

淀生成。该沉淀能完全溶于 6 mol·L^{-1} H$_2$SO$_4$ 溶液,向这个酸性的溶液中加入 6 mol·L^{-1} NH$_3$ 使沉淀呈碱性,可以得到澄清的溶液和深色的沉淀,而这种深色沉淀能完全溶于 6 mol·L^{-1} H$_2$SO$_4$。

存在的离子:_____,不存在的离子:_____,不能确定的离子:_____。

2. 查表,说明还有哪些离子在加入氯离子时能形成沉淀。

3. 写出 Ag$^+$、Cu^{2+} 和 Al^{3+} 离子混合物的分离流程图。

4. 写出并配平下列溶液中发生反应的离子方程式:

(a) AgNO$_3$、KCl 溶液混合

(b) Fe(NO$_3$)$_3$、KOH 溶液混合

(c) NaNO$_3$、KCl 溶液混合

5. 写出具有下列特征的一种物质的分子式:

(a) 蓝色配离子

(b) 无色配离子

(c) 紫红色配离子

(d) 白色沉淀

6. 为什么在分离锰离子的时候要加入 H$_2$O$_2$?

Experiment 10 Designing Scheme for Qualitative Analysis of Cations

Purpose
1. To cultivate comprehensive ability to utilize basic knowledge.
2. To design a scheme for qualitative analysis of cations.

Instructions

You will be provided with a "known" solution that contains the following four cations: ZrO^{2+} (this is a complex ion, similar to NH_4^+), Ni^{2+}, Pb^{2+} and Al^{3+}. You will design your own scheme to separate these cations from each other and then confirm their presence by confirmation tests. You will use the same techniques that you employed in the early experiments.

You will work on this experiment for two lab periods. Analyze the known solution during the first period to confirm that your scheme works. You will be given an unknown solution at the beginning of the second period. This solution contains one or more of the four cations. Analyze your unknown solution, report the ions that you detect and hand it in to your teacher before leaving the laboratory.

You must record *all* observations that you make during both these lab periods in your lab notebook.

1. Exploring the Conditions of Separation of the Ions

Three reagents that are quite useful in the separation of these four cations are hydrochloric acid, sodium hydroxide and ammonia. Begin by determining which cations are soluble in which reagent and which are insoluble. Solutions of each of the individual cations will be provided for you along with 6 mol·L^{-1} solutions of the three reagents. Add 3 drops of the reagent solution to 3 drops of the cation solution and examine the resulting mixture for a precipitate. Record all your observations and summarize them in a table similar to the following in your notebook.

Cation	NaOH	HCl	NH_3
ZrO^{2+}			
Ni^{2+}			
Solubility of nickel precipitate (s)			
Pb^{2+}			
Al^{3+}			

Important Note!

Test the solubility of any precipitates that appear for nickel in the other two reagents. For instance, if you see a precipitate when one of the reagents (NaOH, HCl or NH_3) is added to the nickel solution, centrifuge the test tube, decant the supernatant, wash the precipitate once with water and add 3 or 4 drops of one of the other two reagents to this precipitate. Record your observations in the table you have prepared. Repeat the procedure (you'll need to start with a new sample of the nickel solution to precipitate the solid), this time adding 3 or 4 drops of the other reagent solution to the washed precipitate.

You must develop your own scheme for separating and identifying the cations. There are several different possible procedures which can be used for carrying out this separation, and no one

实验 10　阳离子定性分析方案的设计

实验目的

1. 培养综合应用基础知识的能力。
2. 设计阳离子定性分析方案。

实验指导

实验室将提供一种含有 ZrO^{2+}、Ni^{2+}、Pb^{2+} 和 Al^{3+} 阳离子的已知溶液,要求学生设计一个实验方案分离并确定这些离子的存在,可以用上一个实验的相关技术来进行这个实验。

本实验可以分两部分完成。第一部分,对已知液进行实验,根据实验结果设计分析方案;第二部分,每人将会领到含有四种离子中的一种或多种离子的未知溶液,可以用设计的实验方案对未知液进行分析。实验完成后,向指导教师报告实验结果。

记住,要在实验报告中详细地记录两部分实验的所有实验现象。

1. 离子分离条件的研究

在这四种离子的分离实验中,有三种试剂非常有用:HCl、NaOH 和 NH_3。开始时,要确定这些阳离子在这几种试剂中哪些是可溶的,哪些是不溶的。实验室提供四种单独的阳离子溶液以及浓度为 $6\ mol\cdot L^{-1}$ 的三种试剂供分析测试。

在 3 滴含有某种阳离子的溶液中加入 3 滴某试剂,混合均匀,观察是否有沉淀,并在表中记录下来。

阳离子	NaOH	HCl	NH_3
ZrO^{2+}			
Ni^{2+}			
镍沉淀的溶解性			
Pb^{2+}			
Al^{3+}			

重要提示:当用一种试剂使 Ni^{2+} 沉淀后,用另外两种试剂检查沉淀的溶解性。例如,当将三种试剂(NaOH、HCl 或 NH_3)中的一种加入到镍离子的溶液中,若能观察到沉淀生成,则离心沉降,倾去离心液,用蒸馏水洗涤沉淀一次后,再加入 3~4 滴另外两种试剂中的一种,观察沉淀是否溶解,在预先准备好的表格中记录实验现象。重复一次该过程(重新用镍离子制备沉淀),换一种试剂加入到洗涤过的沉淀中。

每个学生要制定自己的方案分离鉴定这些阳离子。方案不是唯一的,可以按几种不同

way is "correct". You may find a flow chart to be useful in summarizing the steps of your procedure. [Note: Your teacher will take into consideration how much assistance you require in devising your scheme when she or he assigns a grade.]

Begin with 15 drops of either the known or unknown solution. Use 6 to 12 drops of the desired acid or base reagent to precipitate the cations, and remember to check for complete precipitation. Thoroughly wash any precipitates that will be used for further analysis. Use 6 to 12 drops of the desired acid or base reagent to dissolve precipitates. After adding the desired reagent, test the solution with litmus paper to be sure that it is either strongly acidic or basic.

Another important note!

When precipitating the lead cation, add the appropriate reagent, then place the test tube in *ice* for 5 minutes.

2. Exploring the Confirmation Tests for Individual Ion

The following tests should be performed to confirm the presence of the cations *after* they have been completely separated from each other. Each test can also be performed using a known solution of the individual cation in order to observe a positive test result (this is called a *control*). A negative test result can be observed by performing the test on water (this is called a *blank*).

Zirconyl, ZrO^{2+}

Dissolve a precipitate that you suspect contains zirconyl in 6 mol·L^{-1} HCl, *or* add enough HCl to a solution that you suspect contains zirconyl to make it acidic (test with litmus paper). *In a clean test tube*, place 2 drops of Alizarin Red solution. Add 2 drops of the acidic solution of zirconyl to the Alizarin Red solution. Check that the solution is still moderately acidic. Heat the solution for 2 minutes. The appearance of a red-violet color indicates the presence of ZrO^{2+}.

Nickel, Ni^{2+}

Dissolve a precipitate that you suspect contains nickel in 6 mol·L^{-1} HCl, then add 6 mol·L^{-1} NH_3 until strongly basic, *or* add enough NH_3 to a solution that you suspect contains nickel to make it strongly basic (test with litmus paper). Add 1 drop of 1% dimethylglyoxime (DMG) solution. The appearance of a cherry red precipitate indicates the presence of Ni^{2+}. Nickel is toxic and must be disposed of in the appropriate laboratory byproducts jar.

Lead, Pb^{2+}

Add 10 or more drops of water to a precipitate that you suspect contains lead. Heat and dissolve the solid. Add 2 drops of acetic acid and 4 drops of 0.1 mol·L^{-1} potassium chromate K_2CrO_4 solution. The formation of a bright yellow precipitate confirms the presence of Pb^{2+}. This precipitate is toxic and must be disposed of in the appropriate laboratory byproducts jar.

Aluminum, Al^{3+}

Dissolve a precipitate that you suspect contains aluminum in 6 mol·L^{-1} HCl, *or* add enough HCl to a solution that you suspect contains aluminum to make it acidic (test with litmus paper). Add 2 drops of 0.1% aluminon solution, then add 6 mol·L^{-1} NH_3 until the solution tests basic with litmus paper (stir well!). If no precipitate forms (the solution is clear), Al^{3+} is not present. If a red-pink precipitate forms (the solution is cloudy), centrifuge the test tube and decant the supernatant. Add 3 drops of 6 mol·L^{-1} NaOH to the solid. If Al^{3+} is present, the precipitate will dissolve and a clear magenta solution will form. If the precipitate does *not* dissolve in NaOH, ZrO^{2+} is probably present, and possibly Al^{3+} as well.

的过程进行分离,用流程图来表示分离步骤(注意:指导教师在评定成绩时会将所提供的帮助考虑进去,故要尽可能自己独立完成实验方案)。

实验开始时取 15 滴已知或未知液,用 6~12 滴酸或碱试剂使阳离子发生沉淀,记住,要检验沉淀是否完全。沉淀要彻底洗涤,以用于进一步分析。再用 6~12 滴酸或碱试剂溶解沉淀,当加入了所需试剂后,用石蕊试纸检查溶液的酸碱性。

重要提示:当沉淀 Pb^{2+} 时,加入沉淀剂后,要将离心管在冰浴中放 5 分钟。

2. 单个离子的鉴定

当将几种离子完全互相分离后,再做下面的实验确定阳离子的存在。为了观察明确的实验结果,每个实验可用已知的单个离子的溶液进行(这就是对照实验),阴性实验结果则可以用蒸馏水代替离子进行实验(即空白实验)而得到。

ZrO^{2+}

用 6 mol·L^{-1} HCl 将可能含有氧锆离子的沉淀溶解,或者在可能含有氧锆离子的溶液中加入足够的 HCl 使溶液酸化(用石蕊试纸检验)。在一只干净的试管中加入 2 滴茜素红溶液,向该溶液中加入 2 滴酸化了的氧锆离子试液,检查此时溶液是否为中等酸性,加热 2 分钟,若出现紫红色示有 ZrO^{2+}。

Ni^{2+}

用 6 mol·L^{-1} HCl 将可能含有镍离子的沉淀溶解,然后加入 6 mol·L^{-1} NH_3 将溶液调至碱性;或者加入足够的 NH_3 到可能含有镍离子的溶液中,使溶液呈较强的碱性(用石蕊试纸检验)。在试液中加入 1 滴 1% 的丁二酮肟溶液(DMG),樱桃红色的沉淀示有 Ni^{2+}。Ni^{2+} 有毒,须倒入废液缸中。

Pb^{2+}

加入 10 滴以上的蒸馏水至可能含有铅的沉淀中,加热将固体溶解,加入 2 滴 HAc 和 4 滴 0.1 mol·L^{-1} K_2CrO_4,生成亮黄色沉淀示有 Pb^{2+},沉淀有毒,须倒入废液缸中。

Al^{3+}

用 6 mol·L^{-1} HCl 将可能含有铝离子的沉淀溶解,或者在可能含有铝离子的溶液中加入足够的 HCl 使溶液酸化(用石蕊试纸检验)。加入 2 滴 0.1% 的铝试剂,然后用 6 mol·L^{-1} NH_3 将溶液调至碱性,充分搅拌并用石蕊试纸检验,如果溶液是澄清的,没有沉淀,表示没有 Al^{3+}。如果溶液有絮状粉红色沉淀,离心沉降,弃去上层清夜,加入 3 滴 6 mol·L^{-1} NaOH 溶液,如果有 Al^{3+},沉淀溶解,生成澄清的品红色溶液,如果沉淀不溶于 NaOH,则可能有 ZrO^{2+} 存在,可能 Al^{3+} 也存在。

重要提示:铅和铬都有毒,任何可能含有铅或铬的沉淀和溶液都必须倒入标有"铅和铬"的废液缸中,还有,任何可能含有镍的沉淀和溶液都必须倒入标有"镍"的废液缸中。

Part 4 Experiments

More Important Notes!

Both lead and chromium are toxic. Any solutions or precipitates that you suspect contain lead and/or chromium must be disposed of into the Laboratory Byproducts Jar labeled "Lead and Chromium". Also, any solutions or precipitates that you suspect contain nickel must be disposed of into the Laboratory Byproducts Jar labeled "Nickel".

Questions

1. Write out the procedure that you devised for separating the four cations in a step by step manner so that someone else performing the experiment could follow it.

2. Write balanced molecular equations for the formation of any precipitates that appeared upon addition of HCl or NaOH when testing the solubilities of the cations (refer to your solubility table). [Use NO_3^- as the anion in the formulas for the metallic compounds.]

3. A solution is known to contain the following three cations: Ni^{2+}, Pb^{2+} and ZrO^{2+}. How would you *efficiently* separate these three cations (using the least number of steps)?

思 考 题

1. 写出你所设计的分离四种阳离子的详细过程,要求:他人能按这个过程将实验重复出来。

2. 查阅你的溶解性表,写出用 HCl 或 NaOH 检验阳离子溶解性相关的配平的反应方程式。对于金属化合物,用 NO_3^- 作为分子中的阴离子。

3. 怎样用尽可能少的步骤将含有 Ni^{2+}、Pb^{2+} 和 ZrO^{2+} 三种离子的溶液有效地分离?

Experiment 11 Calibration of Volumetric Glassware

Purpose

1. To learn the analytical laboratory techniques, such as burette, volumetric flask and pipette.
2. To understand the significance of glassware calibration and to learn the calibration method.
3. To learn the relative calibration of volumetric flask and pipette.

Principle

As an analytical chemist, it is important to gather the best possible data from the equipment. Thus, it is important to use them correctly, patiently and precisely. A good analytical chemist will calibrate an "instrument" before he or she uses it in order to insure that the data gathered is as accurate as possible.

Three basic items of exact volumetric glassware are used in chemical analysis: volumetric flask, pipette and burette. Their volume has been checked in manufacturing process, and the precision is enough for common analysis, but calibration is recommended for extremely accurate measurement. There are the general methods commonly employed to calibrate glassware. These are as follows: direct (absolute) calibration and relative calibration.

Direct calibration: A volume of water delivered by a buret or pipet, or contained in a volumetric flask, is obtained directly from the weight of the water and its density. To calibrate a piece of glassware, the true volume of distilled water delivered or contained can be obtained by its weight divided by the density, and convert it to the standard volume at 20 ℃. The equation is:

$$V_t = W_t / d_t$$

where V_t(mL) is the volume of water at t℃, W_t(g) is the mass of water at t℃ in the air, d_t (g·mL^{-1}) is the weight of 1 mL water (in glassware) weighed by brass poise at t℃ in the air.

Three factors must be taken into account when calibrating:

(1) The density of water changed with temperature;
(2) The weight changed with air buoyancy;
(3) The volume difference caused by glass expanding or contracting.

After three corrections above, it could be calculated that the volume of the container at 20 ℃ equal the volume of how much water takes at t℃. Table 1 shows the weight of 1 mL water in glassware at different temperature.

Relative calibration: It is often necessary to know only the volumetric relationship between two items of glassware without knowing the absolute volume of either. This situation arises, for example, in taking an aliquot portion of a solution. Suppose that it is desired to titrate *one-fifth* of an unknown sample. The unknown sample might be dissolved and diluted in a 100 mL volumetric flask. A 20 mL pipet would then be used to withdraw an aliquot for titration. For the calculation in this analysis, it would not be necessary to know the exact volume of the flask or the pipet, *but it would be required that the pipet deliver exactly one-fifth of the contents of the flask*. The method used for the relative calibration in this case would be to discharge the pipet five times into the flask and marking the level of the meniscus on the flask.

实验 11　容量仪器的校正

实验目的

1. 学习滴定管、容量瓶、移液管的操作方法。
2. 了解容量器皿校正的意义及学习校正方法。
3. 学会容量瓶与移液管的相对校正。

实验原理

对分析工作者而言,能不能最大限度利用实验仪器的功能,以便得出尽可能好的实验数据是非常重要的。要做到这一点,正确、耐心、细致地使用仪器很关键。为了确保实验数据的准确性,实验前要对仪器进行校正。

化学定量分析中,滴定管、移液管和容量瓶是常用的精密容量仪器。它们的容积在生产过程中已经检定,其所刻容积有一定的精密度,可满足一般分析的要求。但在准确度要求较高的分析工作中,这三种容器的容积应当进行校正。对分析仪器常见的校正有直接(绝对)校正法和相对校正法。

直接校正法:滴定管、移液管或容量瓶所量取的水的体积,是通过水的密度和称量水的质量直接得到的。为了校正一件玻璃器皿,需要称量它所容纳或放出的蒸馏水的质量,按其质量与密度的关系换算成 20℃时的标准容积。

$$V_t = W_t / d_t$$

式中,V_t——t℃时水的容积(mL);

　　W_t——在空气中 t℃时水的质量(g);

　　d_t——t℃时在空气中用黄铜砝码称量 1 mL 水(在玻璃容器中)的校正质量(g/mL)。

校正时必须考虑下述三种因素:

(1) 水的密度随温度的改变;

(2) 空气浮力使重量的改变;

(3) 玻璃的膨胀与收缩引起的体积的不同。

通过上述三项校正,可计算出根据 t℃时容器所装入水的质量求算 20℃时该容器的容积,即 t℃时容器所盛水的计算容积恰好等于 20℃时该容器的容积。为了便于计算,将 20℃容量为 1 mL 的玻璃容器在不同温度时所盛水的重量列于表 1。

Table 1 The weight of 1 mL water in glassware at different temperature

Temperature (℃)	d_t (g/mL)	Temperature (℃)	d_t (g/mL)	Temperature (℃)	d_t (g/mL)
5	0.99853	14	0.99804	23	0.99655
6	0.99853	15	0.99792	24	0.99634
7	0.99852	16	0.99778	25	0.99612
8	0.99849	17	0.99764	26	0.99588
9	0.99845	18	0.99749	27	0.99566
10	0.99839	19	0.99733	28	0.99539
11	0.99833	20	0.99715	29	0.99512
12	0.99824	21	0.99695	30	0.99485
13	0.99815	22	0.99676		

Apparatus and Chemicals

Apparatus: acidic burette, 25 mL; volumetric flask, 100 mL; pipette, 20 mL; thermometer; cone flask, 50 mL; rubber suction bulb.

Chemicals: distilled water.

Procedure

1. Calibration of a pipette (20 mL)

(1) Obtain and weigh a clean and dry 50 mL flask on analytical balance to the nearest milligram (1 mg).

(2) Clean a 20 mL pipette and exercise the operation of it.

(3) Pipette with distilled water using rubber suction bulb, adjust the meniscus to the line, and allow the water to drain into the bottle above, reweigh it to the nearest milligram (1 mg).

(4) Measure the temperature of water and calculate the true volume of the pipette according to the equation.

2. Calibration of burette (25 mL)

(1) Clean and rinse the burette, fill it with distilled water of which the temperature has been measured.

(2) Obtain 50 mL clean and dry flask, weigh it on analytical balance to the nearest milligram (1 mg).

(3) Adjust the meniscus of water to the etched line of 0.00 mL at eye level. Drain the water slowly to the flask till the meniscus touched 5.00 mL etched line. Reweigh the bottle to the nearest milligram (1 mg).

(4) With the weight of water, and d_t from table 1, the true volume between 0.00 mL and 5.00 mL could be calculated according to the equation.

(5) Repeat step (2) to (4) to calibrate the true volume of burette between 0.00 – 10.00 mL, 0.00 – 15.00 mL, 0.00 – 20.00 mL, 0.00 – 25.00 mL.

(6) Calibrate again, and the volumetric difference should be within 0.02 mL.

(7) Calculate the calibration value of each volume (average of two times), calibration line could be done with the volume as x-coordinate and calibration value as y-coordinate.

表 1 在不同温度下 1 mL 的玻璃量器所量得的水在空气中的重量(用黄铜砝码)

温度 t(℃)	d_t(g/mL)	温度 t(℃)	d_t(g/mL)	温度 t(℃)	d_t(g/mL)
5	0.99853	14	0.99804	23	0.99655
6	0.99853	15	0.99792	24	0.99634
7	0.99852	16	0.99778	25	0.99612
8	0.99849	17	0.99764	26	0.99566
9	0.99845	18	0.99749	27	0.99566
10	0.99839	19	0.99733	28	0.99539
11	0.99833	20	0.99715	29	0.99512
12	0.99824	21	0.99695	30	0.99485
13	0.99815	22	0.99676		

相对校正法:实验时常常只需要知道两件仪器的相对体积,而不需要知道仪器的绝对体积,这时就要用到相对校正。这种情况常常用于取溶液的几分之一。例如,需要滴定一个未知样品的五分之一:假设未知试样溶解并稀释在一个体积为 100 mL 的容量瓶中,用一支 20 mL 移液管移取该溶液用于滴定。在这个分析的计算中,并不需要知道容量瓶和移液管的确切体积,而是需要移液管所放出的溶液的体积为容量瓶容积的五分之一。这种情况用相对校正法,用移液管移取五次液体于容量瓶中,并在容量瓶的液体凹面的切线处做上记号。

仪器和试剂

仪器:酸式滴定管,25 mL;容量瓶,100 mL;移液管,20 mL;温度计;磨口塞锥形瓶,50 mL;洗耳球。

试剂:蒸馏水。

实验步骤

1. 移液管的校正(20 mL)

(1) 取洁净且干燥的 50 mL 磨口塞锥形瓶,在分析天平上称重,称准至 1 mg。

(2) 洗净一支移液管,练习移液管的使用。

(3) 用待校正的移液管,移取已测过温度的蒸馏水,将水放入上述磨口锥形瓶中,称重,称准至 1 mg。

(4) 算出水重,测量实验时水温,按公式算出移液管的真实容积。

2. 滴定管的校正(25 mL)

(1) 将 25 mL 滴定管洗净,装入已测过温度的水。

(2) 取已洗净且干燥的 50 mL 具塞磨口锥形瓶,在分析天平上称重,称准至 1 mg。

(3) 将滴定管的液面调节至 0.00 刻度处。按滴定时常用速度(如每秒钟 3 滴)将水放入已称重的锥形瓶中,使其体积至 5.00 刻度处。然后再称重,得瓶加水的质量,称准至 1 mg。

(4) 算出水重,并查出 d_t,即可算出滴定管 0.00～5.00 刻度之间的真实容积。

The method is the same as above to calibrate a 50 mL burette but 10 mL as volume interval. Table 2 shows experiment data of a 25 mL burette calibration at 21 ℃.

3. Relative calibration of volumetric flask and pipette

Fill a clean 20 mL pipette with distilled water and drain into a clean and dry 100 mL volumetric flask, observe weather the meniscus is tangent with the etched line. If it is not tangent, record the position of down-edge of meniscus. Repeat the step above. Make a new mark at the position if the meniscus of water is at the same place. The pipette and the flask can be used together in other experiment.

Table 2 Experiment data of a 25 mL burette calibration at 21 ℃
Weight of 1 mL water is 0.99695 g at 21 ℃

Volume from burette (mL)	Weight of bottle and water (g)	Weight of empty bottle (g)	Weight of water (g)	True volume (mL)	Correction (mL)
0.00 – 5.00	34.14	29.20	4.94	4.96	−0.04
0.00 – 10.00	39.31	29.31	10.00	10.03	+0.03
0.00 – 15.00	44.30	29.35	14.95	15.00	0.00
0.00 – 20.00	49.39	29.43	19.96	20.02	+0.02
0.00 – 25.00	54.28	29.38	24.90	24.98	−0.02

Notes

1. Never assume any glassware is clean unless you washed it yourself.

2. Clean all glassware with soap and tap water. Use distilled water as the final rinse of your glassware, preferably a double or triple rinse.

3. Glassware that will not come clean in soap and water should be cleaned by soaking in a solution of 1 mol·L^{-1} HNO$_3$ or 1 mol·L^{-1} KOH in alcohol, followed by a soap and water wash and distilled water rinse.

4. Never heat any piece of volumetric glassware. Heating will change the volume that will cause it not accurate anymore!

5. All items in this laboratory should be weighed on the analytical balance. These balances read to the nearest 0.1 mg. Thus, all masses recorded must be to the nearest 0.1 mg.

6. You will be assigned a balance to use. You will use the same analytical balance for all measurements.

Questions

1. How to check up whether buret is clean? What are the effects on the experiment results when we use an unclean buret?

2. Why should we weigh it on analytical balance to the nearest milligram 1 mg when calibrating 50 mL or 25 mL buret and 20 mL pipette?

3. How to use a pipette correctly? Why should we wait for 15 seconds after delivering liquid from a pipette? How to deal with a small amount of liquid remaining in the tip of the pipette? Why?

4. Why is the volumetric glassware always calibrated at 20 ℃?

(5) 用上述方法继续校正 0.00～10.00 mL，0.00～15.00 mL，0.00～20.00 mL，0.00～25.00 mL 处的真实容积。

(6) 重复校正一次。两次校正所得同一刻度的体积相差不应大于 0.02 mL。

(7) 算出各个体积处的校正值（两次平均值）。以读数值为横坐标，校正值为纵坐标作出校正曲线，以备滴定时查取。

如校正 50 mL 滴定管，方法同上，只是每隔 10 mL 测一个校正值。现将水温 21 ℃时校正 25 mL 滴定管的实验数据列于表 2 供参考。

3. 移液管与容量瓶的相对校正

用洗净的 20 mL 移液管吸取蒸馏水，放入洗净且沥干的 100 mL 容量瓶中，共放入 5 次，观察容量瓶中弯月面下缘是否与体积刻度线相切。若不相切，记下弯月面下缘的位置。再重复上述实验一次。连续两次实验结果符合后，做出新标志，本实验中所用移液管与容量瓶即可在其他实验中配套使用。

表 2 滴定管的校正（水温＝21 ℃ 1 mL 水重＝0.99695 g）

滴定管读取容积	瓶+水重(g)	空瓶重(g)	水重(g)	真实容积(mL)	校正值
0.00～5.00	34.14	29.20	4.94	4.96	−0.04
0.00～10.00	39.31	29.31	10.00	10.03	+0.03
0.00～15.00	44.30	29.35	14.95	15.00	0.00
0.00～20.00	49.39	29.43	19.96	20.02	+0.02
0.00～25.00	54.28	29.38	24.90	24.98	−0.02

注：如果使用电子天平有除皮键，可以先将空瓶除皮，天平上可以直接读出水重。

注意事项

1. 永远不要想当然地以为仪器是干净的，除非是经过自己亲自洗净的仪器。
2. 实验用的玻璃仪器要用洗涤剂洗涤，再用自来水冲洗，最后用蒸馏水润洗 2～3 次。
3. 用洗涤剂洗不干净的玻璃仪器，可用 1 mol·L^{-1} HNO$_3$ 或 1 mol·L^{-1} KOH 的乙醇溶液浸泡，然后用洗涤剂洗，用自来水冲洗，最后用蒸馏水润洗 2～3 次。
4. 容量仪器绝不能加热！否则其准确度会受到影响。
5. 所有与质量相关的实验数据都是用分析天平称量得到的，分析天平可精确到 0.1 mg，故，所有原始记录的质量也必须精确到 0.1 mg。
6. 每个同学都必须按照学号在固定的位置上使用天平，而且在今后的实验中也固定使用该台天平。

思 考 题

1. 怎样检查滴定管是否洗净？使用未洗净的滴定管滴定，对测定结果有什么影响？
2. 为什么校正 50 mL 或 25 mL 滴定管及 20 mL 移液管时要称准至 1 mg？
3. 使用移液管的操作要领是什么？放完液体后为什么要停留 15 s？最后留在管尖的液体如何处理？为什么？
4. 为什么容量仪器的体积都规定为 20 ℃时的体积？

Experiment 12 Preparation and Standardization of a Standard Sodium Hydroxide Solution

Purpose

1. To learn how to prepare and standardize NaOH solution with a primary standard.
2. To learn usage of basic burette and determination of endpoint in a titration.

Principle

Preparation of a standard NaOH solution is not a simple task. Solid NaOH is hygroscopic, it readily absorbs water from the atmosphere. Another problem in the preparation of standard NaOH solutions is NaOH reacts with CO_2 to produce Na_2CO_3. Thus, solutions of exact NaOH concentrations cannot be prepared by weighing out the calculated amount of NaOH and dissolving it in a given quantity of water. Usually, sodium hydroxide solutions are prepared by diluting a 50% aqueous solution of sodium hydroxide to approximately the desired concentration, followed by standardization of the solution by titration of an acidic primary standard. In 50% solutions of sodium hydroxide, dissolved carbon dioxide precipitates as sodium carbonate.

In this experiment potassium hydrogen phthalate ($C_8H_5O_4K$, called KHP for short) is used as the primary standard. The KHP standard is a stable, non-hygroscopic (non water absorbing) compound obtained in high purity, which slowly dissolves in water to produce K^+ and hydrogen phthalate ($C_8H_5O_4^-$) ions. KHP contains one titratable acidic hydrogen ($pK_a = 4.01$), which reacts according to the equation:

$$\text{(COOH)(COOK)}C_6H_4 + NaOH \longrightarrow \text{(COONa)(COOK)}C_6H_4 + H_2O$$

Apparatus and Chemicals

Apparatus: Erlenmeyer flasks, 3 × 250 mL; basic buret, 25 mL; graduated cylinder, 10 mL, 100 mL; pipet, 20 mL; polyethylene bottles or glass bottles, 2 ×1 L; weighing bottle for drying KHP.

Chemicals: phenolphthalein solution (0.1% in ethanol); potassium hydrogen phthalate solid, primary standard; 50% sodium hydroxide solution (or use 50 g sodium hydroxide in 50 mL of water).

Procedure

1. Preparation of a standard NaOH solution

Add about 6 mL (use a 10 mL graduated cylinder) of the 50% sodium hydroxide solution to a 1 liter bottle. Fill the bottle to its shoulder with deionized water. Place the stopper or the lid on the bottle, and thoroughly mix the solution by shaking the bottle.

2. Standardization of a standard NaOH solution

Weigh three 0.4 to 0.7 g standard primary KHP to the nearest 0.1 mg. Place each in a labeled, 250 mL Erlenmeyer flask. Add about 30 mL of distilled or deionized water to each flask, and dissolve the KHP. Add two or three drops of phenolphthalein solution to each flask, fill the 25 mL buret with sodium hydroxide solution, and titrate the KHP solution in each flask to the end point. At the end point the solution changes from colorless to a faint pink.

实验 12　NaOH 标准溶液的配制与标定

实验目的

1. 学会标准溶液的配制以及用基准物质标定标准溶液浓度的方法。
2. 学会碱式滴定管的使用和滴定终点的判断。

实验原理

配制标准 NaOH 溶液并不简单,因为 NaOH 吸潮,它能够吸收空气中的水,而且还能吸收 CO_2 生成 Na_2CO_3,因此,准确浓度的 NaOH 溶液不能够通过称出计算量的 NaOH 并将其溶于定量水中而制得。通常,可将 50% NaOH 溶液稀释到所需要的近似浓度,然后,用酸性基准物质标定。在 50% NaOH 溶液中,所溶解的 CO_2 可以 Na_2CO_3 的形式沉淀下来。

本实验以邻苯二甲酸氢钾($C_8H_5O_4K$,可用 KHP 表示)作为基准物质。KHP 稳定、不吸湿、易得到纯品,可溶于水中产生钾离子和邻苯二甲酸氢根离子,后者含有一个可滴定的质子($pK_a=4.01$),其反应如下:

$$\underset{\text{COOK}}{\text{COOH}} + \text{NaOH} \longrightarrow \underset{\text{COOK}}{\text{COONa}} + H_2O$$

仪器和试剂

仪器:锥形瓶,3×250 mL;碱式滴定管 25 mL;
　　　量筒,10 mL,100 mL;移液管,20 mL;
　　　聚乙烯或玻璃试剂瓶,2×1 L;称量瓶。
试剂:0.1%酚酞(乙醇)溶液;邻苯二甲酸氢钾,基准物质;
　　　50% NaOH 溶液(即由 50 g NaOH+50 mL 蒸馏水所配)。

实验过程

1. NaOH 标准溶液的配制

用 10 mL 量筒取 6 mL 50% NaOH 溶液于 1 L 试剂瓶中,加入去离子水至溶液为 1 L。盖上瓶塞,充分摇动,混匀。

2. NaOH 标准溶液的标定

准确称取三份 0.4~0.7 g 基准物质邻苯二甲酸氢钾 KHP 于三个已经做了标记的 250 mL 锥形瓶中,加入 30 mL 去离子水或蒸馏水将其溶解,再在每份中加入 2~3 滴酚酞指示剂,用 NaOH 溶液滴定至溶液由无色到浅粉色,即为终点。

Data and Results

Trial number	1	2	3
Wt. of pure KHP/g			
Vol. of NaOH/mL			
Conc. of NaOH/(mol · L^{-1})			
Mean conc. of NaOH/(mol · L^{-1})			
Relative standard deviation (RSD)/%			

Notes

1. Concentrated NaOH is very *corrosive*! Handle this solution very carefully, and avoid contact with your skin or clothes. If you do spill some, wash it off at once. Wash your hands carefully when you complete this experiment.

2. If drops of base fall on the inside wall of the flask, rinse it into the body of the solution with deionized water from your wash bottle.

3. When the solution remains pink for at least 30 seconds, record the burette reading. A good endpoint is when one drop of base turns the solution from colorless to a faint pink.

4. When reading the buret, make sure your eye is level with the bottom of the meniscus.

5. Read the volume to the nearest 0.01 mL by reading between the lines. The buret is marked off in 0.1 mL so you must estimate the uncertain digit.

6. Make sure the tip of the buret is filled with NaOH and no air bubbles are in the tip or drops hanging out of the tip. Air bubbles can be removed from the tip by thumping the tip while the solution is flowing. Dry the tip of the buret and check to see if the buret is leaking.

Questions

1. Why not use NaOH as the primary standard?

2. What is the purpose of boiling the water used to make the NaOH solution?

3. Why is it better to use a 50% NaOH solution, instead of solid NaOH, to prepare the 0.1 mol · L^{-1} NaOH solution?

4. If you had to weigh out solid NaOH on an analytical balance, how would you do it?

5. While you were titrating, some NaOH dripped onto the table instead of into the flask. How would this effect the final concentration of NaOH that you calculate? Why?

6. What is the definition on an end point (or equivalence point)? How do you know this point has been reached?

7. If the endpoint in the titration is surpassed (too pink), what is its influence on the calculated molarity of the NaOH solution? Explain.

数据记录及结果处理

实验编号	1	2	3
W_{KHP}/g			
V_{NaOH}/mL			
$c_{NaOH}/(mol \cdot L^{-1})$			
$\bar{c}_{NaOH}/(mol \cdot L^{-1})$			
相对标准偏差(RSD)/‰			

注意事项

1. 浓 NaOH 具有腐蚀性！接触时要小心，避免浓碱和皮肤及衣服接触，如果不小心弄洒，要立刻冲洗，实验结束后要认真洗手。
2. 滴定时，如果碱溶液滴落到锥形瓶内壁上，用洗瓶中的去离子水吹洗瓶壁。
3. 当溶液的粉红色保持 30 秒以上时，记录滴定体积。好的终点就是滴入一滴 NaOH 溶液就由无色变为淡粉色。
4. 滴定管读数时，要确保视线与半月面的底部在同一水平线上。
5. 体积读数要精确到 0.01 mL，由于滴定管的刻度线是每格 0.1 mL，故必须估读出两个刻度线之间的体积不确定值。
6. 要确保滴定管的尖嘴充满 NaOH 溶液，滴定管的尖嘴里面不能有气泡，也不能有液滴挂在尖嘴上。尖嘴里的气泡可以通过橡皮管向上弯曲挤出溶液而带出。擦干尖嘴，检查滴定管是否漏水。

思 考 题

1. 为什么 NaOH 不能作为基准物质？
2. 为什么要用沸腾蒸馏水配制 NaOH 溶液？
3. 为什么用 50% NaOH 溶液配制 0.1 mol·L⁻¹ NaOH 溶液比用固体 NaOH 配制 0.1 mol·L⁻¹ NaOH 溶液好？
4. 如果不得不在分析天平上称取 NaOH 固体，应该怎样做？
5. 滴定时，若 NaOH 溶液不慎滴到实验台上而没有滴到锥形瓶中，对计算 NaOH 溶液浓度有影响吗？为什么？
6. 什么是滴定终点（或化学计量点）？滴定时如何得知已经到达滴定终点？
7. 请说明如果滴定终点滴过头（粉色太深）对计算 NaOH 溶液的浓度有什么影响。

Experiment 13 Preparation and Standardization of a Standard Solution of Hydrochloric Acid

Purpose

1. To learn how to prepare a standard solution of hydrochloric acid.

2. To learn how to standardize a hydrochloric acid with primary standard anhydrous sodium carbonate.

3. To learn how to detect the end point by methyl orange.

Principle

Molarity of an acidic solution is determined by titration with a standard base solution. Anhydrous sodium carbonate is a suitable chemical as a primary standard for directly preparing a standard solution. The molarity of the given hydrochloric acid can be found by titrating it against the standard sodium carbonate solution prepared. The equation for the complete neutralization of sodium carbonate with dilute hydrochloric acid is.

$$Na_2CO_3(aq) + 2HCl(aq) \rightarrow 2NaCl(aq) + CO_2(g) + H_2O(l)$$

The end-point is marked by using methyl orange as indicator; the color changes from yellow to orange at the end point.

Apparatus and Chemicals

Apparatus: burette pipette, 25 mL; pipet, 20 mL; Erlenmeyer flask, 250 mL.

Chemicals: hydrochloric acid, concentrated, 36.5% to 38%; anhydrous sodium carbonate, solid, primary standard; methyl orange indicator, 0.1%.

Procedure

1. Preparation of 0.1 mol·L^{-1} HCl solution

Obtain about 8 mL of concentrated hydrochloric acid and dilute to about 1 L with deionized water. Mix well, then place in a 1 L glass bottle or in a 1 L polyethylene bottle for storage.

2. Preparation of Na_2CO_3 standard solution

Weigh out about 1.3 g of anhydrous sodium carbonate accurately using the method of "weighing by difference". Put the weighed carbonate to a beaker and add some deionized water to dissolve it completely. Transfer the solution to a 250 mL volumetric flask. Rinse the beaker thoroughly and transfer all the washes into the volumetric flask. Make up the solution to the mark on the neck. Mix well.

3. Standardization of HCl solution against Na_2CO_3 standard solution

Pipette a 20.00 mL aliquot (portion) of sodium carbonate solution to an Erlenmeyer flask. Add 2 drops of methyl orange indicator solution. Titrate with the 0.1 mol·L^{-1} hydrochloric acid. The end-point of the titration is when the solution just changes from yellow to orange. Record the titration volume. Repeat the titration two times.

Notes

1. The titration values of replicates should be within 0.05 mL of each other.

2. If the liquid level inside the buret is changed rapidly, allow a minute for drainage of the upper wall before making a reading.

3. The methyl orange color change is difficult to judge. It is a good idea to first carry out a

实验 13　盐酸标准溶液的配制与标定

实验目的

1. 学会配制盐酸标准溶液。
2. 学会用无水碳酸钠作为基准物质标定盐酸溶液。
3. 学会用甲基橙指示滴定终点。

实验原理

酸溶液的摩尔浓度可以用标准碱溶液滴定而测得。无水碳酸钠可以作为基准物质,能直接配制成标准溶液。本实验中的盐酸溶液可以与所配制的无水碳酸钠标准溶液进行滴定,从而测得其浓度。碳酸钠与稀盐酸完全反应的化学方程式为:

$$Na_2CO_3(aq) + 2HCl(aq) \rightarrow 2NaCl(aq) + CO_2(g) + H_2O(l)$$

反应终点可以用甲基橙指示,终点颜色由黄色变至橙色。

仪器和试剂

仪器:滴定管,25 mL;移液管,20 mL;锥形瓶,250 mL。
试剂:浓盐酸,36.5%~38%;
　　　无水碳酸钠,固体,基准物质;
　　　甲基橙指示剂,0.1%。

实验步骤

1. 配制 0.1 mol·L^{-1} HCl 溶液

取大约 8 mL 浓盐酸用蒸馏水稀释至 1 L,混匀,置入一个 1 L 的玻璃或聚乙烯瓶中。

2. 配制 Na_2CO_3 标准溶液

用减重法准确称取 1.3 g 无水 Na_2CO_3,将其置入一个烧杯中,加入一定量蒸馏水使其完全溶解,溶液转移至一个 250 mL 的容量瓶中,彻底润洗烧杯并将所有的润洗液全部转移到容量瓶中,加水至容量瓶的刻度线,将溶液混合均匀。

3. 用 Na_2CO_3 标准溶液标定 HCl 溶液

移取 20.00 mL 碳酸钠溶液至锥形瓶中,加入 2 滴甲基橙指示剂。用 0.1 mol·L^{-1} HCl 溶液进行滴定,终点为溶液颜色由黄色恰好变为橙色,记录滴定剂的体积。重复滴定两次。

注意事项

1. 重复滴定体积间的差距应在 0.05 mL 之内。
2. 当滴定管内液面快速变化(快速使溶液从滴定管中放出)时,读数之前要等待一会儿,使挂在滴定管内壁的溶液落下之后再读数。

"rough" titration in order to become familiar with the color change at the end point. Fill the other buret with the Na_2CO_3 solution and drain a certain amount of the solution into an Erlenmeyer flask. Add 2 drops of methyl orange indicator. Titrate with the HCl solution until the color of the solution changes from yellow to orange. Then titrate with the Na_2CO_3 solution until the color changes from orange to yellow. Thus, observe the color change by titrating with HCl and Na_2CO_3 solution alternately.

4. Standard solutions may be prepared by two methods. One method involves the direct weighing and dissolution of a high purity "primary standard" to form a solution of known concentration. Alternatively, a known amount of "primary standard" is titrated by a previously prepared solution. The determination of the concentration of the previously prepared solution is called "Standardization".

Questions

1. What is the meaning of "weighing by difference"? Why should "weighing by difference" be used when a portion of primary standard anhydrous sodium carbonate is weighed?

2. What is the effect of CO_2 in the air on the results when sodium carbonate is titrated with hydrochloride acid?

3. If a HCl solution is standardized against a standard solution of NaOH, what indicator can be used to detect the endpoint?

3. 甲基橙指示剂的颜色变化不好判断，可以事先采取"预滴"的办法来熟悉终点时的颜色变化。在另一根滴定管中加入 Na_2CO_3 溶液，放出一定量的 Na_2CO_3 到一个锥形瓶中，加入两滴甲基橙指示剂，用 HCl 滴定至溶液由黄色变橙色，再用 Na_2CO_3 滴定使颜色由橙色变黄色，这样反复交替用 HCl、Na_2CO_3 溶液滴定，仔细观察甲基橙的变色。

4. 标准溶液有两种配制方法。一种是直接称取高纯度的"基准物质"，将其溶解，配制成浓度已知的溶液。另一种方法是先配制一份溶液，然后用已知量的"基准物质"进行滴定，测定其浓度，这种用基准物质测定溶液浓度的过程叫标定过程，称为"标定"。

思 考 题

1. 什么叫"减重称量"？为什么称取无水碳酸钠基准物质时要用"减重称量"方法？
2. 当用盐酸滴定碳酸钠时，空气中的 CO_2 对实验结果有什么影响？
3. 如果用标准 NaOH 溶液标定 HCl 溶液，可以选择什么指示剂指示终点？

Experiment 14 Determination of Aspirin (Acetylsalicylic Acid) using Back Titration

Purpose
1. To know why we use a back titration for this particular analysis.
2. To master the titration end point of phenolphthalein.

Principle
Many reactions are slow or present unfavorable equilibria for direct titration. Aspirin is a weak acid that also undergoes slow hydrolysis; i.e., each aspirin molecule reacts with two hydroxide ions.

$$\text{(COOH, OCOCH}_3\text{)} + OH^- \xrightarrow{\text{快}} \text{(COO}^-\text{, OCOCH}_3\text{)}$$

$$\text{(COO}^-\text{, OCOCH}_3\text{)} + OH^- \xrightarrow{\text{慢}} \text{(COO}^-\text{, O}^-\text{)} + CH_3COOH$$

To overcome this problem, a known excess amount of base is added to the sample solution and an HCl titration is carried out to determine the amount of unreacted base. This is subtracted from the initial amount of base to find the amount of base that actually reacted with the aspirin and hence the quantity of aspirin in the analyte.

This experiment is designed to illustrate techniques used in a typical *indirect* or *back* titration. You will use the NaOH you standardized previous experiment to back titrate an aspirin solution and determine the concentration of aspirin.

Apparatus and Chemicals
Apparatus: two burettes; weighing bottle; water bath.

Chemicals: aspirin; ethanol; phenolphthalein, indicator soln.; hydrochloric acid, standardized soln.; sodium hydroxide, standardized soln.

Procedure
1. Sample Preparation

Using a clean dry weighing bottle, weigh accurately by difference, triplicate 0.2 g samples into labeled 250 mL Erlenmeyer flasks. To each flask, add 20 mL of ethanol (measure by graduated cylinder) and three drops of phenolphthalein indicator. Swirl gently to dissolve.

2. Addition of a known amount of base in excess

Titrate the first aspirin sample with 0.1 mol · L^{-1} NaOH to the first permanent pink colour. The aspirin/NaOH acid-base reaction consumes one mole of hydroxide per mole of aspirin. The slow aspirin/NaOH hydrolysis reaction also consumes one mole of hydroxide per mole of aspirin, and so for a complete titration you will need to use a total of twice the amount of NaOH that you have already used, plus you will add some excess NaOH to ensure that you really have reacted with all of the aspirin in your sample (adding excess reactant drives the equilibrium towards products—Le Chatelier's principle).

实验 14　返滴定法测定阿司匹林(乙酰水杨酸)

实验目的

1. 了解本实验用返滴定法进行分析的原因。
2. 掌握用酚酞指示剂确定滴定终点。

实验原理

许多化学反应反应速率很慢或呈现不利于直接滴定的化学平衡。阿司匹林是一种弱酸,而且发生水解慢反应,即,1 摩尔阿司匹林可以与 2 摩尔的 OH^- 反应。

为了解决阿司匹林的水解问题,在阿司匹林试样中加入已知过量的碱,然后用 HCl 滴定未反应完的碱,样品中阿司匹林的量可以通过最初加入的碱的量减去未和阿司匹林反应的碱的量来计算。

本实验设计说明了典型的间接滴定或返滴定技术的应用。可以用上次实验标定的氢氧化钠返滴定阿司匹林溶液并测定阿司匹林的浓度。

仪器和试剂

仪器:2 支滴定管;称量瓶;水浴锅。
试剂:阿司匹林;乙醇;酚酞指示剂;已标定的 HCl 溶液;已标定的 NaOH 溶液。

实验步骤

1. 样品准备

取一个洁净干燥的称量瓶,用减重法准确称取三份约 0.2 g 的阿司匹林样品于已经做好标记的 250 mL 锥形瓶中。在每个锥形瓶中加入 20 mL 乙醇(用量筒量取)、3 滴酚酞指示剂,旋转轻摇至样品溶解。

2. 加入已知的过量的碱

用约 $0.1\ mol \cdot L^{-1}$ NaOH 溶液滴定第一份阿司匹林样品,淡粉色为终点。

阿司匹林和 NaOH 间的酸碱反应是 1 摩尔阿司匹林消耗 1 摩尔的 NaOH,阿司匹林和 NaOH 间的水解慢反应也是 1 摩尔阿司匹林消耗 1 摩尔的 NaOH,所以整个滴定反应消耗的 NaOH 的量相当于阿司匹林两倍的物质的量,另外还要再加上一些过量的 NaOH,以确

Calculate how much extra NaOH you will need to add, following this reasoning: The volume of base to add for the hydrolysis reaction is equal to the volume of base you have already used to titrate to the acid-base endpoint in previous step plus an additional 10 mL of excess base. (For example: if you used 16 mL of base in the previous step, the volume of base you would add now would be $16+10=26$ mL. Thus, you would have added a total of $16+16+10=42$ mL of base.)

Use your burette (not a graduated cylinder) to add the appropriate amount of extra NaOH to each of your three sample flasks. (Do not to allow the level of base in the burette to fall below the graduated markings; if necessary, record an intermediate volume and refill the burette to continue with your additions.) Record the total volume of NaOH added to each flask within 0.02 mL.

3. Heating the reaction to completion

Put flask in a water bath to speed up the hydrolysis reaction. Avoid boiling, because the sample may decompose. While heating, swirl the flasks occasionally. After 15 minutes, remove samples from the water bath and cool for 5 minutes. If the solution is colourless, add a few more drops of phenolphthalein. If it remains colourless, add 10 mL more of the base and reheat. (Don't forget to add this additional volume of base to the previously recorded total volume.)

4. Back titration with acid

The only base remaining in each flask will be excess base that has not reacted with the aspirin. Using your burette with your $0.1 \text{ mol} \cdot L^{-1}$ HCl solution, titrate the excess base in each flask with HCl until the pink colour just disappears.

Notes

1. Aspirin is not very soluble in water, and the ethanol helps the aspirin dissolve.
2. Aspirin is an ester which is very easily hydrolyzed. So easily that during a normal titration with NaOH, the alkaline conditions break it down leading to errors in analysis.

Questions

1. Why did you use your burette and not a graduated cylinder to add the excess NaOH?
2. Ethanol was used in the solutions to help dissolve the acetylsalicylic acid. Ethanol is slightly acidic, and will react with NaOH. Describe a blank correction experiment; i.e., what experiment you might do to determine how much NaOH reacts with the ethanol in your solution. Once you have determined this, how would you use the results of a blank correction experiment in the data analysis?
3. Why can't we just use the endpoint found from the initial NaOH titration?
4. What are the advantages of a back titration over a standard titration with NaOH for analysing aspirin?

保阿司匹林和 NaOH 反应完全(加入过量的反应物促使平衡向生成物方向移动——理·查得里原理)。

　　计算需加入的 NaOH 的量：水解反应所消耗的碱的体积等于上一步酸碱滴定所用去的碱的体积,再额外加入 10 mL 碱。(例如,上一步滴定用去碱 16 mL 现在要加入的碱的体积是 16+10=26 mL,故,要加入的碱的总体积是：16+16+10=42 mL)

　　用滴定管(而不是用量筒)加入适当过量的 NaOH 于三个装有待测样品的锥形瓶中(所加入的碱的体积不要超过滴定管的最大容量刻度线,如碱溶液的体积比滴定管的最大容量还大,可分两次加,即先放出适当的体积,再将滴定管装满放出另一部分),记录加到每个锥形瓶中的 NaOH 的总体积,准确至 0.02 mL。

　　3. 加热使反应完全

　　将锥形瓶放入水浴锅中以加速水解反应,水浴不要沸腾,因温度过高样品将会分解,加热时,可不时摇动锥形瓶,15 分钟后,从水浴锅中拿出锥形瓶,冷却 5 分钟。如果溶液无色,可加入几滴酚酞,如果仍无色,加入 10 mL NaOH,再加热。(别忘了将这次加入的 NaOH 的体积与前面记录的体积相加)

　　4. 用 HCl 回滴

　　锥形瓶中的碱是与阿司匹林反应后剩下的,用约 0.1 mol·L^{-1} HCl 滴定剩余的碱直到粉红色消失。

注意事项

　　1. 阿司匹林在水中难溶解,乙醇可以助溶。
　　2. 阿司匹林是一种酯,容易水解,故用 NaOH 直接滴定时,在碱性条件下,阿司匹林易水解而导致分析误差。

思 考 题

　　1. 过量的 NaOH 为什么用滴定管而不是用移液管加入？
　　2. 溶液中加入乙醇是为了帮助乙酰水杨酸溶解,乙醇具有微酸性,可以和 NaOH 反应。描述如何做空白实验加以校正,即,什么样的实验可以确定溶液中乙醇所消耗的 NaOH 的体积。一旦做了空白实验,怎样将空白试验结果用在数据分析中？
　　3. 为什么我们不能用最初所滴入的 NaOH 的量来确定反应的终点？
　　4. 与用标准 NaOH 直接滴定阿司匹林相比,返滴定法测定阿司匹林的优点是什么？

Experiment 15 Preparation and Standardization of EDTA Solution

Purpose

1. To learn how to standardize the concentration of EDTA.
2. To Master the principle of complexometric titration.
3. To learn how to detect the end point by Eriochrome Black T.

Principle

EDTA, ethylenediaminetetraacteic acid, $C_{10}H_{16}N_2O_8$, $(HOOCCH_2)_2NCH_2CH_2N(CH_2COOH)_2$, FM=292.24 g/mol, often symbolized by H_4Y, is an excellent chelating agent. It forms very stable complexes with many divalent and trivalent metal ions depending on solution conditions. Ignoring charges for the moment, the reaction is represented as follows:

$$EDTA + M = MEDTA$$

The reaction lies very far to the right because the equilibrium formational constants, K_f, are on the order of $10^8 - 10^{25}$ depending on the metal and other conditions.

EDTA itself is a tetraprotic acid; it has 4 ionizable protons with pK_a's=1.99, 2.67, 6.16, 10.26. In its fully ionized form, Y^{4-}, the four acetate groups and the lone pairs on the two nitrogens make it a *hexidentate* ligand that wraps itself very tightly around a metal ion. Usually, titrations are performed in basic solution, roughly pH 8-11. The fully protonated form, H_4Y, is only sparingly soluble in water, so the standard form of EDTA used analytically is usually the disodium salt $Na_2H_2Y \cdot 2H_2O$ (372.24 g/mol), which is much more soluble and available in sound purity, except for a small (about 0.3%) amount of adsorbed water. The standard EDTA solution is commonly prepared by dissolving disodium edetate in water. In this experiment, the EDTA solution is standardized with primary standard zinc oxide (ZnO) using the indicator Eriochrome Black T at pH≈10.

In complexometric titrations, the indicator (In) must also act as a ligand toward the metal ion to be determined. The end point occurs as the color of the indicator changes from its complexed form (MIn) to its uncomplexed form (In) when the EDTA titrant combines with the last bit of metal ion in solution. In this experiment, this can be represented by the equations:

Before titration: $Zn^{2+} + HIn^{2-} \rightarrow ZnIn^{-} + H^+$
            (pure blue)(red-violet)

Before the end: $Zn^{2+} + H_2Y^{2-} \rightarrow ZnY^{2-} + 2H^+$
            (colorless)

At the endpoint: $ZnIn^- + H_2Y^{2-} \rightarrow ZnY^{2-} + HIn^{2-} + H^+$
            (red-violet)   (colorless) (pure blue)

Calculate the concentration by the formula:

$$c_{EDTA} = \frac{W_{ZnO} \times \frac{20}{100}}{V_{EDTA} \times \frac{M_{ZnO}}{1000}} \quad (M_{ZnO} = 81.38 \text{ g/mol})$$

Apparatus and Chemicals

Apparatus: acidic buret, 25 mL; erlenmeyer flask, 250 mL; volumetric flask, 100 mL; pipet, 20 mL; measuring cylinder, 10 mL, 100 mL.

Chemicals: disodium edetate ($Na_2H_2Y \cdot 2H_2O$); zinc oxide, primary standard, ignited to

实验 15　EDTA 标准溶液的配制和标定

实验目的

1. 学习 EDTA 标准溶液的配制和标定。
2. 掌握配位滴定的原理。
3. 学会用铬黑 T 指示剂判断终点。

实验原理

乙二胺四乙酸[$C_{10}H_{16}N_2O_8$,($HOOCCH_2$)$_2NCH_2CH_2N(CH_2COOH)_2$,FM = 292.24 g/mol,常用符号 H_4Y 表示]简称 EDTA,是一种极好的螯合剂,在溶液中,能与许多二价和三价金属离子形成 1∶1 的稳定螯合物,忽略电荷,可表示为:

$$EDTA + M = MEDTA$$

在一定条件下,不同金属离子和 EDTA 形成螯合物的稳定常数 K_f 在 $10^8 \sim 10^{25}$ 之间,因为平衡常数很大,反应正向进行程度较大。

EDTA 是一个四元酸,其电离常数 pK_{a_1} 至 pK_{a_4} 分别为 1.99、2.67、6.16、10.26。由于 EDTA 完全电离的阴离子 Y^{4-} 的结构具有四个羧基和两个氨基,所以它可以作为六齿配位体与金属离子紧紧络合。通常情况下,滴定要在碱性条件下进行,溶液的 pH 约在 8~11 之间。完全质子化的 EDTA(H_4Y)难溶于水,在分析中常常使用其二钠盐 $Na_2H_2Y \cdot 2H_2O$ (372.24 g/mol)配制标准溶液。$Na_2H_2Y \cdot 2H_2O$ 易溶于水,并且除含少量水(0.3%)外具有一定的纯度。配制 EDTA 标准溶液一般是将其二钠盐溶于水。本实验以 ZnO 为基准物质,用铬黑 T 为指示剂,在 pH≈10 条件下对 EDTA 溶液进行标定。

在配位滴定中,指示剂(In)也是一种配位体,能与被测定金属离子进行配位。当滴定达到终点、溶液中最后所剩的金属离子与 EDTA 反应后,指示剂的颜色由其与金属离子形成的配合物(MIn)的颜色转变为其未配位状态(In)的颜色。本实验以铬黑 T 为指示剂,终点溶液颜色由紫红色变为纯蓝色,滴定反应为:

滴定前：　　$Zn^{2+} + HIn^{2-} \rightarrow ZnIn^- + H^+$
　　　　　　纯蓝色　　紫红色

终点前：　　$Zn^{2+} + H_2Y^{2-} \rightarrow ZnY^{2-} + 2H^+$
　　　　　　　　　　　　　无色

终点时：　　$ZnIn^- + H_2Y^{2-} \rightarrow ZnY^{2-} + HIn^{2-} + H^+$
　　　　　　紫红色　　　　　　无色　　纯蓝色

EDTA 标准溶液浓度的计算公式如下:

$$c_{EDTA} = \frac{W_{ZnO} \times \dfrac{20}{100}}{V_{EDTA} \times \dfrac{M_{ZnO}}{1000}} \quad (M_{ZnO} = 81.38 \text{ g/mol})$$

constant weight at about 800 ℃; dilute HCl; Methyl red IS; ammonia TS; ammonia-ammonium chloride BS (pH≈10.0); Eriochrome Black T indicator.

Procedure

1. Preparation of standard EDTA solution

Dissolve 9.5 g of disodium edetate ($Na_2H_2Y \cdot 2H_2O$) in water to make 500 mL and mix well. Preserve it in a glass bottle or a polythene bottle.

2. Standardization of standard EDTA solution

Dissolve 0.45 g of zinc oxide primary standard, previously ignited to constant weight at about 800 ℃ and accurately weighed, in 10 mL of dilute hydrochloric acid (3 mol $\cdot$ L^{-1}), dilute it with water to make 100 mL and mix it well. Transfer 20.00 mL of it to a flask, add 1 drop of methyl red and add ammonia test solution until a slight yellow color is obtained. Add 25 mL of water, 10 mL of ammonia-ammonium chloride buffer solution (pH≈10.0) and 4 drops of Eriochrome Black T indicator. Titrate with standard disodium edetate solution until the color changes from red-violet to pure blue. Write down the volume of standard disodium edetate solution. Repeat the titration twice and calculate the concentration of standard EDTA solution.

Notes

1. EDTA dissolves slowly in water, so it should be shaken or warmed. If possible, the solution should be allowed to stand over night before using.

2. The solution may be kept in hard glass bottles. A polythene bottle is the most satisfactory for storage because EDTA solution gradually leaches metal ions from glass containers, resulting in a change in the concentration of free EDTA.

3. In order to dissolve ZnO and transfer it into the 100 mL volumetric flask completely, it should be carefully to avoid splashing when adding HCl to ZnO.

4. If the $Zn(OH)_2$ precipitate appears after added ammonia test solution, it can be dissolved by adding some buffer solution.

5. Complexometric titration by EDTA can not be performed too fast because the reaction occurs not very fast like neutralization, especially at lower temperature. Titrate drop by drop with swing the solution thoroughly when the endpoint is coming.

Questions

1. Why should $NH_3 \cdot H_2O - NH_4Cl$ buffer solution be added before the titration?
2. What volume of EDTA solution is required if the weight of ZnO is 0.15 g?
3. What are the differences between acid-base titration and complexometric titration?
4. What is property of $Na_2H_2Y \cdot 2H_2O$? Why do not to use H_4Y for preparing standard EDTA solution?

仪器和试剂

仪器：滴定管，25 mL；锥形瓶，250 mL；容量瓶，100 mL；移液管，20 mL；量筒，10 mL，100 mL。

试剂：乙二胺四乙酸二钠（$Na_2H_2Y \cdot 2H_2O$），分析纯；

　　　ZnO，基准物质，800 ℃下灼烧至恒重；

　　　稀 HCl，3 mol·L^{-1}；氨试液；

　　　甲基红指示剂，0.1%的60%乙醇液；

　　　NH_3—NH_4Cl 缓冲液（pH≈10）；

　　　铬黑 T 指示剂。

实验步骤

1. EDTA 标准溶液（0.05 mol·L^{-1}）的配制

称取 EDTA（$Na_2H_2Y \cdot 2H_2O$）约 9.5 g，加蒸馏水 500 mL 使其溶解，摇匀，贮存于硬质玻璃瓶或聚乙烯瓶中。

2. EDTA 标准溶液的标定

准确称取已在 800 ℃灼烧至恒重的基准物质 ZnO 约 0.45 g 至一小烧杯中，加稀盐酸（3 mol·L^{-1}）10 mL，搅拌使其溶解，并定量转移至 100 mL 容量瓶中，加水稀释至刻度，摇匀。用移液管吸取 20.00 mL 液体至锥形瓶中，加甲基红指示剂 1 滴，用氨试液调至溶液刚呈微黄色，再加蒸馏水 25 mL，加 $NH_3 \cdot H_2O$—NH_4Cl 缓冲液（pH≈10）10 mL，加铬黑 T 指示剂 4 滴，摇匀。用 EDTA 标准溶液滴定至溶液由紫红色转变为纯蓝色，即为终点。记录 EDTA 溶液体积，平行做三次，计算 EDTA 溶液的浓度。

注意事项

1. EDTA 溶解较慢，可先于温水中溶解，再稀释至一定体积，或放置过夜。
2. EDTA 标准溶液可贮存于硬质玻璃瓶中，长期存放时要放在聚乙烯瓶中，因为玻璃瓶壁中的金属离子会缓慢释放出来，从而使 EDTA 浓度发生改变。
3. 样品加稀盐酸溶解时应小心操作，防止溅失，使 ZnO 完全溶解后方可定量转移至 100 mL 容量瓶中。
4. 滴加氨试液后出现 $Zn(OH)_2$ 沉淀，加缓冲液后即可溶解。
5. 配位反应进行的速度较慢（不像酸碱反应能在瞬间完成），故滴加时加入 EDTA 标准溶液的速度不能太快，在室温低时尤其要注意。特别是近终点时，应逐滴加入，并充分振摇。

思 考 题

1. 为什么在滴定前要加 $NH_3 \cdot H_2O$—NH_4Cl 缓冲液？
2. 如果称量 0.15 g ZnO，需要多少毫升 EDTA 标准溶液？
3. 酸碱滴定和配位滴定有什么不同？
4. $Na_2H_2Y \cdot 2H_2O$ 的基本性质怎样？为什么不用 H_4Y 来配制 EDTA 标准溶液？

Experiment 16 Content Assay of Zinc Gluconate

Purpose

1. To learn the method of determining concentration of zinc salts.
2. To Master the principle of complexometric titration.
3. To learn how to detect the end point by Eriochrome Black T.

Principle

The quantitative determination of many metal ions in solution can be achieved by titrating with a standard solution of EDTA by forming very strong 1∶1 complexes with depending on solution conditions. The reaction of zinc ion with EDTA (H_2Y^{2-}) can be represented by the equation:

$$Zn^{2+} + H_2Y^{2-} = ZnY^{2-} + 2H^+$$

The end point of the titration can be obtained by observing the color change of the indicator Eriochrome Black T:

$$ZnIn^- + H_2Y^{2-} = ZnY^{2-} + HIn^{2-} + H^+$$
(red-violet color)　　(pure blue color)

Of course, for the above two works, EDTA must coordinate with the metal somewhat more strongly than the indicator ligand.

Calculate the concentration by the formula:

$$Zn\% = \frac{c_{EDTA \cdot 2Na} \times V_{EDTA \cdot 2Na} \times M_{Zn}}{W_{sample} \times 1000} \times 100\%$$

where $c_{EDTA \cdot 2Na}$ (mol·L^{-1}) and $V_{EDTA \cdot 2Na}$ (mL) represent the concentration and volume of the EDTA solution; W_{sample} stands for the sample determined; M_{Zn} is the relative molecular mass of Zn, 65.409 g·mol^{-1}.

Apparatus and Chemicals

Apparatus: buret, 25 mL; Erlenmeyer flask, 3 × 250 mL; pipet, 20 mL; measuring cylinder, 10 mL, 100 mL.

Chemicals: EDTA standard solution, 0.05 mol·L^{-1}; ammonia-ammonium chloride BS, pH=10.0; Eriochrome Black T indicator.

Procedure

Accurately weigh 0.45 g of zinc gluconate, dissolve it in 20 mL of water (sometimes need to be heated). Add 10 mL ammonia-ammonium chloride buffer solution and 4 drops of Eriochrome Black T solution, and titrate with 0.05 mol·L^{-1} EDTA standard solution, until the color changes from red-violet to pure blue. Repeat the titration twice.

Notes

1. Titrations must be performed swiftly (but carefully) because the NH_3 will evaporate to some degree and thus the pH of the solution will change.

2. It is advantageous to perform a trial titration to locate the approximate endpoint and to observe the color change. In succeeding titrations, titrate very rapidly to within about 1 or 2 mL of

实验 16　葡萄糖酸锌含量的测定

实验目的

1. 学会测定锌盐浓度的方法。
2. 掌握配位滴定的原理。
3. 学会用铬黑 T 指示剂判断终点。

实验原理

由于 EDTA 可以和许多金属生成稳定的 1∶1 的配合物，很多金属离子的浓度可以利用这种配位反应进行定量测定。本实验中，锌离子和 EDTA(H_2Y^{2-})的反应如下：

$$Zn^{2+} + H_2Y^{2-} = ZnY^{2-} + 2H^+$$

滴定终点可以通过指示剂铬黑 T 颜色的变化进行判断：

$$ZnIn^- + H_2Y^{2-} = ZnY^{2-} + HIn^{2-} + H^+$$
$$\text{（紫红色）} \qquad\qquad\qquad \text{（纯蓝色）}$$

上述两个反应中，锌-EDTA 配合物的稳定性要比锌-指示剂配合物的稳定性高。可用下式计算锌的含量：

$$Zn\% = \frac{c_{EDTA \cdot 2Na} \times V_{EDTA \cdot 2Na} \times M_{Zn}}{W_{样品} \times 1000} \times 100\%$$

式中：$c_{EDTA \cdot 2Na}$—EDTA 溶液的浓度，$mol \cdot L^{-1}$；

$V_{EDTA \cdot 2Na}$—EDTA 溶液的体积，mL；

$W_{样品}$—样品的质量，g；

M_{Zn}—Zn 的相对原子质量，65.409 g/mol。

仪器和试剂

仪器：滴定管，25 mL；锥形瓶，3×250 mL；移液管，20 mL；量筒，10 mL，100 mL。

试剂：EDTA 标准溶液，0.05 mol·L^{-1}；氨-氯化铵缓冲溶液，pH=10.0；铬黑 T 指示剂。

实验步骤

准确称取 0.45 g 葡萄糖酸锌于锥形瓶中，加 20 mL 蒸馏水（必要时可适当加热）溶解，加入 10 mL 氨—氯化铵缓冲溶液和 4 滴铬黑 T 指示剂溶液，用 0.05 mol·L^{-1} EDTA 标准溶液进行滴定，溶液的颜色由紫红变为纯蓝，即为终点。平行做三次。

注意事项

1. 由于 NH_3 具有挥发性，为防止溶液 pH 值有变化，滴定时要适当快点（但要小心）进行。

the endpoint, and then titrate very carefully, a drop or half-drop at a time, to the endpoint. Near the endpoint, periodically squirt the sides of the flask and the burette tip and swirl the flask to ensure all the titrant has gotten into the solution in the flask.

3. The color change of Eriochrome Black T at the endpoint is rather subtle, and sometimes it is slow. It is not an abrupt change from violet red to pure blue; but rather it is from purple to pure blue, so you need to be careful at the end.

2. 为了能够准确地判断终点的颜色变化,可先进行预滴。在滴定时,在离终点还需 1~2 mL 滴定剂溶液之前可快速滴定,当接近终点时要非常仔细,可一滴一滴加入,也可半滴半滴加入,并且充分旋转摇动锥形瓶,可以不时地用洗瓶冲洗溅落在锥形瓶壁或悬挂于滴定管尖端的液滴,以确保滴定剂完全进入锥形瓶的溶液中。

3. 在临近终点时要小心滴定,虽然铬黑 T 指示剂在滴定终点时颜色变化很敏锐,但有时变色的速度较慢。终点时颜色的变化并非突然由紫红变到纯蓝,而是从紫色到纯蓝。

Experiment 17 Determination of Water Hardness

Purpose
1. To master the principle of determination of water hardness by complexometric titration.
2. To master the method of determination of water hardness.
3. To understand the way of expressing water hardness.

Principle
Natural water dissolves many inorganic salts as it flows through rocks and soil. Water containing significant amounts of calcium ion (Ca^{2+}), magnesium ion (Mg^{2+}), iron (Ⅱ) ion (Fe^{2+} or ferrous ion), or iron (Ⅲ) ion (Fe^{3+} or ferric ion) is known as hard water. Ca^{2+} and Mg^{2+} ions are the most common sources of hardness in water. Water hardness is an expression for the sum of the calcium and magnesium cation concentration in a water sample. High levels of these ions produce what is called hard water, otherwise is called soft water. Usually the piped water, well water and river water are hard water.

Hardness in water causes problems at home as well as in industry. For example, Ca^{2+} and Mg^{2+} ions react with soap to form water-insoluble salts, as a result, the soap loses some of its cleaning power, and the insoluble Ca^{2+} and Mg^{2+} salts of fatty acids form a scum that sticks to sinks, bathtubs and fabrics to make them even more dirty. When hard water containing relatively large amounts of hydrogen carbonate ion (HCO_3^-) is heated, insoluble calcium, magnesium, and iron (Ⅱ) carbonates ($CaCO_3$, $MgCO_3$, $FeCO_3$) precipitate as "boiler scale" inside pipes and vessels such as hot water heaters, teakettles, and commercial boilers. This may results in a safety hazard. However, hard water is not a health hazard. People regularly take calcium supplements. Drinking hard water contributes a small amount of calcium and magnesium toward the total human dietary needs of calcium and magnesium.

So, water hardness should be determined when water is heated in boilers or is used for preparation of distilled water. The complexometric titration is the common method of determination of water hardness.

Transfer some water, adjust the pH at 10. Titrate the total concentration of calcium and magnesium ions with EDTA using Eriochrome Black T, then we can determine the hardness. The reactions are shown as follows:

Before titration: $Ca^{2+}(Mg^{2+}) + HIn^{2-} = CaIn^-(MgIn^-) + H^+$

Before end: $Ca^{2+}(Mg^{2+}) + H_2Y^{2-} = CaY^{2-}(MgY^{2-}) + 2H^+$

End point: $CaIn^-(MgIn^-) + H_2Y^{2-} = CaY^{2-}(MgY^{2-}) + HIn^{2-} + H^+$

　　　　　(violet-red)　　　　　　　(pure blue)

Two ways of expressing hardness in common:

(1) The hardness of water is expressed in units of parts per million (ppm) of an equivalent amount calcium carbonate ($CaCO_3$). 1 ppm corresponds to about 1 mg of calcium carbonate per liter.

(2) The hardness of water is expressed in units of 10 mg per liter of an equivalent amount CaO is called 1 degree.

Apparatus and Chemicals
Apparatus: acidic buret, 25 mL; Erlenmeyer flask, 250 mL; volumetric flask, 100 mL; pipet,

实验 17　水的硬度的测定

实验目的

1. 掌握用配位滴定法测定水的硬度的原理。
2. 掌握水的硬度的测定方法。
3. 了解水的硬度的表示方法。

实验原理

天然水流过岩石和土壤时溶解了许多无机盐。含有大量钙离子 Ca^{2+}、镁离子 Mg^{2+}、亚铁离子 Fe^{2+} 和铁离子 Fe^{3+} 的水被称为硬水。其中,Ca^{2+} 和 Mg^{2+} 是硬水中最常见的离子,水的硬度可用水样中钙镁离子的总浓度表示,离子浓度高的水叫硬水,离子浓度低的水叫软水。一般管道水、井水和河水都是硬水。

硬水常常给生活和工业生产带来不便。例如,Ca^{2+} 和 Mg^{2+} 与肥皂反应生成不溶于水的盐,其结果使肥皂失去清洁功效,同时,不溶的 Ca^{2+} 和 Mg^{2+} 的脂肪酸盐形成水垢会沾在容器、浴缸、织物或衣服等上,显得更脏。当加热热水器、茶壶或锅炉里的含有较大量的 HCO_3^- 的硬水时,会产生不溶于水的钙、镁、铁的碳酸盐 $CaCO_3$、$MgCO_3$、$FeCO_3$,形成"锅垢"沉积在管道和容器内壁,造成安全隐患。但是硬水对人体没有害处,钙镁是人体必需的营养元素,饮用硬水可以提供人体每天所需的微量钙镁的一部分。

锅炉用水或制备蒸馏水的用水都需要测定其硬度,配位滴定法是最常用的测定方法。取一定量的水样,调节 $pH \approx 10$,以铬黑 T 为指示剂,用 EDTA 标准溶液滴定 Ca^{2+}、Mg^{2+} 的总量,即可计算水的硬度。反应式如下:

滴定前:$Ca^{2+}(Mg^{2+}) + HIn^{2-} = CaIn^-(MgIn^-) + H^+$

终点前:$Ca^{2+}(Mg^{2+}) + H_2Y^{2-} = CaY^{2-}(MgY^{2-}) + 2H^+$

终点时:$CaIn^-(MgIn^-) + H_2Y^{2-} = CaY^{2-}(MgY^{2-}) + HIn^{2-} + H^+$

　　　　　（紫红）　　　　　　　　　（纯蓝）

表示硬度的方法随各国的习惯而有所不同,通用的有以下两种:

(1) 将测得的 Ca^{2+}、Mg^{2+} 折算成 $CaCO_3$ 的质量,以每升水中含有 $CaCO_3$ 的毫克数表示硬度,$1\ mg \cdot L^{-1}$ 可写作 1 ppm。

(2) 将测得的 Ca^{2+}、Mg^{2+} 折算成 CaO 的质量,以每升水中含 10 mg CaO 为 1 度表示水的硬度。

仪器和试剂

仪器:酸式滴定管,25 mL;锥形瓶,250 mL;
　　　容量瓶,100 mL;移液管,20 mL;量筒,100 mL。

试剂:EDTA 标准溶液,$0.05\ mol \cdot L^{-1}$;
　　　NH_3—NH_4Cl 缓冲液,$pH \approx 10$;

20 mL; measuring cylinder, 100 mL.

Chemicals: standard EDTA solution, 0.05 mol·L^{-1}; Ammonia-ammonium chloride BS, pH≈10.0; Eriochrome Black T; Water sample.

Procedure

Transfer 20.00 mL of standard EDTA solution (0.05 mol·L^{-1}) into a 100 mL volumetric flask, and dilute it with water to make 100 mL. The standard solution of EDTA with the concentration 0.01 mol·L^{-1} is obtained. Add a 100 mL of the water sample to a Erlenmeyer flask with a graduated cylinder, add 5 mL of the ammonia-ammonium chloride buffer solution and 5 drops Eriochrome Black T, titration with standard EDTA solution (0.01 mol·L^{-1}) until the color changes from violet-red to pure blue. Repeat the titration twice.

Perform a blank determination and make any necessary correction.

Notes

1. Notice the collection time, collection method and collection apparatus of water sample.

2. In order to correct for any error attributable to the deionized water and the indicator color transition, you will be analyzing a blank solution. The volume of EDTA used to titrate the blank will be subtracted from all other titration volumes.

3. There are two types of water hardness, temporary and permanent. Temporary hardness is due to the bicarbonate ion, HCO_3^-, being present in the water. Bicarbonate hardness is classified as temporary hardness. This type of hardness can be removed by boiling the water to expel the CO_2 and precipitate the $CaCO_3$ as indicated by the following equation:

$$Ca^{2+}(aq) + 2HCO_3^-(aq) \xrightarrow{\triangle} CaCO_3(s) + H_2O + CO_2(g)$$

Permanent hardness is due to the presence of the ions Ca^{2+}, Mg^{+2} and SO_4^{2-} or Cl^-. This type of hardness cannot be eliminated by boiling. The water with this type of hardness is said to be permanently hard. Total hardness is the sum of permanent and temporary hardness.

Questions

1. What is the hardness of water? How are the equations for calculating water hardness deduced?

2. At what pH condition is the titration performed? Why?

3. What indicator is used in the analysis? What is its color change?

4. If a student has a non-zero blank correction, but forgets to include that correction in his/her calculations, would the calculated [Ca] be correct, too high, or too low? Explain your answers briefly.

5. EDTA and other similar chelating agents are frequently used in cleaning products. (a) Briefly explain how a cleaning product containing EDTA can aid in removing water soap scum from a sink. (b) Would the presence of EDTA in a soap powder help prevent soap scum formation in hard water? Briefly explain.

铬黑 T 指示剂；

水样。

实验步骤

用移液管准确量取已标定过的 0.05 mol·L^{-1} EDTA 标准溶液 20.00 mL 至 100 mL 容量瓶中，加水至刻度，即得 0.01 mol·L^{-1} 的 EDTA 标准溶液。

用 100 mL 量筒量取水样 100 mL 置入锥形瓶中，加 NH$_3$·H$_2$O－NH$_4$Cl 缓冲液(pH=10)5 mL 和铬黑 T 指示剂 5 滴，用 EDTA 标准溶液(0.01 mol·L^{-1})滴定，溶液由紫红色转变为纯蓝色，即为终点。平行实验三次。

做空白实验进行校正。

注意事项

1. 应注意水样采集的时间、方式、所用容器等。

2. 为了消除实验所用蒸馏水以及指示剂可能带来的误差，要做空白试验，计算水样硬度时应扣除滴定空白时所消耗的 EDTA 的体积。

3. 水的硬度有两种：暂时硬度和永久硬度。如果水中含有 HCO$_3^-$，水硬度由碳酸氢钙或碳酸氢镁引起，这种水属于暂时性硬水。暂时硬水中含有碳酸氢根，这种类型的硬度可以通过加热水至沸除去 CO$_2$、产生碳酸盐沉淀而解除其硬度。反应如下：

$$Ca^{2+}(aq) + 2HCO_3^-(aq) \xrightarrow{\triangle} CaCO_3(s) + H_2O + CO_2(g)$$

永久硬度则是由于水中含有钙、镁的硫酸盐或氯化物引起的，这种硬度不能通过使水沸腾的方式除去，系永久性硬水。总硬度就是暂时硬度和永久硬度之和。

思 考 题

1. 什么叫水的硬度？水的硬度单位常用哪几种方法表示？试说明硬度计算公式的来源。

2. 滴定要在什么 pH 条件下进行？为什么？

3. 分析时用什么作指示剂？指示剂的颜色是怎样变化的？

4. 如果一个学生做了空白校正，但是计算时忘了扣除空白，那么他计算的未知的 Ca 的浓度正确吗？是太高还是太低？为什么？

5. EDTA 和其他类似的螯合剂常常用于洗涤产品之中。(a) 简要说明含有 EDTA 的洗涤产品是怎样除去容器中的硬水肥皂浮垢的；(b) 简要说明肥皂粉中存在的 EDTA 能否阻止硬水中的肥皂浮垢生成。

Experiment 18 Precipitation Titration: Determination of Chloride by Mohr Method

Purpose
1. To master the principle and operation of Mohr precipitation titrations.
2. To determine the content of chloride in medicinal sodium chloride.

Principle
This method determines the chloride ion concentration of a solution by titration with silver nitrate $AgNO_3$. As the silver nitrate solution is slowly added, a precipitate of silver chloride forms.

$$Ag^+ + Cl^- = AgCl \downarrow \text{ (white)} \qquad K_{sp} = 1.8 \times 10^{-10}$$

The Mohr method enables detection of the end point by formation of red-brown Ag_2CrO_4 in excess Ag^+ when a low concentration of chromate ion is present.

$$2Ag^+ + CrO_4^{2-} = Ag_2CrO_4 \downarrow \text{ (red-brown)} \qquad K_{sp} = 2.0 \times 10^{-12}$$

This method can be used to determine the chloride ion concentration of water samples from many sources or the content of chloride of other sample. The pH of the sample solutions should be between 6.5 and 10. If the solutions are acidic, the gravimetric method or Volhard's method should be used.

Apparatus and Chemicals
Apparatus: volumetric flask, 250 mL; beaker, 500 mL; pipette, 20 mL; buret, 25 mL; Erlenmeyer flask, 250 mL; reagent bottle, brown, 500 mL; Mohr pipet, 2 mL.

Chemicals: NaCl, primary standard; NaCl, Medicinal, solid; K_2CrO_4, 5%; $AgNO_3$, A. R, solid.

Procedure
1. Preparation of standard $AgNO_3$ solution

8.5 g of $AgNO_3$ is weighed, and dissolved and made up to 500 mL with distilled water, then transferred into a brown flask.

2. Standardization of the $AgNO_3$ solution

Reagent-grade NaCl (primary standard) is dried and cooled to room temperature. 1.3 g NaCl is weighed accurately into a beaker and dissolved, transferred quantitatively into a 250 mL volumetric flask and made up to volume. The standard NaCl solution is obtained.

20.00 mL of the standard NaCl solution is transferred into an Erlenmeyer flask and diluted with 25 mL of distilled water. 1.0 mL of 5% K_2CrO_4, the indicator is added. The solution is titrated to the first permanent appearance of red $Ag_2Cr_2O_4$ with the standard $AgNO_3$ solution. The titration should be done parallelly three times. Calculate the concentration of the standard solution of $AgNO_3$ with the consumed volume of $AgNO_3$ and the concentration of the standard NaCl solution.

3. Determination of chloride of a medicinal NaCl sample

1.5 g of the medicinal NaCl is weighed accurately into a beaker and dissolved, transferred quantitatively into a 250 mL volumetric flask and made up to volume.

20.00 mL of the sample solution is transferred into an Erlenmeyer flask and diluted with 25 mL of water. 1.0 mL of 5% K_2CrO_4, the indicator, is added. The solution is titrated to the first

实验 18　Mohr 法测定药用氯化钠的含量(沉淀滴定)

实验目的

1. 掌握沉淀滴定 Mohr 法的基本原理。
2. 学会测定药用氯化钠中氯离子含量。

实验原理

本方法用硝酸银 $AgNO_3$ 溶液进行滴定测定氯离子含量。当硝酸银溶液缓缓滴入含氯离子的溶液中,有白色的氯化银沉淀生成。

$$Ag^+ + Cl^- = AgCl\downarrow(白色) \qquad K_{sp}=1.8\times10^{-10}$$

Mohr 法采用铬酸根离子 CrO_4^{2-} 做指示剂,当滴定终点到达时,过量的 Ag^+ 能与 CrO_4^{2-} 生成红棕色的 Ag_2CrO_4 沉淀。

$$2Ag^+ + CrO_4^{2-} = Ag_2CrO_4\downarrow(红棕色) \qquad K_{sp}=2.0\times10^{-12}$$

该方法可用于测定很多不同水样中的氯离子含量或其他样品中的氯含量。测定样品时的 pH 值应该在 6.5~10 之间,若样品的溶液呈酸性,则要用重量法或 Volhard 法进行测定。

仪器和试剂

仪器:容量瓶,250 mL;烧杯,500 mL;移液管,20 mL;滴定管,25 mL;
　　　锥形瓶,250 mL;棕色试剂瓶,500 mL;吸量管,2 mL。

试剂:氯化钠,基准物质;氯化钠,药用;K_2CrO_4 溶液,5%;$AgNO_3$,A.R,固体。

实验步骤

1. $0.1\ mol \cdot L^{-1}\ AgNO_3$ 标准溶液的配制

称取 8.5 g $AgNO_3$ 溶解于 500 mL 的蒸馏水中,将溶液转入棕色试剂瓶中,置暗处保存,以防止光分解。

2. $0.1\ mol \cdot L^{-1}\ AgNO_3$ 标准溶液的标定

准确称取 1.3 g 经烘干并冷却至室温的基准 NaCl 于小烧杯中,用蒸馏水溶解后,转入 250 mL 容量瓶中,稀释至刻度,制得 NaCl 标准溶液。

准确移取 20.00 mL NaCl 试液于 250 mL 锥形瓶中,加 25 mL 蒸馏水,用吸量管加入 1.0 mL 5% K_2CrO_4 溶液,用 $AgNO_3$ 标准溶液滴定至溶液刚好出现红色,即为终点。平行滴定三份,根据所消耗 $AgNO_3$ 的体积和 NaCl 标准溶液的浓度,计算 $AgNO_3$ 的浓度。

3. 药用氯化钠试样分析

准确称取药用 NaCl 1.5 g 于烧杯中,加水溶解后,转入 250 mL 容量瓶中,定容至刻度。

准确移取 20.00 mL NaCl 试液于 250 mL 锥形瓶中,加 25 mL 蒸馏水,用吸量管加入 1.0 mL 5% K_2CrO_4 溶液,用 $AgNO_3$ 标准溶液滴定至溶液刚好出现红色,即为终点。平行滴定三份,根据试样质量和消耗的 $AgNO_3$ 标准溶液的体积,计算试样中 NaCl 的百分含量。

permanent appearance of red Ag_2CrO_4 with the standard $AgNO_3$ solution. The titration should be done parallelly three times. Calculate the concentration of the percentage of the sample NaCl with the consumed volume of $AgNO_3$ and the mass of the sample.

Notes

1. Silver nitrate solution will stain clothes and skin. Any spills should be rinsed with water immediately.

2. Residues containing silver ions are usually saved for later recovery of silver metal. Wash your burette and flasks immediately when finished the titration.

3. The contents of the flask should be swirled vigorously with each addition of titrant. The end point should be taken as the first permanent red-orange coloration. This end point is a little tricky and is best observed as a change in color that occurs upon the addition of one or two drops of titrant solution and that persists with mixing.

4. It is a good idea to carry out a "rough" titration in order to become familiar with the color change at the end point.

5. The Mohr titration should be carried out under conditions of pH 6.5 – 10. At higher pH silver ions may be removed by precipitation with hydroxide ions, and at low pH chromate ions may be removed by an acid-base reaction to form hydrogen chromate ions or dichromate ions, affecting the accuracy of the end point.

Questions

1. What is the effect on the experimental result when the concentration of indicator K_2CrO_4 is too high or too low?

2. Why should the pH of the Mohr titration be controlled at 6.5 – 10?

3. Can we use $K_2Cr_2O_7$ instead of indicator K_2CrO_4? Why?

注意事项

1. $AgNO_3$ 溶液在光照条件下容易分解生成单质银,若洒落在皮肤、衣服或桌面上会发生变色现象,所以一旦将 $AgNO_3$ 溶液洒落要立即用水冲洗。

2. 实验未用完的含银溶液要保存起来以便日后回收金属银,实验完成后要立即洗涤锥形瓶和滴定管,防止残余银离子、AgCl 沉淀生成金属银。

3. 滴定时,每加入一些滴定剂都必须剧烈摇动锥形瓶(由于 AgCl 沉淀容易吸附溶液中的 Cl^-,以至终点提前而引入误差)。滴定终点应该为刚好出现不再消失的橙红色。滴定终点不太好判断,终点最好以当 1~2 滴滴定剂滴入后,颜色发生突变且振摇之后颜色不发生变化为准。

4. 可以先取一定量的 NaCl 溶液"预滴"来观察终点的颜色变化,以此练习掌握滴定终点判断。

5. Mohr 滴定法测定溶液中氯离子含量必须在 pH 6.5~10 条件下进行:pH 过高会使 Ag^+ 形成 Ag_2O 沉淀;pH 过低导致 CrO_4^{2-} 生成 $Cr_2O_7^{2-}$,从而影响终点的变色。

思 考 题

1. 以 K_2CrO_4 作指示剂时,其浓度太大或太小对测定有什么影响?
2. Mohr 法测定 Cl^- 含量时,为什么溶液的 pH 值需要控制在 6.5~10 之间?
3. 能否用 $K_2Cr_2O_7$ 代替 K_2CrO_4 作指示剂?为什么?

Experiment 19 Preparation and Standardization of Perchloric Acid in Non-aqueous Solvent

Purpose

1. To master the principle and operation of acid-base titrations in non aqueous solvents.
2. To master the method of using micro-burettes.
3. To master the method of preparing and determining of perchloric acid.

Principle

In glacial acetic acid, the acidity of perchloric acid is the strongest. So for the titration in non-aqueous solvents, perchloric acid is adopted as the titrant. For the titration of a basic compound, perchloric acid in glacial acetic is preferred as standard solution. Both perchloric acid and glacial acetic acid contains some amount of water, therefore an amount of acetic anhydride is added to abstract water.

When perchloric acid is standardized, potassium biphthalate is used as primary standard and crystal violet is used as an indicator. The following equation can be used to denote reaction of titrations:

$$\text{C}_6\text{H}_4(\text{COOH})(\text{COOK}) + \text{HClO}_4 \xrightarrow[\text{violet} \to \text{pure blue}]{\text{crystal violet IS}} \text{C}_6\text{H}_4(\text{COOH})_2 + \text{KClO}_4 \downarrow \text{ (white)}$$

Since potassium perchlorate can not be dissolved in the solvent of glacial acetic acid-acetic anhydride, precipitate is produced. The following equation can be used for calculation:

$$c_{\text{HClO}_4} = \frac{W_{\text{KHC}_8\text{H}_4\text{O}_4}}{V_{\text{HClO}_4} \cdot M_{\text{KHC}_8\text{H}_4\text{O}_4}} \times 1000$$

$$M_{\text{KHC}_8\text{H}_4\text{O}_4} = 204.2 \text{ g/mol}$$

where V_{HClO_4} is the volume obtained after making a blank correction.

Apparatus and Chemicals

Apparatus: micro-burette, 10 mL; Erlenmeyer flask, 50 mL; graduated cylinder, 10 mL, 100 mL.

Chemicals: primary standard potassium biphthalate; perchloric acid, A. R, 70%–72%(g/g), specific gravity 1.75; glacial acetic anhydride, A. R; acetic anhydride, 97%, specific gravity 1.08; Crystal violet IS, 0.5% glacial acetic acid.

Procedure

1. Preparation of the solution of perchloric acid (0.1 mol · L^{-1})

To 750 mL of glacial acetic acid, add 8.5 mL of perchloric acid (70%–72%, g/g) and mix well. Add 24 mL of acetic anhydride dropwise with shaking (the amount of acetic anhydride equivalent to 5.22 mL per gram of water present), cool and dilute with dehydrated glacial acetic acid to 1000 mL, mix well and allow it to stand for 24 hours. If the reactant is susceptible to acetylation, the water content of perchloric acid should be determined, and adjusted to 0.01%–0.02% by the addition of water or acetic anhydride.

2. Standardization of the solution of perchloric acid (0.1 mol · L^{-1})

Weigh accurately about 0.16 g of potassium biphthalate primary standard, previously dried to

实验 19 高氯酸标准溶液的配制和标定

实验目的

1. 掌握非水滴定的原理及操作。
2. 掌握微量滴定管的使用方法。
3. 掌握高氯酸标准溶液的配制、标定方法。

实验原理

在冰醋酸中高氯酸的酸性最强。因此常采用高氯酸作滴定剂,以高氯酸的冰醋酸溶液为滴定碱的标准溶液。高氯酸、冰醋酸均含有水分,需加入一定量的醋酐,以除去其中的水分。

标定高氯酸标准溶液时,常用邻苯二甲酸氢钾为基准物质,以结晶紫为指示剂。滴定反应式如下:

[邻苯二甲酸氢钾(COOH, COOK)] + $HClO_4$ —结晶紫 紫色→纯蓝色→ [邻苯二甲酸(COOH, COOH)] + $KClO_4$ ↓ (白色)

生成的 $KClO_4$ 不溶于冰醋酸—醋酐溶剂,因而有沉淀生成。$KClO_4$ 的浓度可按下式计算:

$$c_{HClO_4} = \frac{W_{KHC_8H_4O_4}}{V_{HClO_4} \cdot M_{KHC_8H_4O_4}} \times 1000$$

$$M_{KHC_8H_4O_4} = 204.2 \text{ g} \cdot \text{mol}^{-1}$$

式中,V_{HClO_4}—扣除了空白后的 $HClO_4$ 体积。

仪器和试剂

仪器:微量滴定管,10 mL;锥形瓶,50 mL;量筒,10 mL,100 mL。
试剂:邻苯二甲酸氢钾,基准试剂;
高氯酸,A.R,70%~72%(g/g),比重 1.75;
冰醋酸,A.R;
醋酐,A.R,97%,比重 1.08;
结晶紫指示液,0.5%冰醋酸溶液。

实验步骤

1. 高氯酸标准溶液(0.1 mol·L^{-1})的配制

取无水冰醋酸 750 mL,加入 70%~72%高氯酸 8.5 mL,摇匀,在室温下缓缓加入醋酐 24 mL(每克水相当于要加 5.22 mL 醋酐),边加边摇,加完后再振摇均匀,放冷。加适量的无水冰醋酸使成 1000 mL,摇匀,放置 24 h。若所测样品易乙酰化,则须用水分滴定法测定

constant weight at 105 – 110 ℃, and dissolve in 10 mL of solvent of acetic anhydride-acetic acid (1 : 4). Add 1 drop of crystal violet IS and titrate slowly with perchloric acid (0.1 mol·L^{-1}) to a blue end point. Perform a blank determination and make any necessary correction.

Notes

1. Perchloric acid is easy to explode if it contacts with organic compounds at high temperature. If acetic anhydride is added to perchloric acid directly, it will react violently and give off large amounts of heat. So dilute with glacial acetic acid previously and add acetic anhydride slowly and mix well.

2. Micro-burette should be cleanly washed and inverted to dry previously.

3. Since perchloric acid and glacial acetic acid can corrode skin, pay attention to protect.

4. Perchloric acid should be preserved in well-closed brown glass bottle.

5. Vacuum grease should be used to lubricate the stopcock plug instead of Vaseline.

6. Be careful when using a micro-burette and read correctly the reading of scales. It is measured according to 8 mL when estimating. The volume should be recorded with four digits of significant figure. The last digit of the reading should be '0' or '5'.

7. Wash the inside of conical flask with an amount of solvent when being close to end point.

8. Cover a dry beaker on the top of burette after adding perchloric acid.

9. The solvent is expensive. Please use the solvent by avoiding waste. Recycle the solvent after completing experiments.

Questions

1. How many millilitres of perchloric acid (0.1 mol·L^{-1}) should be used when weighing about 0.16 g of potassium biphthalate primary standard? Which burette should be used?

2. Why must blank determination be done?

3. What solvent is glacial acetic acid as far as perchloric acid, sulfuric acid, hydrochloric acid and nitric acid? What about water?

4. What should we pay attention to when preparing perchloric acid?

5. How about the result of determination if containers and agents contain a little water in acid-base non-aqueous titrations?

本溶液含水量,再用水和醋酐反复调节至本溶液的含水量为 0.01%～0.2%。

2. 高氯酸标准溶液(0.1 mol·L^{-1})的标定

取于 105～110 ℃干燥至恒重的基准物质邻苯二甲酸氢钾约 0.16 g,精密称定。加醋酐—冰醋酸(1∶4)混合溶剂 10 mL 使其溶解,加结晶紫指示液 1 滴,用高氯酸标准溶液(0.1 mol·L^{-1})滴定至蓝色,即为终点。将滴定结果用空白试验校正。

注意事项

1. 高氯酸与有机物遇热接触,易引起爆炸。若将醋酐直接加到高氯酸中,将发生剧烈反应,并放出大量热。因此,配制高氯酸溶液时应先用冰醋酸将高氯酸稀释,然后在不断搅拌下,缓缓加入醋酐。
2. 使用的微量滴定管应预先洗净,倒置沥干;其他玻璃器皿应预先洗净烘干。
3. 高氯酸、冰醋酸能腐蚀皮肤、刺激黏膜,应注意防护。
4. 高氯酸应贮存在密封的棕色玻璃瓶中。
5. 装高氯酸标准溶液的滴定管,其活塞不用凡士林润滑,而应用真空油润滑。
6. 使用微量滴定管时要细心,要正确进行读数。邻苯二甲酸氢钾的估重是按消耗 8 mL 高氯酸进行计算的。微量滴定管的读数为 4 位有效数字,最后一位为"5"或"0"。
7. 近终点时,用少量溶剂荡洗锥形瓶壁。
8. 冰醋酸有挥发性,标准溶液应密闭贮存,防止挥发及水分进入,标准溶液装入滴定管后,其上端应盖上一干燥小烧杯。
9. 溶剂价格昂贵,应注意节约,实验结束后需回收溶剂。

思 考 题

1. 如果称量 0.16 g 的基准邻苯二甲酸氢钾,将会消耗多少毫升的 0.1 mol·L^{-1} HClO$_4$? 滴定时应该使用什么样的滴定管?
2. 为什么要做空白试验?
3. 冰醋酸对于 HClO$_4$、H$_2$SO$_4$、HCl 及 HNO$_3$ 四种酸是什么性质的溶剂? 水对这四种酸是什么溶剂?
4. 在配制高氯酸时应该注意什么问题?
5. 在非水酸碱滴定中,若容器、试剂含有微量水分,对测定结果有什么影响?

Experiment 20 Determination of Sodium Salicylate in Non-aqueous Solvent

Purpose

1. To master the principle and operation of acid-base titrations in non-aqueous solvents on the alkali metal salts of organic acid.

2. To master the detection of the end-point on the assay of sodium salicylate by the color of crystal violet.

Principle

Sodium salicylate is one of the alkali metal alkoxides, in an aqueous solution

$$K_{b_2} = \frac{K_w}{K_{a_1}} = \frac{1.0 \times 10^{-14}}{1.06 \times 10^{-3}} = 9.4 \times 10^{-10}, \qquad cK_{b_2} < 10^{-8}$$

The basicity of sodium salicylate is too weak in an aqueous solution that it can not be titrated with standard acid solution directly. If an appropriate acidic solvent is chosen to increase its basicity, the compound may be titrated with perchloric acid in glacial acetic acid. Choosing the solvent of acetic anhydride-glacial acetic acid (1 : 4), the titrant of perchloric acid, the reactions in titration as follows:

$$HClO_4 + HAc = H_2Ac^+ + ClO_4^-$$
$$C_7H_5O_3Na + HAc = C_7H_5O_3H + Ac^- + Na^+$$
$$H_2Ac^+ + Ac^- = 2HAc$$

Crystal violet is used as the indicator. The total reaction:

$$C_7H_5O_3Na + HClO_4 \xrightarrow[\text{violet} \rightarrow \text{bluish green}]{\text{crystal violet IS}} C_7H_5O_3H + ClO_4^- + Na^+$$

Calculate the content of sodium salicylate as follows:

$$C_7H_5O_3Na(\%) = \frac{(cV)_{HClO_4} \times M_{C_7H_5O_3Na}}{W_{sample} \times 1000} \qquad (M_{C_7H_5O_3Na} = 160.10 \text{ g/mol})$$

where V_{HClO_4} is the volume obtained after making a blank correction.

Apparatus and Chemicals

Apparatus: micro-burette, 10 mL; Erlenmeyer flask, 50 mL; graduated cylinder, 10 mL.

Chemicals: sodium salicylate, for medicine; perchloric acid, $0.1 \text{ mol} \cdot L^{-1}$; glacial acetic acid, A.R; acetic anhydride, A.R, 97%, specific gravity 1.08;

Crystal violet; 0.5% glacial acetic acid.

Procedure

Weigh about 0.13 g of sodium salicylate accurately, previously derided to constant weight at 105 ℃, and dissolve in 10 mL of the mixture of acetic anhydride-glacial acetic acid (1 : 4) in a 50 mL Erlenmeyer flask. Add 1 drop of crystal violet and titrate with the standard perchloric acid solution to a bluish green end point. Perform a blank determination and make any necessary correction.

Notes

1. Glassware which is used must be washed and dried previously.
2. The volume expansion coefficient of glacial acetic aid is relatively high, which makes the

实验 20　非水滴定法测定药用水杨酸钠的含量

实验目的

1. 掌握非水溶液中酸碱滴定测定有机酸的碱金属盐的原理及操作。
2. 掌握用结晶紫作指示剂测定水杨酸钠的终点判断。

实验原理

水杨酸钠为有机酸的碱金属盐,在水溶液中,

$$K_{b_2}=\frac{K_w}{K_{a_1}}=\frac{1.0\times10^{-14}}{1.06\times10^{-3}}=9.4\times10^{-10}, cK_{b_2}<10^{-8}$$

其碱性很弱,因此不能直接用酸标准溶液准确滴定。若选择适当的酸性溶剂,使其碱性增强,则可用高氯酸进行滴定。以醋酐—冰醋酸(1:4)混合液为溶剂,以高氯酸为滴定剂,滴定水杨酸钠的反应是:

$$HClO_4+HAc=H_2Ac^++ClO_4^-$$
$$C_7H_5O_3Na+HAc=C_7H_5O_3H+Ac^-+Na^+$$
$$H_2Ac^++Ac^-=2HAc$$

用结晶紫为指示剂。总反应式为:

$$HClO_4+C_7H_5O_3Na\xrightarrow[\text{紫}\to\text{蓝绿}]{\text{结晶紫}}C_7H_5O_3H+ClO_4^-+Na^+$$

水杨酸钠含量可按下式计算:

$$C_7H_5O_3Na(\%)=\frac{(cV)_{HClO_4}\times M_{C_7H_5O_3Na}}{W_{\text{样品}}\times1000} \quad (M_{C_7H_5O_3Na}=160.10\ \text{g}\cdot\text{mol}^{-1})$$

式中,V_{HClO_4}为空白校正后体积。

仪器和试剂

仪器:微量滴定管,10 mL;锥形瓶,50 mL;量筒,10 mL。
试剂:水杨酸钠,药用品;
　　　高氯酸,0.1 mol·L^{-1};
　　　冰醋酸,A.R;
　　　醋酐,A.R,97%,比重1.08;
　　　结晶紫指示液,0.5%冰醋酸溶液。

实验步骤

取在105 ℃干燥至恒重的水杨酸样品约0.13 g,准确称定。置于50 mL干燥的锥形瓶中,加醋酐—冰醋酸(1:4)混合溶剂10 mL使溶解,加结晶紫指示液1滴,用高氯酸标准溶液(0.1 mol·L^{-1})滴定至蓝绿色,即为终点。将滴定结果用空白试验校正。

volume changes with the temperature greatly. If the temperature at which the titration is performed differs by more than 10 ℃ from the temperature at which the perchloric acid was standardized, the titrant must be standardized again. If the difference does not exceed 10 ℃, the concentration of the titrant can be corrected as follows:

$$c_1 = \frac{c_0}{1 + 0.0011 \cdot (t_1 - t_0)}$$

where 0.0011 is the volume expansion coefficient of glacial acetic acid; t_0 is the temperature at which perchloric acid was standardized; t_1 is the temperature at which the titration is performed; c_0 is the concentration of perchloric acid at t_0; c_1 is the concentration of perchloric acid at t_1.

3. Pay attention to save and recycle the solvent since it is expensive.

Questions

1. Why can't sodium salicylate be titrated with standard acid solution in water directly?

2. Sodium acetate is a weak base in aqueous solution, can it be titrated with hydrochloric acid directly in water? Can it be titrated with hydrochloric acid directly in non-aqueous solutions? If can, please design the simple procedure.

3. Why is the color of the endpoint blue on determining potassium biphthalate, but bluish green on determining sodium salicylate when crystal violet is used as indicator?

4. How to correct the concentration of the perchloric acid when the temperature on the determination and the standardization is different?

注意事项

1. 所用玻璃器皿应预先洗净干燥。
2. 冰醋酸的体积膨胀系数较大,其体积随温度改变较大,故标定高氯酸浓度时应记下室温。如果测定样品时与标定高氯酸时温度超过 10 ℃,要重新标定高氯酸的浓度;如果测定和标定时的温度不超过 10 ℃,则可根据下式将高氯酸的浓度加以校正。

$$c_1 = \frac{c_0}{1+0.0011(t_1-t_0)}$$

式中,0.0011 是冰醋酸的体积膨胀系数;t_0 是标定高氯酸时的温度,t_1 是滴定样品时的温度;c_0 是在温度 t_0 时高氯酸的浓度,c_1 是温度为 t_1 时的高氯酸的浓度。

3. 溶剂价格昂贵,应注意节约,实验结束后需回收溶剂。

思 考 题

1. 为什么水杨酸钠不能在水中直接用酸标准溶液滴定?
2. 醋酸钠在水溶液中为一弱碱,是否可用盐酸标准溶液直接滴定?能否用非水酸碱滴定法测定其含量?若能测定,试设计一简单的操作步骤。
3. 以结晶紫为指示剂,为什么测定邻苯二甲酸氢钾时终点颜色为蓝色,而测定水杨酸钠时终点颜色为蓝绿色?
4. 如果测定温度和标定温度不同,应该怎样对高氯酸浓度进行校正?

Experiment 21 Preparation and Standardization of Potassium Permanganate Solution

Purpose

1. To learn the preparation and preservation of $KMnO_4$ solution.
2. To grasp the theory, method and titration condition for the standardization of $KMnO_4$ solution using $Na_2C_2O_4$.
3. To practice the use of self-indicator in the direction of endpoint.
4. To learn using fine-porosity and sintered-glass funnel for the filtration of $KMnO_4$ solution.

Principle

$KMnO_4$ is a strong oxidant and its concentration can generally be standardized by $Na_2C_2O_4$. The following is the reaction of $KMnO_4$ with $Na_2C_2O_4$.

$$2MnO_4^- + 5C_2O_4^{2-} + 16H^+ \xrightarrow{75-85\ ℃} 2Mn^{2+} + 10CO_2 \uparrow + 8H_2O$$

Heat is necessary because the reaction progresses slowly at room temperature. Even so, the reaction rate is still slow and the color of $KMnO_4$ can't fade rapidly at the beginning of the titration. But the reaction is accelerated once Mn^{2+}, catalyzer of the action, is formed during the reaction. The endpoint can be indicated by the color of MnO_4^-.

The concentration of $KMnO_4$ can be calculated by the following formula.

$$c_{KMnO_4} = \frac{2 \times W_{Na_2C_2O_4}}{5 \times V_{KMnO_4} \times \frac{M_{Na_2C_2O_4}}{1000}}$$

Apparatus and Chemicals

Apparatus: acidic burette, 25 mL; conical flask, 250 mL; graduated cylinder, 10 mL, 100 mL; regent bottle, brown, 500 mL; fine-porosity and sintered-glass funnel; equipment of suction filtration; water bath; hotplate; thermometer.

Chemicals: $KMnO_4$, A. R; $Na_2C_2O_4$, primary standard; H_2SO_4, conc., relative density 1.83 – 1.84 g/mL.

Procedure

1. Preparation of $KMnO_4$ solution (0.02 mol·L^{-1})

Dissolve about 1.6 g $KMnO_4$ in 500 mL of distilled water. Heat the solution at a gentle boil for about 1 h. After the solution turn cold, pour it into a brown bottle. Let the solution stand in dark place for 7 – 10 days. Remove MnO_2 by filtering through a fine-porosity, sintered-glass funnel. Transfer the solution to a clean brown glass-stoppered bottle; store in the dark when not in use.

2. Standardization of $KMnO_4$ solution

About 0.15 g $Na_2C_2O_4$, previously dried at 105 ℃, is weighed accurately and dissolved in 125 mL of water and 5 mL of H_2SO_4. The solution is then heated to 75 – 85 ℃ using water bath and about 15 mL of $KMnO_4$ solution is added to the above solution from a burette. After the color of $KMnO_4$ fades, titration will be continued until a pale pink color is observed and persists for 0.5 min. The temperature of the solution at the endpoint should not be lower than 55 ℃.

实验 21　高锰酸钾溶液的配制与标定

实验目的

1. 学习 $KMnO_4$ 标准溶液的配制方法和保存方法。
2. 掌握用 $Na_2C_2O_4$ 作基准物标定高锰酸钾溶液的原理、方法及滴定条件。
3. 练习使用自身指示剂确定反应终点。
4. 学会使用垂熔玻璃漏斗进行 $KMnO_4$ 溶液的过滤。

实验原理

$KMnO_4$ 是一强氧化剂,其溶液的浓度常用草酸钠($Na_2C_2O_4$)做基准物来标定。$KMnO_4$ 与 $Na_2C_2O_4$ 的反应如下:

$$2MnO_4^- + 5C_2O_4^{2-} + 16H^+ \xrightarrow{75\sim85℃} 2Mn^{2+} + 10CO_2 \uparrow + 8H_2O$$

由于室温下 $KMnO_4$ 与 $Na_2C_2O_4$ 反应较慢,需要加热以加快反应速率。即便如此,反应开始时仍然较慢,开始滴定时加入的 $KMnO_4$ 颜色不能立即褪去,但一经反应生成 Mn^{2+} 后,Mn^{2+} 对反应有催化作用,反应速率加快。滴定时利用 MnO_4^- 本身的颜色指示滴定终点。

高锰酸钾 $KMnO_4$ 的浓度可以根据以下公式计算:

$$c_{KMnO_4} = \frac{2 \times W_{Na_2C_2O_4}}{5 \times V_{KMnO_4} \times \dfrac{M_{Na_2C_2O_4}}{1000}}$$

仪器和试剂

仪器:酸式滴定管,25 mL;锥形瓶,250 mL;量筒,10 mL、100 mL;棕色试剂瓶,500 mL;微孔玻璃漏斗;抽滤装置;水浴锅;电炉;温度计。

试剂:$KMnO_4$,A.R;$Na_2C_2O_4$,基准物质;浓 H_2SO_4,A.R,密度 1.83~1.84 g/mL。

实验步骤

1. $KMnO_4$ 标准溶液(0.02 mol·L^{-1})的配制

称取 $KMnO_4$ 1.6 g,溶于 500 mL 蒸馏水中,加热煮沸约 1 h,放冷后置棕色玻璃瓶中,于暗处放置了 7~10 天,用垂熔玻璃漏斗过滤除去 MnO_2,保存于另一干净的具塞棕色玻璃瓶中,不用时溶液要放于暗处。

2. $KMnO_4$ 标准溶液(0.02 mol·L^{-1})的标定

取 105 ℃ 干燥至恒重的 $Na_2C_2O_4$ 基准物约 0.15 g,准确称定,置 250 mL 锥形瓶中,加新鲜蒸馏水 125 mL 与浓 H_2SO_4 5 mL,旋摇使其溶解,置水浴锅加热至 75~85 ℃,取出。迅速自滴定管中加入 $KMnO_4$ 溶液约 15 mL,待褪色后,继续滴定至溶液显淡红色并保持半分钟不褪,滴定终点时,溶液温度不低于 55 ℃。

Notes

1. Commercial $KMnO_4$ can't be used directly to prepare a standard solution due to the existence of MnO_2 as an impurity, which can accelerate the decomposition of $KMnO_4$. MnO_2 can be eliminated by filtration, but filter paper can not be used.

2. Distilled water generally contains some organic compounds, which can deoxidize $KMnO_4$, so the solution must be boiled to destroy these reducing substances.

3. Light may accelerate the decomposition of $KMnO_4$, so $KMnO_4$ solution must be preserved in brown bottle and kept in dark place for 7 - 10 days.

4. The reaction is fairly slow, so the titration shouldn't be too fast. The temperature of the solution at the endpoint should not be lower than 55 ℃.

5. $Na_2C_2O_4$ should be heated in water bath at 75 - 85 ℃ and can't be heated on a hotplate directly. $Na_2C_2O_4$ decomposes at high temperature.

6. The shape of the meniscus in burette is difficult to discern since $KMnO_4$ is a colorful solution, the liquid level is read from the edge of the liquid.

Questions

1. Why can the standard $KMnO_4$ solution not be prepared directly from solid $KMnO_4$? What should be noticed when a $KMnO_4$ solution is prepared? Why?

2. Why should the solution be adjusted to acidic with sulfuric acid? Can HCl or HNO_3 be used instead of sulfuric acid?

3. Under what kind of conditions should titration be performed when standardizing $KMnO_4$ solution against $Na_2C_2O_4$?

4. What is the residue on the sintered-glass funnel when the $KMnO_4$ solution is filtered? What reagent can be used to remove it?

5. Why should we use an acid burette rather than a basic burette to keep the titrant, $KMnO_4$ solution in this experiment? What is the brown residue on the tip of the burette? and How to remove it?

注意事项

1. 市售 $KMnO_4$ 不可直接配制标准溶液,因其常含有少量 MnO_2 杂质。配成溶液后,MnO_2 起催化剂作用促使 $KMnO_4$ 逐渐分解,故必须过滤除去 MnO_2 杂质。过滤不可用滤纸。

2. 蒸馏水中常含有少量有机杂质,能还原 $KMnO_4$ 产生 MnO_2,而生成的 MnO_2 又促进 $KMnO_4$ 的进一步分解,因此溶液必须煮沸,以便破坏这些还原性物质。

3. 光线能促使 $KMnO_4$ 分解,故 $KMnO_4$ 溶液应贮于棕色玻璃瓶中并在暗处放置 7~10 天。

4. 由于氧化还原反应速度较慢,滴定速度不宜过快。滴定终点时,溶液温度不应低于 55 ℃。

5. 不可将 $Na_2C_2O_4$ 溶液在电炉上直接加热,而应在水浴上加热至 75~85 ℃。温度过高会引起草酸($H_2C_2O_4$)的分解。

6. $KMnO_4$ 溶液为有色溶液,滴定管内的弯月面很模糊,读数时读液面的边沿位置。

思 考 题

1. 为什么不能用 $KMnO_4$ 固体直接配制 $KMnO_4$ 标准溶液?在配制 $KMnO_4$ 标准溶液时,应注意哪些问题?为什么?

2. 为什么要用 H_2SO_4 调节溶液呈酸性?用 HCl 或 HNO_3 可以吗?

3. 用 $Na_2C_2O_4$ 标定 $KMnO_4$ 溶液时,应在什么反应条件下进行?

4. 配制 $KMnO_4$ 溶液时,过滤后的滤器上留下的污物是什么?应选用什么物质将污垢清洗干净?

5. 本实验的滴定剂 $KMnO_4$ 溶液为什么一定要装在酸式滴定管内而不是装在碱式滴定管内?在装 $KMnO_4$ 溶液的滴定管尖部残存的棕色物质是什么?应如何洗净?

Experiment 22 Assay of Ferrous Ammonium Sulfate

Purpose

1. To be familiar with the method and principle to assay Ferrous ammonium sulfate by the standard solution of potassium permanganate.

2. To master the use of self-indicator in the direction of endpoint.

Principle

In acidic solution, Fe^{2+} is oxidized to Fe^{3+} by potassium permanganate. The titration endpoint can be indicated by the color of MnO_4^-. The reaction is as follows:

$$MnO_4^- + 5Fe^{2+} + 8H^+ = Mn^{2+} + 5Fe^{3+} + 4H_2O$$

$$E^{\ominus}_{MnO_4^-/Mn^{2+}} = 1.51\ V, \qquad E^{\ominus}_{Fe^{3+}/Fe^{2+}} = 0.771\ V$$

The percentage of FAS is calculated as follows:

$$FAS\% = \frac{(cV)_{KMnO_4} \times 5 \times \frac{M_{FAS}}{1000}}{W_{sample}(g)} \times 100\%$$

$$M_{FAS} = 394.14\ (g/mol)$$

The color of endpoint may be confused by yellow color of Fe^{3+} ion produced during the titration. Fe^{3+} ion forms a colorless complex of $FeHPO_4^+$ ion when appropriate amount of H_3PO_4 is added. This simplifies detection of endpoint. Also, the concentration of Fe^{3+} in the solution will decrease because of formation of the complex ion, as a result, the conditional potential decreases and the potential jump of titration increases. This is helpful for the reaction forward completed and sensitizes the endpoint.

Apparatus and Chemicals

Apparatus: acidic burette, 25 mL; Erlenmeyer flask, 250 mL; graduated cylinder, 100 mL.

Chemicals: standard solution of potassium permanganate, 0.02 mol · L^{-1}; FAS; concentrated sulfate acid solution; H_3PO_4, 85%.

Procedure

0.8 - 0.9 g FAS, accurately weighed, is dissolved in 10 mL of 1 : 3 H_2SO_4 and 2 mL of 85% H_3PO_4. Then 50 mL oxygen-free distilled water is added to make the solid dissolved and the solution is mixed well, and is immediately titrated with the standardized solution of potassium permanganate with the concentration of 0.02 mol · L^{-1} until the first permanent pale pink color is observed (not disappears in 30 s). Calculate the percentage of $(NH_4)_2SO_4 \cdot Fe(SO_4) \cdot 6H_2O$ in the sample from the volume of $KMnO_4$ consumed (V_{KMnO_4}).

Notes

1. Be attention to the pH of the reaction solution. The sample should be dissolved with sulfate acid solution first, then diluted with distilled water.

2. Pay attention to how to read the volume of colorful solution in a burette.

3. Recover the rest solution of potassium permanganate after experiment.

4. If the brown manganese dioxide remains at the tip of burette, clean the burette with the mixture of sodium oxalate and sulfate acid.

实验 22 硫酸亚铁铵含量的测定

实验目的

1. 熟悉用 $KMnO_4$ 法测定硫酸亚铁铵含量的方法和原理。
2. 掌握 $KMnO_4$ 自身指示剂确定反应终点的方法。

实验原理

在硫酸酸性溶液中，$KMnO_4$ 能将亚铁(Fe^{2+})氧化成三价铁(Fe^{3+})，利用 $KMnO_4$ 自身作指示剂指示终点，滴定反应如下：

$$MnO_4^- + 5Fe^{2+} + 8H^+ = Mn^{2+} + 5Fe^{3+} + 4H_2O$$

$$E^{\ominus}_{MnO_4^-/Mn^{2+}} = 1.51\ V, \quad E^{\ominus}_{Fe^{3+}/Fe^{2+}} = 0.771\ V$$

计算公式如下：

$$FAS\% = \frac{(cV)_{KMnO_4} \times 5 \times \frac{M_{FAS}}{1000}}{W_{样品}(g)} \times 100\%$$

$$M_{FAS} = 394.14(g/mol)$$

由于在滴定过程中生成黄色的 Fe^{3+} 对终点颜色有干扰，可加入适量 H_3PO_4，使之与 Fe^{3+} 形成无色的配合物 $FeHPO_4^+$ 而得到掩蔽，消除 Fe^{3+} 颜色的影响，有助于终点的观察。同时，由于配合物的生成减小了 Fe^{3+} 的浓度，降低了 Fe^{3+}/Fe^{2+} 的条件电位，使滴定的突跃增大，有利于反应进行完全，也增加了滴定终点的敏锐度。

仪器和试剂

仪器：酸式滴定管，25 mL；锥形瓶，250 mL；量筒，100 mL。
试剂：标准 KMO_4 溶液，0.02 mol·L^{-1}；FAS；浓硫酸；H_3PO_4，85%。

实验步骤

分别准确称取 0.8～0.9 g 硫酸亚铁铵 FAS 置于三只 250 mL 锥形瓶中，分别加入 1∶3 H_2SO_4(8 mL 浓硫酸溶解于 24 mL 水中)10 mL、85%磷酸 2 mL，再加入 50 mL 无氧蒸馏水，摇匀，使固体溶解，用已标定好的 0.02 mol·L^{-1} 标准 $KMnO_4$ 溶液滴定至刚好出现粉红色(半分钟之内不褪色)。根据所消耗的 $KMnO_4$ 标准溶液体积(V_{KMnO_4})计算样品中的 $(NH_4)_2SO_4 \cdot Fe(SO_4) \cdot 6H_2O$ 的百分含量。

注意事项

1. 注意反应溶液的酸度。样品先用硫酸溶解，再用蒸馏水稀释。
2. 注意有色溶液在滴定管中的读数方法。
3. 实验完毕回收没有用完的高锰酸钾溶液。

Questions

1. Can a solution of hydrochloric acid or nitric acid be used as the acidic medium instead of a sulfate acid solution when Fe^{2+} ion is titrated with the potassium permanganate solution?

2. Why should H_3PO_4 be added into the Erlenmeyer flask before titrating Fe^{2+} against $KMnO_4$ solution? How many millilitres of H_3PO_4 should be added?

3. Why should the burette be cleaned immediately after using $KMnO_4$ solution?

4. 如果棕色二氧化锰残留在滴定管尖,可用草酸钠和硫酸的混合物进行清洗。

思 考 题

1. 用 $KMnO_4$ 法测定亚铁离子时,能否用盐酸或硝酸代替硫酸作酸性介质?
2. 为什么用 $KMnO_4$ 溶液滴定 Fe^{2+} 之前,要加入 H_3PO_4?加入多少量合适?
3. 装过 $KMnO_4$ 溶液的滴定管,为什么应立即洗净?

Experiment 23 Determination of Medical Hydrogen Peroxide

Purpose
1. To grasp the theory, method for determination of medical hydrogen peroxide.
2. To learns expression of the liquid sample content.

Principle
The following reaction occurs when potassium permanganate solution is added to hydrogen peroxide solution acidified with dilute sulphuric acid:

$$2MnO_4^- + 5H_2O_2 + 6H^+ \rightleftharpoons 2Mn^{2+} + 5O_2 \uparrow + 8H_2O$$

The endpoint can be indicated by the color of MnO_4^-.
The concentration of H_2O_2 can be calculated by the following formula.

$$H_2O_2 \%(g/g) = \frac{(cV)_{KMnO_4} \times \frac{5}{2} \times \frac{M_{H_2O_2}}{1000}}{W_{样品}(g) \times \frac{10}{100}} \times 100\%$$

$$H_2O_2 \%(g/mL) = \frac{(cV)_{KMnO_4} \times \frac{5}{2} \times \frac{M_{H_2O_2}}{1000}}{W_{样品}(mL)} \times 100\%$$

$$M_{H_2O_2} = 34.02 \text{ g/mol}$$

Apparatus and Chemicals
Apparatus: acidic burette, 25 mL; Erlenmeyer flask, 250 mL; beaker, 50 mL; measuring cylinder, 10 mL, 100 mL; volumetric flask, 250 mL; pipette, 1 mL, 20 mL.

Chemicals: medical H_2O_2, 30%; $KMnO_4$, A.R; $Na_2C_2O_4$, primary standard; H_2SO_4, A.R, relative density 1.83 – 1.84 g/mL; H_2SO_4, 6 mol·L^{-1}.

Procedure
Add 5 mL distilled water into a 50 mL beaker and weigh it accurately. By using a pipette, add 1.00 mL of the sample 30% H_2O_2 to the 50 mL beaker. Immediately reweigh the beaker accurately. Record the weight as W. Transfer the sample to a 250 mL volumetric flask. Thoroughly rinse the beaker into the volumetric flask. Dilute to volume with distilled water and mix well.

Add 60 mL of distilled water and 15 mL of H_2SO_4 (6 mol·L^{-1}) into a 250 mL Erlenmeyer flask. Then, pipette 20.00 mL of the diluted sample solution into the 250 mL Erlenmeyer flask, titrate it against the standardized potassium permanganate solution with concentration of 0.02 mol·L^{-1} until the first appearance of a pale pink color that persists for 30 seconds. Record the volume delivered as V. Calculated the content of H_2O_2 in the sample.

Notes

1. Hydrogen Peroxide solutions are strong oxidizes and corrosive to the eyes, mucous membranes and skin. In case of contact with the eyes, skin or clothing, flush with large amounts of water for 15 minutes. In case of ingestion, sit upright, drink large quantities of water to dilute the stomach contents and seek immediate medical attention.

2. Hydrogen peroxide in contact with combustible materials may cause fires.

实验 23　医用双氧水的含量测定

实验目的

1. 掌握医用双氧水含量的测定方法及原理。
2. 学会液体含量的表示方法。

实验原理

过氧化氢溶液(即双氧水)用稀硫酸溶液酸化后,加入 $KMnO_4$,发生如下反应:
$$2MnO_4^- + 5H_2O_2 + 6H^+ \rightleftharpoons 2Mn^{2+} + 5O_2\uparrow + 8H_2O$$
滴定时利用 MnO_4^- 本身的颜色指示终点。

其中 H_2O_2 的浓度可按下式计算:

$$H_2O_2\%(g/g) = \frac{(cV)_{KMnO_4} \times \frac{5}{2} \times \frac{M_{H_2O_2}}{1000}}{W_{样品}(g) \times \frac{10}{100}} \times 100\%$$

$$H_2O_2\%(g/mL) = \frac{(cV)_{KMnO_4} \times \frac{5}{2} \times \frac{M_{H_2O_2}}{1000}}{W_{样品}(mL)} \times 100\%$$

$$M_{H_2O_2} = 34.02 \text{ g/mol}$$

仪器和试剂

仪器:酸式滴定管,25 mL;锥形瓶,250 mL;烧杯,50 mL;量筒,10 mL、100 mL;容量瓶,250 mL;移液管,1 mL、20 mL。

试剂:医用双氧水,30%水溶液(市售品);

　　　$KMnO_4$,A.R;

　　　$Na_2C_2O_4$,基准物质;

　　　H_2SO_4,A.R,密度 1.83~1.84 g/mL;

　　　H_2SO_4,6 mol·L^{-1}。

实验步骤

向一个干净的 50 mL 烧杯中加入 5 mL 水,准确称其质量。用移液管精密量取 1.00 mL 30%的 H_2O_2 试样置入小烧杯中,立即再准确称重。记录所加入样品的质量(记作 W)。然后将溶液转移至 250 mL 的容量瓶中,用水彻底荡洗小烧杯,并将洗涤液转移至容量瓶中,加水至刻度,混匀备用。

向锥形瓶中加水 60 mL、6 mol·L^{-1} H_2SO_4 溶液 15 mL,再用移液管吸取 20.00 mL 已稀释的样品溶液置入该 250 mL 锥形瓶中,用 0.02 mol·L^{-1} $KMnO_4$ 标准溶液滴定至溶液刚好呈粉红色(在 30 秒内不褪色)即为终点。记录所消耗的高锰酸钾的体积(记作 V)。计

Questions

1. In determining H_2O_2, besides $KMnO_4$ titration, are there any other methods?
2. When determinating H_2O_2 by using $KMnO_4$, can we use HNO_3 or HCl or HAc instead of H_2SO_4? Why?

算试样中的 H_2O_2 的含量。

注意事项

1. 双氧水溶液是强氧化剂，对眼睛、黏膜和皮肤都有腐蚀性。一旦不小心弄到眼睛、皮肤或衣服上，要用大量的水冲洗15分钟；如果不小心进入口腔，坐直了喝大量的水以稀释胃液，并马上去医院。
2. 双氧水与易燃物接触会着火。

思 考 题

1. 测定 H_2O_2 含量，除 $KMnO_4$ 法外，还可用其他什么方法测定？
2. 用 $KMnO_4$ 法测定 H_2O_2 含量时，能否用 HNO_3 或 HCl、HAc 来控制酸度？为什么？

Experiment 24 Iodimetric Titration of Ascorbic Acid in Vitamin C Tablets

Purpose

To determine the ascorbic acid content of Vitamin C tablets using an iodimetric titration.

Principle

Ascorbic acid (vit-C) is a mild reducing agent that reacts rapidly and quantitatively with iodine to reduce it to iodide ion.

$$C_6H_8O_6 + 2H_2O + I_2 = C_6H_6O_6 + 2I^- + 2H_3O^+ \qquad (1)$$

In this experiment, a known excess of iodine is formed by the reaction of an accurately weighed amount of iodate ion in the presence of excess iodide ion. (Eq. 2)

$$IO_3^- + 5I^- + 6H^+ = 3I_2 + 3H_2O \qquad (2)$$

Once the ascorbic acid has quantitatively reacted with the iodine (Eq. 1), the remaining iodine is back-titrated with thiosulfate (Eq. 3).

$$I_2 + 2S_2O_3^{2-} = 2I^- + S_4O_6^{2-} \qquad (3)$$

Thus, the ascorbic acid content of the Vitamin C tablet is calculated according to the amounts of the solution of iodate ion used and the solution of thiosulfate consumed in a titration:

$$W_{C_6H_8O_6} = \left[3(cV)_{IO_3^-} - \frac{1}{2}(cV)_{S_2O_3^{2-}}\right] \times M_{C_6H_8O_6}$$

The titrant, thiosulfate, is standardized against iodine using the same iodate and iodide ion reaction. To indicate the end point, the disappearance of a blue starch color is used.

Apparatus and Chemicals

Apparatus: volumetric flask, 500 mL; beaker, 250 mL; transfer pipette, 20 mL; Erlenmeyer flask, 250 mL; burette, 25 mL; measuring cylinder, 10 mL.

Chemicals: sodium thiosulfate, $Na_2S_2O_3 \cdot 5H_2O$, FW 158.11, solid; sodium carbonate, Na_2CO_3 A.R, solid; potassium iodide, KI, solid; potassium iodate, KIO_3, FW 214.0, solid; sulfuric acid, H_2SO_4, 0.5 mol $\cdot$ L^{-1}; starch indicator, containing HgI_2 as the preservative; Vitamin C tablets, containing 250 mg of vitamin C, $C_6H_8O_6$, FW 176.12.

Procedure

1. Preparation of the Thiosulfate Titrant (0.1 mol $\cdot$ L^{-1})

Prepare a solution of approximate 0.1 mol $\cdot$ L^{-1} sodium thiosulfate by dissolving about 12.40 g of solid $Na_2S_2O_3 \cdot 5H_2O$ in 500 mL of freshly boiled, deionized water containing about 0.05 grams of Na_2CO_3. Store this solution in the brown bottle which is rinsed with the boiled (sterile) deionized water. Keep the bottle tightly capped.

2. Preparation of the Potassium Iodate Solution (0.013 mol $\cdot$ L^{-1})

Carefully prepare a 0.013 mol $\cdot$ L^{-1} potassium iodate solution by accurately weighing out 1.39 g of predried, primary standard KIO_3. Carefully dissolve the KIO_3 and quantitatively dilute in a volumetric flask of 500 mL.

3. Standardization of the Thiosulfate Solution

Carefully pipet 20.00 mL of the KIO_3 solution into a 250 mL Erlenmeyer flask. Add approximately 1.0 grams of solid KI and 5 mL of 0.5 mol $\cdot$ L^{-1} sulfuric acid. Swirl and

实验 24　碘量法测定维生素 C 的含量

实验目的

用碘量法测定维生素 C 片中抗坏血酸的含量。

实验原理

抗坏血酸(维生素 C)是中等强度的还原剂,能与碘迅速发生定量反应,使碘单质还原为碘离子:

$$C_6H_8O_6 + 2H_2O + I_2 = C_6H_6O_6 + 2I^- + 2H_3O^+ \tag{1}$$

在本实验中,精确称重的碘酸钾与过量的碘离子反应生成过量的且数量已知的碘单质:

$$IO_3^- + 5I^- + 6H^+ = 3I_2 + 3H_2O \tag{2}$$

然后,抗坏血酸再与其中生成的部分碘单质定量发生反应(反应 1),剩余的碘单质则用标准硫代硫酸根离子溶液进行回滴(反应 3):

$$I_2 + 2S_2O_3^{2-} = 2I^- + S_4O_6^{2-} \tag{3}$$

因此,抗坏血酸含量可以根据碘酸根离子的用量和所消耗的硫代硫酸根离子用量求算:

$$W_{C_6H_8O_6} = \left[3(cV)_{IO_3^-} - \frac{1}{2}(cV)_{S_2O_3^{2-}} \right] \times M_{C_6H_8O_6}$$

在本实验中,硫代硫酸根离子溶液是滴定剂,其浓度采用与测定抗坏血酸相同的方法进行标定,即碘酸盐溶液与过量碘离子生成的碘单质反应。以淀粉作指示剂,滴定终点时,碘—淀粉复合物的蓝色消失。

仪器和试剂

仪器:容量瓶,500 mL;烧杯,250 mL;移液管,20 mL;锥形瓶,250 mL;
　　　滴定管,25 mL;量筒,10 mL。
试剂:硫代硫酸钠,$Na_2S_2O_3 \cdot 5H_2O$,FW 158.11,固体;碳酸钠,A.R,固体;
　　　碘化钾,KI,固体;碘酸钾,KIO_3,FW 214.0,固体;硫酸,0.5 mol·L^{-1};
　　　淀粉指示剂;维生素 C 片,250 mg/片,FW 176.12。

实验步骤

1. 0.1 mol·L^{-1}硫代硫酸钠溶液的配制

在 500 mL 新煮沸的蒸馏水中加入 0.05 g Na_2CO_3,将 12.40 g $Na_2S_2O_3 \cdot 5H_2O$ 固体溶解其中,制成硫代硫酸钠溶液;事先用煮沸过的蒸馏水将一个棕色试剂瓶仔细荡洗干净,将配制好的硫代硫酸钠溶液装入洗净的棕色瓶中,并将瓶子盖紧。

2. 0.013 mol·L^{-1}碘酸钾溶液的配制

准确称取 1.39 g 事先经过干燥恒重的基准物质 KIO_3 固体,小心将其溶解并定量转移

immediately titrate with the thiosulfate, titrant solution, until the solution has lost almost all of its color (has a straw yellow color). At this point add 2 mL of the starch indicator. If the solution is clear and no blue color is seen after addition of the starch, the trial must be discarded and redone. Carefully complete the titration until one drop of titrant removes the blue color.

4. Analysis of ascorbic acid in the Vitamin C Tablets

In a 250 mL Erlenmeyer flask, dissolve one of the 250 mg ascorbic acid tablets in 20 mL of cold, recently boiled distilled water; add 3 mL of $0.5 \text{ mol} \cdot \text{L}^{-1}$ sulfuric acid. Use your glass rod to help break-up the tablet (Some solid binding material may not dissolve). Add about 1.0 g of KI and exactly 20.00 mL of the standard KIO_3 solution. Swirl to mix and then carefully titrate with the standardized thiosulfate as done above. Add 2 mL of starch only when the solution is light yellow in color. Continue the titration until one drop of titrant removes the blue color.

Notes

1. If a standard iodine solution is used as a titrant for an oxidizable analyte, the technique is iodimetry. If an excess of iodide is used to quantitatively reduce a chemical species while simultaneously forming iodine, and if the iodine is subsequently titrated with a reducing reagent such as thiosulfate, the technique is iodometry. Iodometry is an example of an indirect determination since a product of a preliminary reaction is titrated.

2. I_2 exists as I_3^-, tri-iodide ion, in the presence of I^-:

$$I_2 + I^- \rightleftharpoons I_3^-$$

The formation of the triiodide ion increases the solubility of the iodine. The resulting solution has a much lower vapor pressure than a solution of iodine in pure water, consequently the loss by volatilization is considerably diminished.

3. Amylose coils into a helical secondary structure resembling a tube with a hollow core. Iodine can lodge inside the core. The complex of iodine stuck inside the amylose coil produces a characteristic blue color. If iodine is attached the surface, the iodine can be desorbed from amylose rapidly. If iodine is deep inside the hollow, the iodine can be desorbed from amylose much slowly.

4. Since starch rapidly degrades, the solution must be fresh or stored with a preservative.

5. Thiosulfate ions are not stable and form the sulfur precipitate when the solution is acidic. The microbes in the water may decompose thiosulfate ions. So, thiosulfate solutions are prepared with the boiled water containing sodium carbonate.

6. The pH of solutions is critical for those chemical equations in the titration.

7. The relative standard deviation should be less than 3 parts per thousand. Note all of these titrations should be performed carefully but rapidly to minimize air oxidation of the iodide ion.

Questions

1. Why should an excess of KI be used in the experiment?

2. Why should the thiosulfate solution be preserved in a tightly capped container?

3. What is the effect when the starch solution, the indicator, is added in earlier time in the titration?

4. Why is it necessary to dissolve the vitamin C tablet with cold, recently boiled distilled water?

至一个 500 mL 的容量瓶中,稀释至刻度。

3. 硫代硫酸钠溶液的标定

用移液管小心移取 20.00 mL KIO_3 溶液于 250 mL 的锥形瓶中,加入 1.0 g 固体 KI 以及 5 mL 0.5 mol·L^{-1} 的硫酸溶液,摇动使之溶解,立即用硫代硫酸钠溶液滴定至溶液的深棕色大部分消失(溶液呈草黄色);此时,向溶液加入 2 mL 淀粉指示剂,溶液显蓝色(如果此时溶液不显蓝色,无色透明,则说明此次滴定失败,需要重做)。继续完成滴定,直至最后一滴滴定剂的加入使溶液的蓝色刚好消失。

4. 维生素 C 片中抗坏血酸的测定

取一片维生素 C 片置于 250 mL 锥形瓶中,加入 20 mL 新煮沸放冷的蒸馏水,并加入 3 mL 0.5 mol·L^{-1} 硫酸,用玻璃棒将维生素 C 片捣碎(维生素 C 片中有些固体辅料不能溶解)。加入 1.0 g KI 固体,用移液管移取 20.00 mL KIO_3 溶液加入其中;振摇锥形瓶使其充分混合,用硫代硫酸钠标准溶液小心进行滴定,滴定过程与前面的操作相同,当溶液颜色变成亮黄色时,加入 2 mL 淀粉指示剂,继续滴定至蓝色消失。

注意事项

1. 用标准碘溶液为滴定剂分析还原性物质的方法称为"碘滴定法"(直接碘量法);先用过量碘离子还原某氧化剂,定量生成碘单质,再用还原剂(如硫代硫酸钠溶液)作为滴定剂滴定生成的碘单质,这种方法称作"滴定碘法"(间接碘量法)。在滴定碘法中,被滴定的物质(I_2)是由先前一步反应生成的,因此属于间接测定。

2. 碘单质 I_2 在过量 I^- 存在的情况下以 I_3^- 形式存在:

$$I_2 + I^- \rightleftharpoons I_3^-$$

I_3^- 的形成不仅能加大碘单质 I_2 的溶解度,并且,因为 I_2 在 KI 溶液中(以 I_3^- 存在)的蒸汽压与其在纯水中的蒸汽压相比要低得多,使得 I_2 由蒸发造成的损失大大减小。

3. 在淀粉链的二级结构中,螺旋卷曲的管子形成一个空洞,碘分子能钻入这个空洞里形成碘—淀粉复合物,显示特征的蓝色。如果碘分子处于复合物的表面,则淀粉对碘能快速解吸附,如果碘分子处在空洞的深处,则淀粉对碘的解吸附速度就较慢。

4. 由于淀粉容易变质,因此淀粉溶液必须是新配制或者加有防腐剂保存的。

5. 硫代硫酸根离子在酸性溶液中不稳定,容易生成单质 S 沉淀。另外,水中的微生物也可能分解硫代硫酸根离子。因此,在配制硫代硫酸钠溶液时,需要使用经过煮沸的蒸馏水,并且应向水中加入碳酸钠将溶液调成碱性。

6. 在本实验中,pH 值是相关化学反应的重要条件。

7. 本实验的相对标准偏差应该小于千分之三。为了减少空气中氧气氧化碘离子对结果的不良影响,所有的滴定应该快速而小心。

思 考 题

1. 在实验中为什么要加入过量的 KI?
2. 为什么储存硫代硫酸钠溶液时要盖紧容器的盖子?
3. 滴定时如果过早加入淀粉指示剂对实验结果有什么影响?

5. If you were working for the manufacturer of the vitamin tablets, would your be satisfied with your results? Yes or no? Why? Comment about the use of iodimetric titrations and the starch indicator used.

4. 为什么溶解维生素 C 片需要冷的新煮沸的蒸馏水？
5. 如果你在一家维生素 C 片生产厂家工作，你对该实验结果满意吗？为什么？请对碘量法和淀粉指示剂作出评论。

Experiment 25 Determination of Dissociation Constant of Weak Acid

Purpose

1. To learn how to determine the dissociation constant of a weak acid by pH-electrode potential titration.

2. To learn how to measure the pH of a solution using a pHS-25 acidometer.

3. To understand the concept of the dissociation equilibrium.

Principle

The dissociation constant of a weak acid can be determined by the method of pH-electrode potential titration. The acid is titrated with a NaOH solution with known concentration. The values of pH are measured with a pH meter after each addition of base, and a "Titration Curve" is created as shown as Figure 1. The two ways to obtain the dissociation constant according to the "Titration Curve" are described as follows.

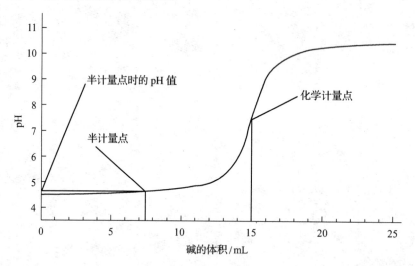

Figure 1 Titration Curve

Acetic acid is a weak monoproton acid. A dissociation equilibrium in aqueous solution can be represented by

$$HAc \rightleftharpoons H^+ + Ac^- \tag{1}$$

The dissociation constant is generally written as:

$$K_a = \frac{[H^+][Ac^-]}{[HAc]} \tag{2}$$

Taking the negative logarithm of above formation Henderson equation is obtained:

$$pH = pK_a + \lg\frac{[Ac^-]}{[HAc]} \tag{3}$$

1. Method of half-neutralization point

Since acetic acid is neutralized with the solution of NaOH as $HAc + OH^- \rightleftharpoons H_2O + Ac^-$, at the half-neutralization point ($V_{NaOH,\frac{1}{2}eq}$), the approximate concentrations of Ac^- and HAc are equal to each other, $[Ac^-] = [HAc]$. Based on Eq 3., there is

$$\lg K_{HAc} = \lg[H^+]_{\frac{1}{2}eq} = -pH_{\frac{1}{2}eq} \tag{4}$$

实验 25 弱酸电离常数的测定

实验目的

1. 学会用 pH 电位滴定法测定弱酸离解常数。
2. 学会 pHS-25 酸度计的使用。
3. 通过实验进一步理解电离平衡的概念。

实验原理

弱酸的电离常数可以采用 pH 电位滴定法进行测定。用浓度已知的 NaOH 溶液对弱酸进行滴定,记录每一次滴加的 NaOH 溶液的体积及其相应的 pH 值,并绘制如图 1 所示的滴定曲线。根据滴定曲线,有两种求 pK_a 的方法。

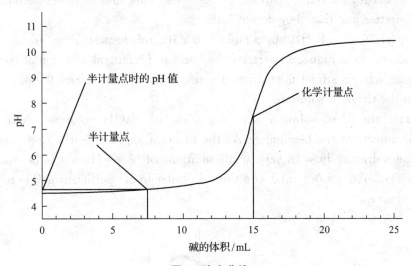

图 1　滴定曲线

醋酸是一元弱酸,在水溶液中的电离平衡可表示如下:

$$HAc \rightleftharpoons H^+ + Ac^- \tag{1}$$

其电离平衡常数的表达式为:

$$K_a = \frac{[H^+][Ac^-]}{[HAc]} \tag{2}$$

对上式取负对数,得到 Henderson 方程:

$$pH = pK_a + \lg\frac{[Ac^-]}{[HAc]} \tag{3}$$

1. 半中和点法

当 NaOH 溶液加入到醋酸溶液中,醋酸被不断地中和,$HAc + OH^- \rightleftharpoons H_2O + Ac^-$。当中和进行到半中和点($V_{NaOH, \frac{1}{2}eq}$)时,$Ac^-$ 与 HAc 的浓度相等,即$[Ac^-] = [HAc]$。根据方程(3),有:

So the dissociation constant K_a is estimated by the value of pH at the half-neutralization point.

2. Method of linear regress

The dissociation constant is also determined using a linear regress by dealing with the data from the titration curve according Eq. 3. Make a linear plot with pH as the y-axis and with $\lg \frac{c(Ac^-)}{c(HAc^-)}$ as x-axis. The intercept of the plot is the value of pK_a.

Apparatus and Chemicals

Apparatus: pHS-25 acidometer; pH electrode; beaker, 100 mL; burette, 25 mL; pipette, 20 mL; magnetic stirrer with a stirbar.

Chemicals: NaOH standard solution, $0.1\ mol \cdot L^{-1}$; HAc, $0.1\ mol \cdot L^{-1}$; standard buffer solution, pH=4.00, 6.86, 9.18.

Procedure

1. Calibrate the acidometer with the standard buffer solutions.

2. Rinse and fill the burette with the base. Make sure that there are no bubbles in the tips of the burettes and that they do not leak.

3. Transfer 20.00 mL HAc by a pipette to a 100 mL beaker.

4. As shown as in Figure 2, arrange the buret to permit the addition of NaOH to the 100 mL beaker while a stirrer is positioned underneath the beaker and the pH electrode is immersed in the titration solution.

5. Titrate the HAc solution with the standard NaOH solution. Add NaOH in 2.00 mL increments at the beginning. As the titration approaches the *equivalence point*, the pH changes drastically with very small additions of base. Here it is necessary to add small volumes of base (≈ 0.1 mL) at a time, in order to get sufficient points to define the shape of the curve.

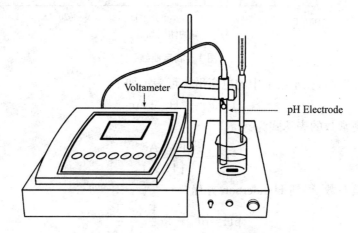

Figure 2 Experimental Setup

$$\lg K_{HAc} = \lg[H^+]_{\frac{1}{2}eq} = -pH_{\frac{1}{2}eq} \tag{4}$$

因此 K_a 可以由半中和点的 pH 测定出来。

2. 线性回归法

平衡常数也可以根据方程(3)，通过对处理滴定曲线的数据进行线性回归而得到。以 pH 为 y 轴、以 $\lg \dfrac{c(Ac^-)}{c(HAc)}$ 为 x 轴作图，线性回归得到一条直线，直线的截距就是 pK_a。

仪器和试剂

仪器：pHS-25 酸度计；pH 电极；烧杯，100 mL；
　　　滴定管，25 mL；移液管，20 mL；
　　　磁搅拌器；搅拌子。
试剂：标准 NaOH 溶液，0.1 mol·L^{-1}；HAc 溶液，0.1 mol·L^{-1}；
　　　标准缓冲溶液，pH=4.00、6.86、9.18。

实验步骤

1. 用缓冲溶液校正酸度计；
2. 用碱溶液润洗并盛装在滴定管中，要确定滴定管嘴没有气泡并且保证不渗漏。
3. 用移液管移取 20.00 mL HAc 溶液于 100 mL 烧杯中。
4. 安装滴定装置如图 2 所示：要使 NaOH 溶液能滴到 100 mL 烧杯中去，搅拌器位于烧杯的下方，pH 电极浸入滴定液中。
5. 用标准 NaOH 溶液滴定 HAc 溶液。开始时每次加入 2.00 mL NaOH，接近滴定化学计量点时，每加入少量 NaOH 都能使 pH 发生很大的变化，为了得到较精确的曲线，注意此时每次加入的 NaOH 体积要少(≈0.1 mL)。

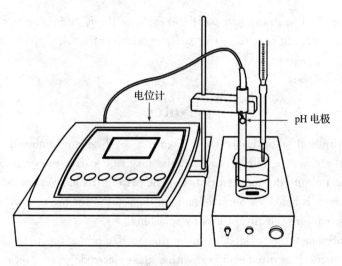

图 2　实验装置

Data and Results

1. Data record

$c_{NaOH} = $ _____ ; V^0_{HAc} _____ ;

Table 1 pH Titration of HAc with NaOH

No.	V_{NaOH}/mL	$\lg \dfrac{c_{NaOH} V_{NaOH}}{c_{HAc} V_{HAc} - c_{NaOH} V_{NaOH}}$	pH
1	0.00		
2	2.00		
3	4.00		
4	6.00		
...	...		
...	...		
...	...		
...	...		
...	...		
...	...		
...	...		
...	25.00		

2. Make the titration curve at the point of pH vs. volume of NaOH added.

3. Pick out the end point from the titration curve and calculate the concentration of the solution of HAc: c_{HAc} _____

4. Measure the half-neutralization point and evaluate the K_a according to the point.

5. Draw the graph at $\lg (c_{HAc}/c_{NaAc})$ vs. pH and calculate the regressive equation and evaluate the K_a from the linear regress.

Compare two methods.

Notes

1. The experimental setup is shown in Figure 2. A beaker contains the acid dissolved in distilled water and a magnetic stirbar, so that the magnetic stirrer motor below can be used to continually stir the mixture during the titration. The pH electrode is supported by a clamp on the left, while the burette is held by a burette clamp in the right of the beaker, with its tip well below the lip of the beaker to prevent titrant loss by splashing.

2. pHS-25 acidometer is a kind of apparatus to determine the pH value with electrode potential measurement. It is often equipped with a glass electrode as an indicative electrode and a calomel electrode or a silver chloride electrode as a primary reference electrode. Also these two electrodes can form a compound electrode.

3. A new glass electrode should be immersed into distilled water no less than 8 hours before being used or after being used. But a compound electrode should be coated with a plastic cup

数据记录及结果

1. 实验记录

$c_{NaOH}=$_____;V^0_{HAc}_____。

表 1 NaOH 溶液对 HAc 溶液的 pH 滴定

No.	V_{NaOH}/mL	$\lg \dfrac{c_{NaOH}V_{NaOH}}{c_{HAc}V_{HAc}-c_{NaOH}V_{NaOH}}$	pH
1	0.00		
2	2.00		
3	4.00		
4	6.00		
...	...		
...	...		
...	...		
...	...		
...	...		
...	...		
...	25.00		

2. 以 NaOH 的体积为横坐标、pH 值为纵坐标绘制滴定曲线。

3. 从滴定曲线上找出终点的位置,并根据终点所消耗的 NaOH 体积计算 HAc 溶液的浓度 c_{HAc}。

4. 在滴定曲线上找到半中和点,用半中和点法求算 HAc 的 pK_a 和 K_a。

5. 根据实验原始数据计算 $\lg(c_{HAc}/c_{NaAc})$。以 $\lg(c_{HAc}/c_{NaAc})$ 为横坐标、pH 值为纵坐标作图,得到线性回归方程,根据直线方程得到 pK_a 和 K_a。

比较由半终点法和线性回归法得到的结果。

注意事项

1. 实验装置图如图 2 所示。烧杯中盛放蒸馏水配制的酸溶液和搅拌子,其下面的磁搅拌器能在滴定时连续搅拌。烧杯左边放置 pH 电极,右边放置滴定管。滴定管的尖嘴要伸进烧杯中,以免滴定时滴定剂溅失。

2. pHS-25 酸度计是一种通过测定电极电势来测定 pH 值的仪器,它常常以玻璃电极作为指示电极,以甘汞电极或氯化银电极作参比电极,有时也可把这两种电极合二为一做成复合电极。

containing 3 mol·L^{-1} KCl solution.

4. There exists saturated KCl solution inside the calomel electrode (as a salt bridge), so there should be excessive KCl crystals to keep KCl solution saturated.

5. After the experiment, rinse the electrodes with distilled water, then reserve these electrodes correctly.

Questions

1. Is the dissociation constant dependent on the concentration of HAc?. What is the dependence of the dissociation degree of HAc upon its concentration?

2. Is it right that the stronger an acid is, its dissociation degree is larger?

3. Discuss the difference between the precise Henderson equation and the approximate Henderson equation.

4. If the concentration of HAc is constant, and its temperature is various, is there any changes in the dissociation constant and dissociation degree of HAc?

5. Is the precise concentration of HAc necessary in determing the K_a of HAc? If to determine a diluted acid with unknown concentration by a NaOH solution, how about the K_a value?

3. 新启用玻璃电极,使用前要在蒸馏水中浸泡 8 小时以上,使用后也要浸泡在蒸馏水中。但复合电极则应该浸泡在 3 mol·L^{-1} KCl 溶液中。

4. 甘汞电极中有饱和 KCl 溶液(作为盐桥),因此 KCl 溶液中应该有 KCl 晶体以维持溶液的饱和状态。

5. 实验完毕,用蒸馏水将电极冲洗干净后正确保存。

思 考 题

1. 实测的电离常数 K_a 值与 HAc 的浓度有关吗?为什么 HAc 的电离度与其浓度有关?
2. "一种酸的酸性越强,其电离度越大",这句话正确吗?
3. 讨论精确 Henderson 方程和近似的 Henderson 方程间的差别。
4. 如果 HAc 的浓度不变,温度是变化的,HAc 的电离常数和电离度有变化吗?
5. 测定 HAc 的 K_a 值一定要知道 HAc 的准确浓度吗?用 NaOH 溶液测定一种未知浓度的稀酸,其 K_a 值又会如何?

Experiment 26 Buffer Action and Buffer Solution

Purpose

1. To explore quantitative formulation of buffer solution from experiment.
2. To design a buffer solution, prepare and determine buffer solution.
3. To master how to use acidometer.

Principle

When an equilibrium system of a weak electrolyte is added with a strong electrolyte containing common ion and sequentially the concentration of the ion increases, the equilibrium shifts to the direction of association. The phenomenon is called the common ion effect. For instance, the chemical equilibrium shifts in the direction to form HAc when the solution of HAc is added with NaAc to increase the concentration of Ac^-. Buffers are solutions that resist large changes in pH when small quantities of strong acid or strong base are added to them. These solutions contain a weak acid and its conjugate base in nearly equal quantities. When small amounts of strong acid are added, according to the principle of the common ion effect, it reacts with the conjugate base present to produce weak acid; small additions of strong base react with weak acid to form conjugate base. This mechanism is termed as buffer action.

The pH value of a buffer solution can be affected by the factors such as concentrations, K_a and temperature, *etc*. A multi-factor equation is needed for estimation of the pH of a buffer solution. To get the equation, we can split it into several single-factor experiments. Firstly, try to find out the quantitative relationship between pH and concentrations when keeping pK_a and temperature fixed. Then, change the kind of the acid and its conjugated base (changing pK_a) by keeping the concentrations and temperature fixed, and find out the dependence of pH upon pK_a. The buffer equation can be obtained after integrated with the two relationships under certain temperature.

Before the empirical formulation obtained in such way is accepted as a theoretical equation, it needs reasoning theoretically to prove the relationships between the factors established. In the solution of a weak acid HA, the ionization of HA occurs,

$$HA \rightleftharpoons H^+ + A^-$$

According to the chemical equilibrium,

$$K_a = \frac{[H^+][A^-]}{[HA]}$$

And

$$[H^+] = K_a \frac{[HA]}{[A^-]}$$

Then

$$pH = pK_a + \lg \frac{[A^-]}{[HA]}$$

This is so called Henderson-Hasselbalch equation (or buffer equation). The formulation can be established when it is conformed by empirical formulation from experiment result.

Apparatus and Chemicals

Apparatus: pHS-25 acidometer; transfer pipette; burette.

实验 26　缓冲溶液与缓冲作用

实验目的

1. 从实验中探索缓冲公式的定量关系。
2. 设计缓冲溶液配方,配制缓冲溶液,测定缓冲容量。
3. 掌握 pH 计的使用方法。

实验原理

在弱电解质的电离平衡体系中加入与弱电解质有共同离子的强电解质时,由于该离子浓度的增加,而使电离平衡向生成分子的方向移动,这种现象称为"同离子效应"。例如,向 HAc 溶液中加入 NaAc 可以增加溶液中 Ac^- 浓度,使平衡向生成 HAc 方向移动。缓冲溶液是一类向其加入少量强酸或强碱而不能引起其 pH 较大变化的溶液,这些溶液由含量相近的弱酸及其共轭碱所组成。根据同离子效应,当向缓冲溶液中加入少量强酸时,强酸能与其中的共轭碱反应产生弱酸;而加入少量强碱时,强碱能与弱酸反应生成共轭碱。这就是缓冲溶液的缓冲原理。

决定和影响缓冲溶液 pH 值的因素从理论上是可以估计的,包括浓度、酸的电离常数以及温度等。因此描述缓冲溶液 pH 的数学关系是一个多因素方程。为得到这个方程,我们可以将其分解成几个单因素实验:可先考察 pH 与浓度的关系,这时要把酸(即 pK_a)和温度固定不变,找出 pH 与 $\lg(c_{HA}/c_{A^-})$ 的数学关系;再固定共轭酸碱对浓度和温度不变,改变不同酸碱对的种类(pK_a),找出 pH 与 pK_a 的关系。将这两个关系式合并,应能得到一定温度下的缓冲公式。

这样得到的公式是经验式,只有从理论上证明它的正确,证明它确实反映几种因素的内在关系,才能作为一个理论关系而成立。在弱酸 HA 溶液中,HA 发生电离反应:

$$HA \rightleftharpoons H^+ + A^-$$

根据化学平衡原理:

$$K_a = \frac{[H^+][A^-]}{[HA]}$$

则:

$$[H^+] = K_a \cdot \frac{[HA]}{[A^-]}$$

有:

$$pH = pK_a + \lg\frac{[A^-]}{[HA]}$$

这就是 Henderson-Hasselbalch 方程(缓冲公式)。实验结果所得的经验式应与之互相验证,才能成立。

Chemicals: NaH_2PO_4, 0.1 mol·L^{-1}; Na_2HPO_4, 0.1 mol·L^{-1}; $NH_3·H_2O$, 0.1 mol·L^{-1}; NH_4Cl, 0.1 mol·L^{-1}; HAc, 0.1 mol·L^{-1}; NaAc, 0.1 mol·L^{-1}; NaOH, 0.1 mol·L^{-1}; HCl, 0.1 mol·L^{-1}; $NaHCO_3$, 0.1 mol·L^{-1}.

Procedure

1. Relationship between pH and concentration ratio

Measure different volumes of HAc and NaOH solution respectively, mix them to get five combinations at least and adjust the ratios of c_{NaAc}/c_{HAc} in the range of 10 : 1 to 1 : 10 in the total volume of 20 mL and test their pH values respectively.

No.	V_{HAc}	V_{NaOH}	c_{NaAc}/c_{HAc}	$lg(c_{NaAc}/c_{HAc})$	pH
1					
2					
3					
4					
5					

Draw the coordinates with pH as vertical axis and $lg(c_{NaAc}/c_{HAc})$ as parallel axis respectively, analyze the lineal relationships between pH and $lg(c_{NaAc}/c_{HAc})$.

2. Relationship between pH and pK_a

Mix the solutions of acids, such as HCl, HAc, NH_4Cl, NaH_2PO_4 and $NaHCO_3$, etc. with the solution of NaOH with the volume ratios of 2 : 1 (V/V) to make the ratio of the concentration c_{salt}/c_{acid} be 1 : 1 in the mixture, measure the pH values. Plot the graph by assigning pH to y-coordinate and pK_a to x-coordinate, and analyze the relationship between pH and pK_a.

Acid/Base Pair	HCl/Cl^-	HAc/Ac^-	NH_4^+/NH_3	$H_2PO_4^-/HPO_4^{2-}$	HCO_3^-/CO_3^{2-}
pK_a					
pH					

Discuss how to integrate the two formulations derived from the above sections into a unified one indicating the relationship of the pH of buffer solutions with the concentration ratios of c_{salts}/c_{acids} and pK_a respectively. Compare this empirical formulation and the result deduced upon the equilibrium of a weak acid. Analyze the reasons for the differences between them if there are any.

3. Designing and Preparing a Buffer Solution with pH=7.50

Design and prepare 100 mL of buffer solution with pH=7.50, and measure its pH value. Compare the theoretical value and the measured value and explain the deviation if there is any.

4. Measuring buffer capacity

Buffer capacity is the ability of the buffer to resist changes of pH. There are usually two ways to describe the buffer capacity. One is called calculus buffer capacity which is defined as quantity of strong acid or base that must be added to change the pH of one liter of solution by one pH unit.

仪器和试剂

仪器：pHS-25 型酸度计；移液管；滴定管。

试剂：NaH_2PO_4，$0.1\ mol\cdot L^{-1}$；Na_2HPO_4，$0.1\ mol\cdot L^{-1}$；$NH_3\cdot H_2O$，$0.1\ mol\cdot L^{-1}$；NH_4Cl，$0.1\ mol\cdot L^{-1}$；HAc，$0.1\ mol\cdot L^{-1}$；NaAc，$0.1\ mol\cdot L^{-1}$；NaOH，$0.1\ mol\cdot L^{-1}$；HCl，$0.1\ mol\cdot L^{-1}$；$NaHCO_3$，$0.1\ mol\cdot L^{-1}$。

实验步骤

1. 测定 pH 值与浓度比的关系

准确量取已知准确浓度的 $0.1\ mol\cdot L^{-1}$ HAc 和 $0.1\ mol\cdot L^{-1}$ NaOH 溶液不同体积（V_{HAc} 及 V_{NaOH}，mL），混合后。溶液总体积为 $V_{HAc}+V_{NaOH}$，则：

$$c_{NaAc}=\frac{0.1\times V_{NaOH}}{V_{HAc}+V_{NaOH}} \quad c_{HAc}=\frac{0.1\times V_{HAc}-0.1\times V_{NaOH}}{V_{HAc}+V_{NaOH}}$$

可近似认为 c_{HAc} 和 c_{NaAc} 分别等于其平衡浓度 $[HAc]$ 和 $[NaAc]$。

改变 V_{HAc} 及 V_{NaOH} 以制备 c_{NaAc}/c_{HAc} 比值从 10∶1 到 1∶10 范围内的混合溶液 5 份以上，每份混合液的总体积为 20 mL，分别测出其 pH 值。

编号	V_{HAc}	V_{NaOH}	c_{NaAc}/c_{HAc}	$\lg(c_{NaAc}/c_{HAc})$	pH
1					
2					
3					
4					
5					

以 pH 对 c_{NaAc}/c_{HAc} 作图。再用 pH 对 $\lg(c_{NaAc}/c_{HAc})$ 作图，分析其线性关系。讨论恒温时，同一缓冲对的 pH 与共轭酸碱对的浓度比的对数值之间的数学关系，并用数学式表达。

2. 测定 pH 与酸电离常数 pK_a 的关系

取 $0.1\ mol\cdot L^{-1}$ HCl、HAc、NH_4Cl、NaH_2PO_4 及 $NaHCO_3$ 的溶液，分别与 $0.1\ mol\cdot L^{-1}$ NaOH 按体积比 2∶1 混合，组成 $c_{盐}/c_{酸}$（或 $c_{盐}/c_{碱}$）=1∶1 的缓冲溶液，并测定其 pH 值。画出 pH 对 pK_a 的关系曲线，讨论恒温时相同浓度比时 pH 与 pK_a 之间关系，用数学关系式表达。

共轭酸碱对	HCl/Cl^-	HAc/Ac^-	NH_4^+/NH_3	$H_2PO_4^-/HPO_4^{2-}$	HCO_3^-/CO_3^{2-}
pK_a					
pH					

Part 4 Experiments

Such definition, although has its practical applications, gives different values of buffer capacity for acid addition and for base addition. This contradicts intuition that for a given buffer solution its resistance should be identical regardless of whether acid or base is added. So, the other buffer capacity definition that takes this intuition into account is given by

$$\beta = \frac{dn}{d(\text{pH})}$$

where dn is an infinitesimal amount of added base and $d(\text{pH})$ is the resulting infinitesimal change in pH.

Although the Henderson-Hasselbalch equation indicates that the pH of the buffer solution is determined by the ratio of concentrations of buffer pair but not by the absolute concentration of the solution, it is obvious that the lower concentration of buffer solution results in the weaker buffer capacity.

Prepare the following buffer solutions and measure the pH values of them. Then add some quantity of NaOH solution with 1/20 of the volume of the buffer solution, and measure the pH values again. Calculate the capacity of the buffer solution based on $\frac{\Delta V \times c_{\text{NaOH}}}{\Delta \text{pH} \times V}$. Summarize the factors that affect the buffer capacity of the solution.

No.	V_{HAc} (0.1 mol·L^{-1})	V_{NaOH} (0.1 mol·L^{-1})	V_{H_2O}	pH	pH (after NaOH added)	capacity
1	10	1	9			
2	10	5	5			
3	5	2.5	12.5			
4	10	9	1			

Henderson-Hasselbalch Equation

American biochemist, Henderson [Henderson, Lawrence Joseph (1878 – 1942)] is remembered for his discovery of the chemical means by which acid-base equilibria are maintained in nature in 1908. In his investigations of body fluids, he discovered that the formation of carbonic acid from carbon dioxide and water in the presence of the salt of the acid (bicarbonates) is the only system in nature that maintains acid-base equilibrium. Henderson developed a chemical equation used to describe these systems, known as physiological buffers. This equation is of fundamental importance to biochemistry. In 1916, Danish biochemist, Hasselbalch [Hasselbalch, Karl Albert (1874 – 1962)] took the equation developed by Henderson and converted it into logarithmic form resulting in the Henderson-Hasselbalch equation. Although the Henderson-Hasselbalch equation is only approximately true, it still remains the most useful mathematical model for treating problems when dealing with buffer solutions. The equation expresses the pH of a buffer solution as a function of the concentration of the weak acid or base and the salt components of the buffer. The Henderson-Hasselbalch equation is one of the fundamental principles of chemistry. You can calculate the pH of a buffer solution or the concentration of the acid and base using the H-H Equation.

讨论如何把上面两个实验结果的数学式合并起来,成为一个同时表达浓度比和 pK_a 决定缓冲溶液 pH 值的总方程式。再由电离常数关系式推导以验证其是否正确。仔细分析实际情况与理论是否存在差异,如果存在说明原因。

3. 设计并配制 pH＝7.50 的缓冲溶液

设计并配制 pH＝7.50 的缓冲溶液 100 mL,测其 pH 值。比较设计的理论值与实测的 pH 有无差距,从理论上加以解释。

4. 缓冲容量的测定

缓冲容量是缓冲溶液抵抗 pH 变化能力的大小。通常有两种表示缓冲容量的方法。一种称为积分缓冲容量,是指每一升缓冲溶液的 pH 值改变 1 个 pH 单位所需要加入的强酸或强碱的量。这种定义的缓冲容量虽然有应用价值,但是由于强酸、强碱的不断加入,缓冲溶液的缓冲容量会不断发生变化,这与我们的直觉相矛盾——缓冲容量大小应该与加入的酸碱量无关。因此,另一种被称为微分缓冲容量的表示方法考虑到了这一点,其定义为:

$$\beta = \frac{dn}{d(\text{pH})}$$

其中,dn 是指加入的极微量的强酸或强碱,$d(\text{pH})$ 是因此引起的极微小的 pH 变化。

虽然从 Henderson-Hasselbalch 方程(缓冲公式)可见,pH 由缓冲对的浓度比,而不是浓度的绝对值决定,但显然,缓冲物质越少,缓冲能力越小,也就是缓冲容量越低。本实验要研究缓冲容量与浓度比的关系。

取下列缓冲溶液测定 pH 值,再分别加入体积为溶液总体积 1/20 的 0.1 mol·L^{-1} NaOH,测定溶液的 pH 值。由根据 $\dfrac{\Delta V \times c_{\text{NaOH}}}{\Delta \text{pH} \times V}$ 计算各溶液的缓冲容量,并小结影响缓冲容量的因素。

序号	V(0.1 mol·L^{-1} HAc)/mL	V(0.1 mol·L^{-1} NaOH)/mL	$V_{\text{H}_2\text{O}}$	pH(加入 NaOH 前)	pH(加入 1.0 mL 0.1 mol·L^{-1} NaOH 后)	缓冲容量 (mol·L^{-1})
1	10	1	9			
2	10	5	5			
3	5	2.5	12.5			
4	10	9	1			

亨德森-哈塞尔巴尔赫方程

1908 年,美国生物化学家亨德森[Henderson, Lawrence Joseph(1878—1942)]因发现本质上维持酸碱平衡的化学方法而闻名。在他的有关人体体液的研究中,他发现在碳酸盐存在下,由二氧化碳和水形成的碳酸在本质上是维持酸碱平衡的唯一体系,亨德森创建了描述这些体系(称作生理学上的缓冲器)的化学方程,这个方程对于生物化学是一重要的基本

Questions

1. There are two methods to prepare the buffer solutions with the required pH. Is there any difference between the method introduced in the handbook and that designed according to the buffer equation?

2. Why does the difference of the pH value exist between theoretical and actual when the buffer with the designed pH 7.5 is prepared?

3. Discuss the approximation in Henderson-Hasselbalch equation and the limitation for its application.

方程。1916年，丹麦生物化学家哈塞尔巴尔赫[Hasselbalch, Karl Albert(1874—1962)]发展了亨德森的方程，将其转换为对数形式，产生了亨德森－哈塞尔巴尔赫 Henderson-Hasselbalch 方程。虽然 Henderson-Hasselbalch 方程只是一个近似方程，方程仍然保留了处理缓冲溶液的最有用的数学模型。方程表达了缓冲组分中弱酸/碱及其盐的浓度对溶液 pH 值的影响。Henderson-Hasselbalch 方程是化学的基本方程之一，用该方程可以计算缓冲溶液的 pH 值或酸和碱的浓度。

思 考 题

1. 配制缓冲溶液有两种方法，一是按手册上的配方配制，一是按要求的 pH 设计。两种方法有什么区别？
2. 为什么配制的 pH＝7.50 的缓冲溶液与实测值有差距？
3. 讨论缓冲公式的 Henderson-Hasselbalch 方程的近似性和适用范围。

Experiment 27 Determining Coordination Number of $[Ag(NH_3)_n]^+$ Complex Ion

Purpose

1. To determine the coordination number (n) and the stability constant of $[Ag(NH_3)_n]^+$ on the basis of coordination equilibrium principle and solubility product principle.

2. To learn how to deal with the data and how to draw the graph.

Principle

The stable complex ion consisting of Ag^+ and NH_3 can be formed when excess ammonia water is added into silver nitrate solution. Nevertheless, silver bromide (AgBr) is precipitated when a potassium bromide solution is added into the solution of the complex. The coordination equilibrium and the precipitation equilibrium coexist in this system.

The coordination Equilibrium: $Ag^+ + nNH_3 = [Ag(NH_3)_n]^+$

$$K_{st} = \frac{[Ag(NH_3)_n^+]}{[Ag^+][NH_3]^n} \tag{1}$$

The precipitation Equilibrium: $Ag^+ + Br^- = AgBr(s)$

$$K_{sp} = [Ag^+][Br^-] \tag{2}$$

(1) × (2):

$$K = K_{sp} K_{st} = \frac{[Ag(NH_3)_n^+][Br^-]}{[NH_3]^n} \tag{3}$$

Then:

$$[Br^-] = \frac{K[NH_3]^n}{[Ag(NH_3)_n^+]} \tag{4}$$

$[NH_3]$, $[Br^-]$ and $[Ag(NH_3)_n^+]$ represent the equilibrium concentrations and can be approximately calculated by the following method:

In each portion of the mixture, if the volume of silver nitrate solution we initially took is the same as V_{Ag^+} and the concentration is represented as $[Ag^+]_0$. If the volume of ammonia water (substantial excess) we added sequentially into each portion is V_{NH_3} and the concentration is $[NH_3]_0$. If the volume of potassium bromide solution we titrate into each portion is V_{Br^-} and the concentration is $[Br^-]_0$. And if the total volume of mixture is V_t, then, when equilibrium is reached in the mixture, we have:

$$[Br^-] = [Br^-]_0 \times \frac{V_{Br^-}}{V_t} \tag{5}$$

$$[Ag(NH_3)_n^+] = [Ag^+]_0 \times \frac{V_{Ag^+}}{V_t} \tag{6}$$

$$[NH_3] = [NH_3]_0 \times \frac{V_{Ag^+}}{V_t} \tag{7}$$

Put (5), (6) and (7) into Equation (4), then we have:

$$V_{Br^-} = V_{NH_3}^n \cdot K \cdot \left(\frac{[NH_3]_0}{V_t}\right)^n \cdot \frac{V_t}{[Br^-]_0} \cdot \frac{V_t}{[Ag^+]_0 V_{Ag^+}} \tag{8}$$

Except for $V_{NH_3}^n$, all the terms are constants in the right side of Equation (8), so Equation (8) can be transformed to:

$$V_{Br^-} = V_{NH_3}^n \cdot K' \tag{9}$$

Make the logarithms at both sides of the equation (9), we can obtain a linear equation:

实验 27　银氨配离子配位数的测定

目的要求

1. 应用配位平衡及溶度积原理,测定 $[Ag(NH_3)_n]^+$ 配离子的配位数 n,并计算稳定常数 K_{st}。
2. 学习如何处理实验数据和作图。

实验原理

在硝酸银溶液中加入过量的氨水,即生成稳定的银氨配离子 $[Ag(NH_3)_n]^+$。再往溶液中加入溴化钾溶液,直至刚刚开始有 AgBr 沉淀(浑浊)出现为止。这时混合溶液中同时存在着配位平衡和沉淀平衡。

配位平衡:$Ag^+ + nNH_3 = [Ag(NH_3)_n]^+$

$$K_{st} = \frac{[Ag(NH_3)_n^+]}{[Ag^+][NH_3]^n} \tag{1}$$

沉淀平衡:$Ag^+ + Br^- = AgBr(s)$

$$K_{sp} = [Ag^+][Br^-] \tag{2}$$

(1)×(2)得:

$$K = K_{sp} K_{st} = \frac{[Ag(NH_3)_n^+][Br^-]}{[NH_3]^n} \tag{3}$$

则:

$$[Br^-] = \frac{K[NH_3]^n}{[Ag(NH_3)_n^+]} \tag{4}$$

$[Br^-]$、$[NH_3]$ 和 $[Ag(NH_3)_n^+]$ 皆指平衡时的浓度,它们可以近似地计算如下:设每份混合溶液最初取用的 $AgNO_3$ 溶液的体积为 V_{Ag^+}(各份相同),浓度为 $[Ag^+]_0$,每份加入的氨水(大量过量)和溴化钾溶液的体积分别为 V_{NH_3} 和 V_{Br^-},其浓度为 $[NH_3]_0$ 和 $[Br^-]_0$,混合溶液总体积为 V_t,则混合后并达到平衡时:

$$[Br^-] = [Br^-]_0 \times \frac{V_{Br^-}}{V_t} \tag{5}$$

$$[Ag(NH_3)_n^+] = [Ag^+]_0 \times \frac{V_{Ag^+}}{V_t} \tag{6}$$

$$[NH_3] = [NH_3]_0 \times \frac{V_{Ag^+}}{V_t} \tag{7}$$

将(5)、(6) 和(7)代入(4)式并整理后得

$$V_{Br^-} = V_{NH_3}^n \cdot K \cdot \left(\frac{[NH_3]_0}{V_t}\right)^n \cdot \frac{V_t}{[Br^-]_0} \cdot \frac{V_t}{[Ag^+]_0 V_{Ag^+}} \tag{8}$$

因此上式等号右边除 $V_{NH_3}^n$ 外,其他皆为常数,故(8)式可写为:

$$V_{Br^-} = V_{NH_3}^n \cdot K' \tag{9}$$

$$\lg V_{Br^-} = n \cdot \lg V_{NH_3} + \lg K' \tag{10}$$

Make the graph (assign y-coordinate to $\lg V_{Br^-}$, x-coordinate to $\lg V_{NH_3}$ and intercept to $\lg K'$) and then we can get the value of the slope (n) which is also the value of the coordination number of $[Ag(NH_3)_n]^+$ (take the closest integer).

Apparatus and Chemicals

Apparatus: transfer pipette, 20 mL; Erlenmeyer flask, 250 mL; graduated cylinder, 100 mL; burettes, 2×25 mL.

Chemicals: silver nitrate, $0.01\ mol \cdot L^{-1}$; potassium bromide, $0.01\ mol \cdot L^{-1}$; ammonia solution, $2\ mol \cdot L^{-1}$.

Procedure

1. Pipet 20.00 mL aliquots of the silver nitrate solution with the concentration $0.01\ mol \cdot L^{-1}$ into each 250 mL Erlenmeyer flask.

2. Add an aliquot 40.00 mL, 35.00 mL, 30.00 mL, 25.00 mL and 20.00 mL of the $2.0\ mol \cdot L^{-1}$ ammonia water into each Erlenmeyer flask through the basic burette respectively.

3. Add the respective volumes of distilled water by the graduated cylinder to make the total volumes of the mixtures equal in each of the Erlenmeyer flasks.

4. Titrate the complex ion solutions in the Erlenmeyer flasks by a $0.01\ mol \cdot L^{-1}$ potassium bromide solution until the precipitate of silver bromide appears respectively. Then stop the titration and write down the volume V_{Br^-} of potassium bromide solution consumed and the ultimate volume V_t of the solution in that flask. In order to keep V_t in every titration as the same volume as that of the first, make up some distilled water when the end point is approaching.

Data and Results

1. Records

No.	V_{Ag^+}/mL ($0.01\ mol \cdot L^{-1}$)	V_{NH_3}/mL ($2.0\ mol \cdot L^{-1}$)	V_{Br^-}/mL ($0.01\ mol \cdot L^{-1}$)	V_{H_2O}/ mL	V_t/mL	$\lg V_{NH_3}$	$\lg V_{Br^-}$
1	20.00	40.00		40			
2	20.00	35.00		45+			
3	20.00	30.00		50+			
4	20.00	25.00		55+			
5	20.00	20.00		60+			

2. Data Process

(1) Please make the graph with y-coordinate assigned to $\lg V_{Br^-}$ and x-coordinate assigned to $\lg V_{NH_3}$.

(2) Get the value of n from the graph and calculate the value of K' from Equation (10).

(3) Calculate the value of K from Equation (8).

(4) On the basis of the equation $K = K_{sp} K_{st}$, we can finally get the value of K_{st}. The reference value: $K_{sp,\ AgBr} = 4.1 \times 10^{-13}$.

将(9)式两边取对数,得直线方程

$$\lg V_{Br^-} = n \cdot \lg V_{NH_3} + \lg K' \tag{10}$$

作图(以 $\lg V_{Br^-}$ 为纵坐标, $\lg V_{NH_3}$ 为横坐标,以 $\lg K'$ 为截距),求出直线的斜率 n,即得 $[Ag(NH_3)_n]^+$ 的配位数 n(取最接近的整数)。

仪器和试剂

仪器:移液管,20 mL;锥形瓶,250 mL;量筒,100 mL;滴定管,2×25 mL。

试剂:$AgNO_3$,0.01 mol·L^{-1};KBr,0.01 mol·L^{-1};氨水,2 mol·L^{-1},应新鲜配制。

实验步骤

1. 用移液管吸取 20.00 mL 0.01 mol·L^{-1} $AgNO_3$ 溶液放入每个 250 mL 锥形瓶中。

2. 用滴定管分别量取 40.00 mL、35.00 mL、30.00 mL、25.00 mL 和 20.00 mL 2.0 mol·L^{-1} 氨水溶液分别加入上述锥形瓶中。

3. 并用量筒量取相应体积的蒸馏水(如,第一个锥形瓶中加入 40 mL 水),使每个锥形瓶中溶液的体积保持一致。

4. 分别从滴定管中向锥形瓶逐滴加 0.01 mol·L^{-1} KBr,直至开始产生的 AgBr 浑浊不再消失时,停止滴定,记下所加入的 KBr 溶液的体积 V_{Br^-} 和溶液的总体积 V_t。在进行重复操作中,为了使溶液的总体积(V_t)与第一个滴定的(V_t)大致相同,当接近终点时应加入适量的蒸馏水。

数据记录及结果处理

1. 数据记录

溶液编号	V_{Ag^+}/mL (0.01 mol·L^{-1})	V_{NH_3}/mL (2.0 mol·L^{-1})	V_{Br^-}/mL (0.01 mol·L^{-1})	V_{H_2O}/mL	V_t/mL	$\lg V_{NH_3}$	$\lg V_{Br^-}$
1	20.00	40.00		40			
2	20.00	35.00		45$^+$			
3	20.00	30.00		50$^+$			
4	20.00	25.00		55$^+$			
5	20.00	20.00		60$^+$			

2. 结果处理

(1) 以 $\lg V_{Br^-}$ 为纵坐标,以 $\lg V_{NH_3}$ 为横坐标作图。

(2) 从图求得 n,并从公式 $\lg V_{Br^-} = n \cdot \lg V_{NH_3} + \lg K'$ 求出 K'。

(3) 利用(8)式计算 K 值。

(4) 利用 $K = K_{sp} K_{st}$,求出 K_{st}。(文献值:$K_{sp,AgBr} = 4.1 \times 10^{-13}$)

Notes

1. During the concoction for each mixture, ammonia solution is added finally for it is volatile.

2. It is preferable that the turbidity degrees at the endpoints are the same in each titration with respect of the parallel principle.

3. At last, pipettes and Erlenmeyer flasks contaminated with silver nitrate should be washed with the left ammonia solution because silver ion is converted to metal silver easily.

Questions

1. Why do the coordination equilibrium and precipitation equilibrium coexist in the system of this experiment?

2. Why could the following situations be ignored when the equilibrium concentrations of $[NH_3]$, $[Br^-]$ and $[Ag(NH_3)_n^+]$ are calculated?

(1) The amount of Br^- and Ag^+ used to produce the precipitate of silver bromide.

(2) The amount of Ag^+ dissociated from $[Ag(NH_3)_n^+]$.

(3) The amount of ammonia used to produce the complex ion of $[Ag(NH_3)_n^+]$.

3. If potassium bromide solution was added in excess during titration, is it the best disposal that the solution in the Erlenmeyer flask was thrown away for a turn-over-a-new-leaf?

4. Is V_{Br^-} always the same in each titration? And how about $[Br^-]$?

注意事项

1. 配制每一份混合溶液时,最后加氨水(防止氨水挥发掉)。
2. 反应一定要达平衡(振摇后沉淀不消失)后再观察终点,且每次浑浊度要一致。
3. 实验完毕将取 $AgNO_3$ 溶液用的移液管、锥形瓶用实验剩余的氨水洗涤以防 Ag^+ 转化为 Ag。

思 考 题

1. 为什么本实验的溶液体系中配位平衡与沉淀平衡共存?
2. 在计算平衡浓度$[NH_3]$、$[Br^-]$ 和 $[Ag(NH_3)_n^+]$时,为什么可以忽略以下情况?
(1) 生成 AgBr 沉淀时消耗掉的 Br^- 和 Ag^+。
(2) 配离子$[Ag(NH_3)_n]^+$离解出的 Ag^+。
(3) 生成配离子$[Ag(NH_3)_n]^+$时消耗的 NH_3。
3. 滴定时,若 KBr 溶液加过量了,有无必要弃去锥形瓶中的溶液重新开始?
4. 每次滴定时 V_{Br^-} 是否一样?$[Br^-]$是否一样?

Experiment 28 Determination of Reaction Rate and Activation Energy

Purpose

1. To understand the effect of concentration, temperature and catalyst on the reaction rate.

2. To determine the rate of ammonia peroxydisulfurite [$(NH_4)_2S_2O_8$] reacting with potassium iodide (KI), and calculate the reaction order, rate constant and activation energy.

3. To learn how to control the reaction temperature with ice water bath and hot water bath.

Principle

In the homogeneous phase reactions, the reaction rate is determined by the native of the reactant, the temperature, the concentration and catalyst. The rate can be formulated by the change in concentration of one of the reactants or products which occurs in the small time interval after the reagents are mixed.

In this experiment, when $(NH_4)_2S_2O_8$ oxidizes KI into I_2, I_2 then reacts with starch solution and a blue complex is formed, which can be regarded as the end of the reaction. Shorter period of the appearance of blue means quicker the reaction is.

1. In aqueous solution, the reaction between $(NH_4)_2S_2O_8$ and KI is:

$$S_2O_8^{2-} + 2I^- = 2SO_4^{2-} + I_2 \tag{A}$$

where the average reaction rate law has the form:

$$\bar{v} = -\frac{\Delta[S_2O_8^{2-}]}{\Delta t} = k[S_2O_8^{2-}]^m[I^-]^n \tag{1}$$

$\bar{v}$ is the average reaction rate; $\Delta[S_2O_8^{2-}]$ is the change in concentration of $S_2O_8^{2-}$ during Δt time interval; $[S_2O_8^{2-}]$ and $[I^-]$ are the initial concentration of $S_2O_8^{2-}$ and I^- respectively; k is reaction rate constant; m and n are reactant orders.

2. In order to determine the value of $\Delta[S_2O_8^{2-}]$ during Δt time interval, when mixing the solution of $(NH_4)_2S_2O_8$ and KI, add starch solution and definite amount of $Na_2S_2O_3$ with definite concentration value. So when Reaction (A) is going on, the other reaction as the following also exists:

$$2S_2O_3^{2-} + I_2 = S_4O_6^{2-} + 2I^- \tag{B}$$

The rate of Reaction (B) is much more quickly than Reaction (A), so the product of I_2 in Reaction (A) will react with $S_2O_3^{2-}$ immediately forming $S_4O_6^{2-}$ (colorless) and I^-. As soon as the reactant $S_2O_3^{2-}$ is exhausted, the product of I_2 produced in Reaction (A) will react with starch solution at once, then blue color appears.

From the Reaction (A) and Reaction (B), 1 mol of $S_2O_8^{2-}$ consumed equal to 2 mol of $S_2O_3^{2-}$ consumed, so we can write:

$$\Delta[S_2O_8^{2-}] = \frac{\Delta[S_2O_3^{2-}]}{2} \tag{2}$$

Record the time interval Δt from the beginning of the reaction to the appearance of blue color. Because $S_2O_3^{2-}$ has been completely exhausted in Δt time, $-\Delta[S_2O_3^{2-}]$ equals to the initial concentration of $Na_2S_2O_3$. And the reaction rate has the form $-\frac{\Delta[S_2O_8^{2-}]}{\Delta t}$.

3. Taking the logarithm of reaction rate formulation $\bar{v} = k[S_2O_8^{2-}]^m[I^-]^n$ gives:

实验 28　化学反应速率与活化能的测定

实验目的

1. 了解浓度、温度和催化剂对反应速率的影响。
2. 测定过二硫酸铵 $(NH_4)_2S_2O_8$ 与碘化钾 KI 反应的反应速率,并计算反应级数、反应速率常数及反应的活化能。
3. 学会使用冰浴、热水浴来控制所需要的反应温度。

实验原理

在均相反应中,反应速率决定于反应物的本性、浓度、温度和催化剂。当反应试剂混合后,其反应速率可以根据一小段时间间隔内某种反应物或生成物浓度发生的变化量来求得。

在本实验中,$(NH_4)_2S_2O_8$ 氧化 KI 生成 I_2,以 I_2 与淀粉生成蓝色加合物作为反应完成的标志,蓝色出现的时间越快意味着反应的速率越快。

1. 在水溶液中,$(NH_4)_2S_2O_8$ 与 KI 发生以下反应(A):

$$S_2O_8^{2-} + 2I^- = 2SO_4^{2-} + I_2 \tag{A}$$

该反应的平均反应速率可用下式表示:

$$\bar{v} = -\frac{\Delta[S_2O_8^{2-}]}{\Delta t} = k[S_2O_8^{2-}]^m[I^-]^n \tag{1}$$

式中,$\bar{v}$ 为平均反应速率,$\Delta[S_2O_8^{2-}]$ 为 Δt 时间内 $S_2O_8^{2-}$ 浓度的变化;$[S_2O_8^{2-}]$ 和 $[I^-]$ 分别为 $S_2O_8^{2-}$ 与 I^- 的起始浓度;k 为反应速率常数;m 和 n 则为反应级数。

2. 为了测 Δt 时间内 $S_2O_8^{2-}$ 的浓度变化,在将 $(NH_4)_2S_2O_8$ 溶液与 KI 溶液混合的同时,加入一定体积的已知浓度的 $Na_2S_2O_3$ 溶液和淀粉溶液。这样在以上反应进行的同时,还发生以下反应(B)(称记时反应):

$$2S_2O_3^{2-} + I_2 = S_4O_6^{2-} + 2I^- \tag{B}$$

反应(B)的速率比反应(A)快得多,所以由反应(A)生成的 I_2 立即与 $S_2O_3^{2-}$ 作用生成了无色的 $S_4O_6^{2-}$ 和 I^-。当 $Na_2S_2O_3$ 耗尽,反应(A)生成微量的 I_2 就立即与淀粉作用,使溶液显蓝色。从反应(A)和反应(B)可以看出,$S_2O_8^{2-}$ 减少 1 mol 时,$S_2O_3^{2-}$ 则减少 2 mol。

即:

$$\Delta[S_2O_8^{2-}] = \frac{\Delta[S_2O_3^{2-}]}{2} \tag{2}$$

记录从反应开始到溶液出现蓝色所需要的时间 Δt。由于在 Δt 时间内 $S_2O_3^{2-}$ 全部耗尽,所以 $\Delta[S_2O_3^{2-}]$ 实际上就是反应开始时 $Na_2S_2O_3$ 的浓度,进而可以计算反应速率。

3. 对反应速率表示式 $\bar{v} = k[S_2O_8^{2-}]^m[I^-]^n$ 两边取对数,

得:

$$\lg\bar{v} = m\lg[S_2O_8^{2-}] + n\lg[I^-] + \lg k \tag{3}$$

当 $[I^-]$ 不变时,以 $\lg\bar{v}$ 对 $\lg[S_2O_8^{2-}]$ 作图,可得一直线,斜率即为 m。同理,当 $[S_2O_8^{2-}]$ 不变时,以 $\lg\bar{v}$ 对 $\lg[I^-]$ 作图,可求得 n。

求出 m 和 n 后,可根据下式求得反应速率常数 k。

$$\lg \bar{v} = m\lg[S_2O_8^{2-}] + n\lg[I^-] + \lg k \tag{3}$$

When $[I^-]$ is constant, make the graph (assign y-coordinate to $\lg\bar{v}$, x-coordinate to $\lg[S_2O_8^{2-}]$ and intercept to $n\lg[I^-] + \lg k$) and we can get the value of the slope m; when $[S_2O_8^{2-}]$ is constant, make the graph (assign y-coordinate to $\lg\bar{v}$, x-coordinate to $\lg[I^-]$ and intercept to $m\lg[S_2O_8^{2-}] + \lg k$) and then we can get the value of the slope n.

Then reaction rate constant k can be obtained as following form:

$$k = \frac{\bar{v}}{[S_2O_8^{2-}]^m [I^-]^n} \tag{4}$$

4. With reaction rate constant k and reaction temperature T, we can predict the activation energy of chemical reaction, then the following form is obtained:

$$\lg k = A - \frac{E_a}{2.303RT} \tag{5}$$

E_a represents the activation energy of the reaction, R is universal gas constant, T is absolute temperature (or the temperature on the Kelvin scale). A plot of $\lg k$ as a function of $1/T$ should be a straight line with the slope of $-\frac{E_a}{2.303R}$. So the value of E_a is obtained.

Apparatus and Chemicals

Apparatus: stopwatch; thermometer; water bath; beaker, 100 mL, 250 mL; measuring cylinder, 10 mL, 100 mL.

Chemicals: $(NH_4)_2S_2O_8$, 0.2 mol·L^{-1}; KI, 0.2 mol·L^{-1}; $Na_2S_2O_3$, 0.01 mol·L^{-1}; $(NH_4)_2SO_4$, 0.2 mol·L^{-1}; KNO_3, 0.2 mol·L^{-1}; starch solution, 0.2%.

Procedure

1. Concentration effects on the reaction rate and calculation of the reaction order

At room temperature, measure 20 mL of 0.2 mol·L^{-1} KI solution, 8 mL of 0.01 mol·L^{-1} $Na_2S_2O_3$ solution and 4 mL of 0.2% starch solution with three measuring cylinders. Then put them to a beaker (250 mL) and mix well. Add 20 mL of 0.2 mol·L^{-1} $(NH_4)_2S_2O_8$ solution with measuring cylinder to the beaker quickly, turning on stopwatch at the same time, and stirring continuously. As soon as blue appears, please turn off the stopwatch, record the passing time interval Δt and the room temperature.

Do the other four experiments according to the given amount as shown in Table 1. in the same way. In order to keep the ionic strength and the total volume constant in each portion of the mixture, add 0.2 mol·L^{-1} $(NH_4)_2SO_4$ solution and 0.2 mol·L^{-1} KNO_3 solution respectively to supplement the inadequate part.

Calculate the value of the reaction rate $\bar{v}$, and fill in Table 1. Make the graph ($\lg\bar{v}$ to $\lg[S_2O_8^{2-}]$) with the data in test Ⅰ, Ⅱ, Ⅲ to calculate the value of m; make the graph ($\lg\bar{v}$ to $\lg[I^-]$) with the data in test Ⅰ, Ⅳ, Ⅴ to calculate the value of n. With the value of m, n, we can get value of reaction rate constant k in each test.

$$k = \frac{\bar{v}}{[S_2O_8^{2-}]^m [I^-]^n} \tag{4}$$

4. 反应速率常数 k 与反应温度 T 有以下关系：

$$\lg k = A - \frac{E_a}{2.303RT} \tag{5}$$

式中，E_a 为反应活化能，R 为摩尔气体常数，T 为热力学温度。测出不同温度时的 k 值，以 $\lg k$ 对 $\frac{1}{T}$ 作图，可得一条直线，由直线斜率（等于 $-\frac{E_a}{2.303R}$）可求得反应的活化能 E_a。

仪器和试剂

仪器：秒表；温度计，100 ℃；水浴锅；烧杯，100 mL，250 mL；量筒，10 mL，100 mL。

试剂：$(NH_4)_2S_2O_8$，0.2 mol·L^{-1}；KI，0.2 mol·L^{-1}；$Na_2S_2O_3$，0.01 mol·L^{-1}；$(NH_4)_2SO_4$，0.2 mol·L^{-1}；KNO_3，0.2 mol·L^{-1}；淀粉，0.2%。

实验步骤

1. 测定浓度对化学反应速率的影响，求反应级数

在室温下，用 3 只量筒分别量取 20 mL 0.2 mol·L^{-1} KI 溶液，8 mL 0.01 mol·L^{-1} $Na_2S_2O_3$ 溶液和 4 mL 0.2% 淀粉溶液，加到同一烧杯中，混合均匀。再用另 1 只量筒量取 20 mL 0.2 mol·L^{-1} $(NH_4)_2S_2O_8$ 溶液，快速加到烧杯中，同时开动秒表，并不断搅拌。当溶液刚开始出现蓝色时，立即停秒表。记下时间及室温。

用同样的方法按照表 1 中的用量进行另外四次实验。为了使每次实验中的溶液的离子强度和总体积保持不变，不足的量分别用 0.2 mol·L^{-1} KNO_3 溶液和 0.2 mol·L^{-1} $(NH_4)_2SO_4$ 溶液补足。

算出各实验中的反应速率 $\bar{v}$，并填入表 1 中。用表中 Ⅰ、Ⅱ、Ⅲ 实验的数据以 $\lg \bar{v}$ 对 $\lg[S_2O_8^{2-}]$ 作图，求出 m；用实验 Ⅰ、Ⅳ、Ⅴ 的数据以 $\lg \bar{v}$ 的数据对 $\lg[I^-]$ 作图，求得 n。求出 m 和 n 后，再算出各实验的反应速率常数 k，把计算结果填入下表 1 中。

表 1　浓度对化学反应速率的影响

	反应温度(℃)					
	实验序号	Ⅰ	Ⅱ	Ⅲ	Ⅳ	Ⅴ
试剂用量 (mL)	0.2 mol·L^{-1} $(NH_4)_2S_2O_8$	20	10	5	20	20
	0.2 mol·L^{-1} KI	20	20	20	10	5
	0.01 mol·L^{-1} $Na_2S_2O_3$	8	8	8	8	8
	0.2% 淀粉	4	4	4	4	4
	0.2 mol·L^{-1} KNO_3	—	—	—	10	15
	0.2 mol·L^{-1} $(NH_4)_2SO_4$	—	10	15	—	—
起始浓度 (mol·L^{-1})	$(NH_4)_2S_2O_8$					
	KI					
	$Na_2S_2O_3$					

Part 4 Experiments

Table 1 Concentration effects on the chemical reaction rate

			I	II	III	IV	V
	Experiment temperature(℃)						
	Experiment number		I	II	III	IV	V
Reagents (mL)		0.2 mol·L^{-1} (NH$_4$)$_2$S$_2$O$_8$	20	10	5	20	20
		0.2 mol·L^{-1} KI	20	20	20	10	5
		0.01 mol·L^{-1} Na$_2$S$_2$O$_3$	8	8	8	8	8
		0.2% starch solution	4	4	4	4	4
		0.2 mol·L^{-1} KNO$_3$	—	—	—	10	15
		0.2 mol·L^{-1} (NH$_4$)$_2$SO$_4$	—	10	15	—	—
Initial concentration		(NH$_4$)$_2$S$_2$O$_8$					
		KI					
		Na$_2$S$_2$O$_3$					
Δt (s)							
$\Delta[S_2O_8^{2-}]$							
$\bar{v} = -\dfrac{\Delta[S_2O_8^{2-}]}{\Delta t}$							
$k = \dfrac{\bar{v}}{[S_2O_8^{2-}]^m [I^-]^n}$							

2. Temperature effects on the reaction rate and calculation of the activation energy

Add 10 mL of KI solution, 8 mL of Na$_2$S$_2$O$_3$ solution, 4 mL of starch solution and 10 mL of KNO$_3$ solution to a beaker (250 mL), mix them well. Add 20 mL of (NH$_4$)$_2$S$_2$O$_8$ solution to another small beaker, then put the two beakers into a water bath at 0 ℃. When the solution in the beakers is cooled to 0 ℃, mix the solutions and turn on the stopwatch, stirring continuously. Record the passing time interval Δt when the solution just turns blue.

At the temperature of about 10 ℃, 20 ℃, 30 ℃, do the above experiment again. Finally we can get four values of Δt at four different temperatures. Calculate the reaction rate and reaction rate constant at four temperatures, and fill in Table 2. Make the graph (lg$k$ to $1/T$) with the data got in the above experiments and obtain the value of E_a in reaction (A).

Table 2 Temperature effects on the chemical reaction rate

Experiment Number	I	II	III	IV
Reaction temperature, T				
Reaction time interval, t				
Reaction rate, $\bar{v}$				
Reaction rate constant, k				
lg$k$				
$\dfrac{1}{T}$				

续表

反应温度(℃) 实验序号	Ⅰ	Ⅱ	Ⅲ	Ⅳ	Ⅴ
反应时间 $\Delta t/s$					
$S_2O_8^{2-}$ 的浓度变化，$\Delta[S_2O_8^{2-}]$					
反应的平均速率，$\bar{v}=-\dfrac{\Delta[S_2O_8^{2-}]}{\Delta t}$					
反应速率常数，$k=\dfrac{\bar{v}}{[S_2O_8^{2-}]^m[I^-]^n}$					

2. 测定温度对化学反应速率的影响，求活化能

在 250 mL 烧杯中加入 10 mL KI 溶液、8 mL $Na_2S_2O_3$ 溶液、4 mL 淀粉溶液和 10 mL KNO_3 溶液，混合均匀。在另一个小烧杯中加入 20 mL $(NH_4)_2S_2O_8$ 溶液，并把它们同时放在冰水浴中冷却（即按表 1 中Ⅳ号的用量进行反应）。等烧杯中的溶液都冷却到 0 ℃时，把 $(NH_4)_2S_2O_8$ 溶液加到 KI 等混合溶液中，同时开动秒表，并不断搅拌。当溶液刚出现蓝色时，立即停秒表，记下反应时间。

在约 10 ℃、20 ℃、30 ℃的条件下，重复以上实验，得到 4 个温度（0 ℃、10 ℃、20 ℃、30 ℃）下的反应时间，计算 4 个温度下的反应速率及反应速率常数，把数据及计算结果填入下表 2 中。用表中各次实验的 $\lg k$ 对 $\dfrac{1}{T}$ 作图，求出反应(A)的活化能 E_a。

表 2　温度对化学反应速率的影响

实验序号	Ⅰ	Ⅱ	Ⅲ	Ⅳ
反应温度(℃)				
反应时间 t				
反应速率 $\bar{v}$				
反应速率常数 k				
$\lg k$				
$\dfrac{1}{T}$				

注：T 为热力学温度。

注意事项

1. 因本实验是利用 $Na_2S_2O_3$ 的浓度来衡量反应产生的 I_2 的浓度，从而计算消耗的 $(NH_4)_2S_2O_8$ 的浓度，所以准确添加 $Na_2S_2O_3$ 的量是本实验成败的关键。

2. 本实验的目的是测定常数，对时间、浓度等数值读取必须符合仪器精密度要求。

3. 经计算得出的 5 个 k 值其最大和最小数之间的系数差值不得超过 0.5。

Notes

1. In this experiment we use $[S_2O_3^{2-}]$ to determine the concentration of the product I_2, then calculate the consumed concentration of $S_2O_8^{2-}$. So adding $Na_2S_2O_3$ accurately is the key step of the experiment.

2. The purpose of the experiment is to determine the constants, so the value of the time and the concentration should be recorded according to the precision of the apparatus.

3. The difference between the maximum value and the minimum value among the five values of k should be no more than 0.5.

4. Make the graph ($\lg k$ to $1/T$) on the graph paper, and the proportion and configuration should be apposite.

Questions

1. What is chemical reaction rate? What factors will effect on the reaction rate? How to validate the effects of concentration, temperature on the reaction rate?

2. Please explain the law of mass action on Le Chatelier's principle.

3. Illustrate the process if you can get the reaction order according to the reaction equations.

4. The purpose of this experiment is to determine the rate constant of the reaction, $S_2O_8^{2-} + 2I^- = 2SO_4^{2-} + I_2$, please tell the reason why we should add a definite amount of $Na_2S_2O_3$ when mix the reagents.

5. Which items in Table 1 and Table 2 can be calculated before hand, which ones can be calculated after the experiment?

6. Why should we add KNO_3 solution and $(NH_4)_2SO_4$ solution to supplement the inadequate volume? Can we use water?

4. $\lg k$ 对 $\frac{1}{T}$ 作图时,比例、布局必须合适。

思 考 题

1. 何谓化学反应速率？影响化学反应速率的因素有哪些？本实验中如何试验浓度、温度对反应速率的影响？

2. 试说明质量作用定律和理·查得里原理。

3. 根据反应方程式,是否能确定反应级数？举例说明。

4. 本实验是测定 $S_2O_8^{2-} + 2I^- = 2SO_4^{2-} + I_2$ 的反应速率常数,为什么反应物混合时,同时要加一定量的 $Na_2S_2O_3$？

5. 实验的两个表格中哪些项目可以预先计算好填入,哪些需要实验后经计算才能填入？

6. 为什么用 KNO_3 和 $(NH_4)_2SO_4$ 溶液补足溶液的体积？能否用水补充？

Experiment 29 Determination of K_{sp} of Silver Acetate

Purpose

1. To determine the solubility product constant K_{sp} of silver acetate.
2. To learn the Volhard's method of determining the concentration of silver ion Ag^+.

Principle

The equilibrium between the sparingly soluble salt, $AgC_2H_3O_2(s)$, and its saturated solution is represented by Equation (1).

$$AgC_2H_3O_2(s) = Ag^+(aq) + C_2H_3O_2^-(aq) \tag{1}$$

The solubility product constant expression for the above reaction is given by Equation (2) in which $[Ag^+]$ and $[C_2H_3O_2^-]$ are the concentration of Ag^+ and $C_2H_3O_2^-$ ions in the saturated solution.

$$K_{sp} = [Ag^+][C_2H_3O_2^-] \tag{2}$$

1. The above equilibrium system can be established in the way to prepare a saturated silver acetate solution by mixing solutions of two salts, one containing Ag^+ as the cation and the other containing $C_2H_3O_2^-$ as the anion, for example, aqueous solutions of $AgNO_3$ and $NaC_2H_3O_2$.

2. The concentration of Ag^+ at equilibrium in the saturated $AgC_2H_3O_2$ solutions will be determined by titration with a standard solution of potassium thiocyanate, KSCN. Addition of potassium thiocyanate to a saturated solution of silver acetate will cause the formation of solid silver thiocyanate, AgSCN:

$$Ag^+(aq) + SCN^-(aq) = AgSCN(s)$$

The indicator used in the titration is a solution of ferric ammonium sulfate, $NH_4Fe(SO_4)_2$. When the precipitation of AgSCN is complete, Fe^{3+} ions from the ferric ammonium sulfate indicator will react with excess thiocyanate ions to form the red $Fe(SCN)^{2+}$ complex ion. For the end point of the titration the color of the solution changes from colorless to a pale orange (salmon orange).

$$Fe^{3+}(aq) + SCN^-(aq) = Fe(SCN)^{2+}(aq)$$

3. The concentration of $C_2H_3O_2^-$ at equilibrium in the saturated $AgC_2H_3O_2$ solutions can be calculated based on the determined concentration of Ag^+. So, the solubility product constant K_{sp} is obtained according to Equation (2).

Apparatus and Chemicals

Apparatus: test tube with stopper, 20 mL; filter; Erlenmeyer flask, 25 mL; transfer pipet, 5 mL; buret, 25 mL.

Chemicals: silver nitrate ($AgNO_3$), 0.200 mol·L^{-1}; sodium acetate ($NaC_2H_3O_2$), 0.200 mol·L^{-1}; potassium thiocyanate (KSCN), 0.0400 mol·L^{-1}; ferric ammonium sulfate $[NH_4Fe(SO_4)_2·12H_2O]$, saturated solution (Add 8 g of ferric ammonium sulfate to 20 mL of distilled water and add a few drops of concentrated nitric acid); Diluted nitric acid solution.

Procedure

1. Preparation of Saturated Silver Acetate Solutions by Precipitation

Prepare the three mixtures listed below using the transfer pipet. Place each mixture in a clean, dry, labeled 20 mL test tube. Stopper the test tubes, and then mix well by swirling, but carefully! Let the test tubes sit overnight or longer.

实验29 醋酸银的 K_{sp} 测定

实验目的

1. 测定醋酸银的溶度积常数 K_{sp}。
2. 学习佛尔哈德法测定溶液中银离子(Ag^+)的浓度。

实验原理

难溶盐醋酸银 $AgC_2H_3O_2(s)$ 与其饱和溶液间的沉淀溶解平衡可以用方程(1)表示：

$$AgC_2H_3O_2(s) = Ag^+(aq) + C_2H_3O_2^-(aq) \tag{1}$$

上述反应的溶度积常数由方程(2)给出，其中$[Ag^+]$ 和 $[C_2H_3O_2^-]$ 分别代表饱和溶液中 Ag^+ 和 $C_2H_3O_2^-$ 的浓度。

$$K_{sp} = [Ag^+][C_2H_3O_2^-] \tag{2}$$

1. 以上平衡体系可以通过如下方法建立：将含阳离子(Ag^+)与含有阴离子($C_2H_3O_2^-$)的溶液，如 $AgNO_3$ 和 $NaC_2H_3O_2$ 的水溶液，进行混合，生成醋酸银沉淀和其饱和溶液。

2. 饱和溶液中 Ag^+ 的平衡浓度可用硫氰化钾 KSCN 标准溶液进行滴定而测得。向饱和的醋酸银溶液中加入硫氰化钾会生成硫氰化银沉淀：

$$Ag^+(aq) + SCN^-(aq) = AgSCN(s)$$

滴定采用硫酸铁铵 $NH_4Fe(SO_4)_2$ 溶液做指示剂。当 AgSCN 沉淀完全时，硫酸铁铵指示剂中的 Fe^{3+} 将与过量的硫氰酸根 SCN^- 反应，生成红色的 $Fe(SCN)^{2+}$ 离子，滴定终点可以观察到溶液由无色变为浅橙色。

$$Fe^{3+}(aq) + SCN^-(aq) = Fe(SCN)^{2+}(aq)$$

3. 饱和溶液中的 $C_2H_3O_2^-$ 平衡浓度可根据已测得的 Ag^+ 浓度计算出来，这样，溶度积常数就可根据方程(2)计算得到。

仪器和试剂

仪器：具塞试管，20 mL；漏斗；锥形瓶，25 mL；移液管，5 mL；滴定管，25 mL。

试剂：硝酸银($AgNO_3$)，0.200 mol·L^{-1}；

醋酸钠($NaC_2H_3O_2$)，0.200 mol·L^{-1}；

硫氰化钾(KSCN)，0.0400 mol·L^{-1}；

硫酸铁铵 $[NH_4Fe(SO_4)_2·12H_2O]$，饱和溶液 [将 8 g $NH_4Fe(SO_4)_2·12H_2O$ 溶解在 20 mL 加有数滴浓硝酸的蒸馏水中]；

稀硝酸 HNO_3 溶液。

实验步骤

1. 沉淀法制备醋酸银饱和溶液

按照下表所列，用移液管制备 3 份混合物。将每份混合物分别置入一个干净、干燥的

	$V(AgNO_3$ of 0.200 mol·$L^{-1})$	$V(NaC_2H_3O_2$ of 0.200 mol·$L^{-1})$	V(Mixture)
Mixture 1	5.00 mL	5.00 mL	10.00 mL
Mixture 2	6.00 mL	4.00 mL	10.00 mL
Mixture 3	3.00 mL	7.00 mL	10.00 mL

2. Determination of [Ag^+] by titration with Standard KSCN Solution

(1) Put three dry and clean funnels into the funnel support and put filter paper into each one, but do not wet the filter paper with deionized water. Filter the three mixtures prepared above into separate clean, dry, labeled 20 mL test tubes.

(2) Use a clean, dry transfer pipet to measure exactly 5.00 mL of filtrate #1 into a dry 25 mL Erlenmeyer flask for titration. Add about 6 drops of ferric alum indicator and about 6 drops of dilute nitric acid solution to it. (The HNO_3 prevents hydrolysis by ferric ion)

(3) Add standard KSCN solution to the buret. Titrate the mixture with standard KSCN. A white precipitate of AgSCN will form but it will not interfere with the endpoint. Titrate carefully because there is not enough filtrate to repeat the titration. At the endpoint the aqueous solution above the AgSCN (s) will change from colorless to a pale salmon color.

(4) Repeat steps (2) and (3) for sample #2 and 3. Before using the 5 mL transfer pipet to measure 5.00 mL of the solution, rinse it with two small portions of the solution.

(5) Calculate the K_{sp} for each sample, and calculate the average K_{sp}.

Data and Results

$c_{KSCN}/(mol·L^{-1})=$

	Mixture 1	Mixture 2	Mixture 3
V_{KSCN}/mL			
$c_{Ag^+}/(mol·L^{-1})$			
$c_{NaC_2H_3O_2}/(mol·L^{-1})$			
K_{sp}			
$K_{sp, average}$			
RSD			

Notes

1. The saturated silver acetate solutions must be prepared at least one day before the titration is performed.

2. Silver solutions cause stains. Handle carefully!

3. To avoid diluting the concentrations of the saturated silver acetate solution, all the equipment should be dry and clean when the mixture is filtered. For the same reason, the Erlenmeyer flasks should be cleaned and dried when titration.

20 mL 试管中,并做好标记,将试管的塞子盖好,然后小心旋摇试管,使混合物充分混匀。让混合物静置过夜,或者更长时间。

	0.200 mol·L^{-1} AgNO$_3$ 的体积	0.200 mol·L^{-1} NaC$_2$H$_3$O$_2$ 的体积	混合物体积
混合物 1	5.00 mL	5.00 mL	10.00 mL
混合物 2	6.00 mL	4.00 mL	10.00 mL
混合物 3	3.00 mL	7.00 mL	10.00 mL

2. 用 KSCN 标准溶液测定银离子的浓度[Ag$^+$]

(1) 将 3 个干净、干燥的漏斗放在漏斗架上,在每个漏斗中放好滤纸,但不要用蒸馏水湿润滤纸;将上述 3 份混合物分别过滤至干净、干燥的做好标记的 20 mL 试管中。

(2) 用一根干燥、干净的移液管精确移取 1 号滤液 5.00 mL 至一个干燥的 25 mL 锥形瓶中,以备滴定。向锥形瓶中加入 6 滴铁铵矾指示剂,并加入大约 6 滴稀硝酸。(硝酸用于防止铁离子的水解)

(3) 在滴定管中装入 KSCN 标准溶液。用 KSCN 标准溶液滴定锥形瓶中的混合液,有白色沉淀生成,但沉淀并不干扰终点的观察。鉴于没有足够的溶液重复滴定,一定要小心滴定。到达终点时,AgSCN 沉淀之上的溶液颜色由无色变为浅橙红色。

(4) 重复步骤(2)和(3)的操作,对 2 号和 3 号滤液进行滴定。注意:在用 5 mL 移液管移取 5.00 mL 滤液时,要用滤液润洗移液管两次。

(5) 计算每份样品的 K_{sp},并计算出平均 K_{sp}。

数据记录及结果处理

$c_{KSCN}/(\text{mol}\cdot\text{L}^{-1})=$

	混合物 1	混合物 2	混合物 3
V_{KSCN}/mL			
$c_{Ag^+}/(\text{mol}\cdot\text{L}^{-1})$			
$c_{NaC_2H_3O_2}/(\text{mol}\cdot\text{L}^{-1})$			
K_{sp}			
$\overline{K}_{sp}$			
RSD			

注意事项

1. 饱和醋酸银溶液的制备必须比滴定至少提早一天进行。

2. 银离子溶液滴在物体表面会形成银斑,使用的时候要小心。

3. 为了避免饱和醋酸银溶液被稀释,所有用于混合物过滤的实验装置都应该是干燥、洁净的;同理,滴定时锥形瓶也必须干燥、洁净。

Questions

1. Why must the saturated silver acetate solutions be prepared in advance at least one night?

2. Why should the precipitate of silver acetate be filtrated out before the titration of Ag^+ is performed?

3. The filter paper can not be wet with deionized water when the mixture is filtrated. Why? What should be used if the filter paper need to be wet?

4. How to calculate the concentration of $C_2H_3O_2^-$ at equilibrium in the saturated $AgC_2H_3O_2$ solutions when determining the concentration of Ag^+?

5. The K_{sp} of $AgC_2H_3O_2$ is 1.94×10^{-3} at the temperature 25 ℃. Discuss the precision and accuracy of the value of K_{sp} determined and the factors which affect the precision and accuracy.

思 考 题

1. 为什么饱和醋酸银溶液必须至少提前一个晚上配制？
2. 在滴定 Ag^+ 时，为什么醋酸银沉淀应该过滤除去？
3. 过滤混合物时，滤纸不能用蒸馏水润湿。为什么？如果滤纸需要润湿，可以用什么进行润湿？
4. 如何根据测得的银离子浓度计算饱和醋酸银溶液中 $C_2H_3O_2^-$ 的平衡浓度？
5. 在 25 ℃时，$AgC_2H_3O_2$ 的 K_{sp} 为 1.94×10^{-3}。讨论你的 K_{sp} 测定结果的精密度与准确度，并讨论影响精密度与准确度的因素。

Part 5   Appendix

5.1   Using Excel for Graphing

The following are guidelines for acquiring a graph using *Excel*, including a best-fit line (called a trendline in the program). Be judicious about the data included. If a points lies far outside a reasonable value, provide two plots for grading: one with the data point and one without. Also, you should perform a Q-Test and report the confidence level for rejecting the data point.

Instructions for graphing with 97-2003 *Excel*

(1) Enter data points to be graphed in two adjacent columns. Data for the x axis should be in the left column and data for the y axis should be in the right column.

(2) Highlight the two columns of data by holding down the left mouse button and dragging the mouse over the column heading letters.

(3) Go to the Insert menu and select Chart. The Chart Wizard window should appear.

(4) In the first window, Step 1 of 4, select XY Scatter for Chart Type and Scatter for Sub Chart Type (the choice without any lines connecting the data points). Click the Next button at the bottom of the Chart Wizard window.

(5) In the second window, Step 2 of 4, make sure Series in Columns is chosen and then click the Next button.

(6) In the third window, Step 3 of 4, there are five tabs: titles, axes, gridlines, legend, and data labels. In the Title tab enter the Chart Title and the Labels for the x and y axes (make sure that you include units). In the Gridlines tab check Major Gridlines for both x and y axes. And, in the Legend tab uncheck Show Legend. Click the Next button.

(7) In the fourth window, Step 4 of 4, determine whether you want your graph on the same sheet with your data, or on a new sheet. Click the Finish button.

(8) Click on the data points so that only they are highlighted. Go to the Chart menu and choose Add Trendline. In the Add Trendline window, under the Type tab, choose the correct line type (which is usually linear for General chem. lab). Under the Options tab, check Display Equation on Chart. Click OK.

Note: You may change locations, sizes, gridline units, fonts, et cetera of your chart by clicking to highlight the areas you wish to change.

Instructions for graphing with 2007 *Excel*

(1) Enter data points to be graphed in two adjacent columns. Data for the x axis should be in the left column and data for the y axis should be in the right column.

(2) Highlight the two columns of data by holding down the left mouse button and dragging the mouse over the column heading letters.

(3) Click on the Insert tab and select Scatter icon in the Charts section. Choose the Scatter

第五部分 附 录

5.1 Excel 作图

Excel 软件可以用于处理实验数据和作图,其中包括曲线的拟合(Excel 软件称其为曲线趋势)以及数据判断。如果某个数据点远离合理值的范围,程序能提供两种曲线:一个包括该数据点,另一个不包括该数据点。当然,你应当用 Q 检验法判断并报告被去除的数据点的置信度。

Excel 97-2003 作图指南

(1) 在相邻的两个数据栏中输入作图所需要的数据点。x-轴的数据应该在左边一栏,y-轴数据在右边一栏。

(2) 点击鼠标左键选定这两个数据栏,并用鼠标拖拽至数据栏的标题字母。

(3) 在插入菜单中选定图表,图表向导窗口被打开。

(4) 图表向导的第一个窗口显示"4 步骤之 1",在标准类型栏中选定 xy 散点图,在子图表类型中选定散点图(即没有任何连线的数据点图)。然后点击图表向导窗口的下一步按钮。

(5) 在第二个向导窗口"4 步骤之 2"中,查看数据区域是否确定,点击下一步按钮。

(6) 在第三个向导窗口"4 步骤之 3"中,有五个选项卡:标题、坐标轴、网格线、图例和数据标志。在标题中输入图表的标题和 x-轴、y-轴的标志(坐标轴不要忘记带单位);在网格线中选定 x-轴、y-轴的主要网格线;在图例中去掉显示图例。点击下一步。

(7) 在第四个向导窗口"4 步骤之 4"中,选择将所作图和数据表放在一起(作为其中之一插入 sheet)还是作为新的工作表存在(作为新工作表插入 chart),然后点击"完成"。

(8) 点击图上的数据点,数据点被选定,在图表菜单中选择添加趋势线选项,打开添加趋势线窗口,在类型选项卡栏中,选取合适的回归分析类型(基础化学实验所设计的曲线通常是线性的)。在选项中,选择显示公式。点击确定按钮。

注意事项:当需要对图的位置、大小、网格线单位、字体等进行修改时,点击需要修改的部分,将其选定后即可进行修改。

Excel 2007 作图指南

(1) 在相邻的两个数据栏中输入作图所需的数据点。x-轴的数据应该在左边一栏,y-轴

with Only Markers chart type (the top left choice). The plot of data will appear.

(4) Click on the plot that just appeared. Chart Tools should appear at the very top of the screen in the middle. Choose the Layout tab.

a. In the Analysis section choose Trendline. Choose More Trendline Options at the bottom of the pull down window. Choose the Trend/Regression Type and click on the box next to Display Equation on chart.

b. In the Label section choose Chart Title. Choose Above Chart and enter a title in the field that appears at the top of the plot. In the same section choose Axis Titles. Enter the titles.

c. If the initial and final values of either axis need to be changed, click on the Axis button in the Axis section and make the necessary changes. Gridlines can be altered by clicking on the Gridline button in the Axis section.

数据在右边一栏。

(2) 点击鼠标左键选定这两个数据栏,并用鼠标拖拽至数据栏的标题字母。

(3) 点击插入,选择图表选项中的散点图。选择图表类型中的仅带数据标记的散点图(左上角的选项),数据图随即显现。

(4) 点击随即显现的数据图,图表工具就会在屏幕中间的最上方显现,选择布局选项。

  a. 在分析选项中选择趋势线,选取下拉窗口底部的其他曲线趋势线选项,在趋势预测/回归分析类型项目中,点击方框下面的显示公式。

  b. 在标签选项中选择图表标题,选择图表上方,并在曲线上方位置的区域输入标题。同样,在标签选项中选择坐标轴标题并输入坐标轴标题。

  c. 如果需要修改坐标轴上的数据范围,则点击坐标轴选项中的坐标轴按钮进行必要的修改。点击坐标轴选项中的网格线按钮可以对网格线加以修改。

5.2 Identification Tests of Common Inorganic Ions

Cation ions

Aluminum Salts

1. Add sodium hydroxide TS to a solution of the substance being examined; a gelatinous white precipitate appears which is soluble in an excess of sodium hydroxide TS.

2. Add ammonia TS to a solution of the substance being examined until a gelatinous white precipitate is formed. Add a few drops of sodium alizarin sulfonate IS; the precipitate becomes cherry red in colour.

Ammonium Salts

1. Heat a quantity of the substance being examined with an excess of sodium hydroxide TS; the characteristic odour of ammonia is perceived the vapour turns moistened red litmus paper to blue and blackens a strip of filter paper moistened with mercurous nitrate TS.

2. To a solution of the substance being examined add 1 drop of alkaline mercuric potassium iodide TS; a reddish brown precipitate is produced.

Calcium Salts

1. Moisten the substance being examined with hydrochloric acid on a platinum wire, it imparts a brick red colour to a nonluminous flame.

2. Add 2 drops of methyl red IS to a solution of the substance being examined $(1 \rightarrow 20)$. Neutralize with ammonia TS and then acidify with hydrochloric acid. Add ammonium oxalate TS; a white precipitate is produced which is soluble in hydrochloric acid but insoluble in acetic acid.

Copper Salts

1. Add a few drops of ammonia TS to a solution of the substance being examined; a light blue precipitate is produced which is soluble in an excess of the reagent, forming a dark blue solution.

2. Add potassium ferrocyanide TS to a solution of the substance being examined; a reddish brown colour or precipitate is produced.

Ferric Salts

1. Add potassium ferrocyanide TS to a solution of the substance being examined; a dark blue precipitate is formed which is insoluble in dilute hydrochloric acid, it decomposes to form a brown precipitate on addition of sodium hydroxide TS.

2. Add ammonium thiocyanate TS to a solution of the substance being examined; a red colour is produced.

Ferrous Salts

1. Add potassium ferricyanide TS to a solution of the substance being examined; a dark blue precipitate is formed which is insoluble in dilute hydrochloric acid, it decomposes to form a brown precipitate on addition of sodium hydroxide TS.

2. To a solution of the substance being examined, add a few drops of a 1% solution of o-phenanthroline in ethanol; a deep red colour is produced.

Barium Salts

1. Moisten the substance being examined with hydrochloric acid on a platinum wire. it imparts an yellowish green colour to a nonluminous flame, or a blue colour when viewed through a green glass plate.

2. Add dilute sulfuric acid to a solution of the substance being examined; a white precipitate

5.2 药典中常见无机离子的鉴别

阳离子

铝盐

1. 取供试品溶液,加氢氧化钠试液,即生成白色胶状沉淀;分离,沉淀能在过量的氢氧化钠试液中溶解。
2. 取供试品溶液,加氨试液至生成白色胶状沉淀,滴加茜素磺酸钠指示液数滴,沉淀即显樱红色。

铵盐

1. 取供试品,加过量的氢氧化钠试液后,加热,即分解,发生氨臭;遇湿润的红色石蕊试纸,能使之变蓝色,并能使硝酸亚汞试液湿润的滤纸显黑色。
2. 取供试品溶液,加碱性碘化汞钾试液 1 滴,即生成红棕色沉淀。

钙盐

1. 取铂丝,用盐酸湿润后,蘸取供试品,在无色火焰中燃烧,火焰即显砖红色。
2. 取供试品溶液(1→20),加甲基红指示液 2 滴,用氨试液中和,再滴加盐酸至恰呈酸性,加草酸铵试液,即生成白色沉淀;分离,沉淀不溶于醋酸,但可溶于盐酸。

铜盐

1. 取供试品溶液,滴加氨试液,即生成淡蓝色沉淀;再加过量的氨试液,沉淀即溶解,生成深蓝色溶液。
2. 取供试品溶液,加亚铁氰化钾试液,即显红棕色或生成红棕色沉淀。

铁盐

1. 取供试品溶液,加亚铁氰化钾试液,即生成深蓝色沉淀;分离,沉淀在稀盐酸中不溶,但加氢氧化钠试液,即分解成棕色沉淀。
2. 取供试品溶液,加硫氰酸铵试液,即显血红色。

亚铁盐

1. 取供试品溶液,加铁氰化钾试液即生成深蓝色沉淀;分离,沉淀在稀盐酸中不溶,但加氢氧化钠试液,即分解成棕色沉淀。
2. 取供试品溶液,加 1%邻二氮菲的乙醇溶液数滴,即显深红色。

钡盐

1. 取铂丝,用盐酸湿润后,蘸取供试品,在无色火焰中燃烧,火焰即显黄绿色;通过绿色玻璃透视,火焰显蓝色。
2. 取供试品溶液,滴加稀硫酸,即生成白色沉淀;分离,沉淀在盐酸或硝酸中均不溶解。

锂盐

1. 取供试品溶液,加氢氧化钠试液碱化后,加入碳酸钠试液,煮沸,即生成白色沉淀;分离,沉淀能在氯化铵试液中溶解。
2. 取铂丝,用盐酸湿润后,蘸取供试品,在无色火焰中燃烧,火焰显胭脂红色。

is produced which is insoluble in hydrochloric acid or nitric acid.

Lithium Salts

1. Alkalize the solution of the substance being examined with sodium hydroxide TS, add sodium carbonate TS, a white precipitate is produced on boiling which is soluble in ammonium chloride TS.

2. Moisten the substance being examined with hydrochloric acid on a platinum wire, it imparts a crimson colour to a nonluminous flame.

3. To a quantity of the substance being examined add dilute sulfuric acid or soluble sulfates solution, no precipitate is produced (distinction from strontium).

Magnesium Salts

1. Add ammonia TS to a solution of the substance being examined; a white precipitate is produced which redissolves on addition of ammonium chloride TS. Add 1 drop of disodium hydrogen phosphate TS and shake; a white precipitate insoluble in ammonia TS is produced.

2. Add sodium hydroxide TS to a solution of the substance being examined; a white precipitate is produced, filter the precipitate is insoluble in an excess of sodium hydroxide TS, but is coloured reddish brown on addition of iodine TS.

Mercuric Salts

1. Add sodium hydroxide TS to a solution of the substance being examined; a yellow precipitate is produced.

2. Add potassium iodide TS to a neutral solution of the substance being examined; a scarlet precipitate is produced which is soluble in an excess of the reagent. To the solution add sodium hydroxide TS and an ammonium salt; a reddish brown precipitate is produced.

3. When applied to bright copper foil, solutions of mercury salts, free from an excess of nitric acid, yield a deposit that upon rubbing, becomes bright and silvery in appearance.

Mercurous Salts

1. To a quantity of the substance being examined add ammonia TS or sodium hydroxide TS; a black colour is developed.

2. To a quantity of the substance being examined add potassium iodide TS and shake; a yellowish green precipitate is produced, changing to grayish green rapidly and then to grayish black gradually.

Potassium Salts

1. Moisten the substance being examined with hydrochloric acid on a platinum wire, it imparts a violet colour to a nonluminous flame. If sodium is also present, the yellow colour can be screened out by viewing through a blue glass plate.

2. Ignite the substance being examined to remove any ammonium salt contaminated, cool, dissolve it in water, add acetic acid and a 0.1% solution of sodium tetraphenylborate; a white precipitate is produced.

Sodium Salts

1. Moisten the substance being examined with hydrochloric acid on a platinum wire, it imparts an intense yellow colour to a nonluminous flame.

2. Add zinc uranylacetate TS to a neutral solution of the substance being examined; a yellow precipitate is produced.

3. 取供试品适量,加入稀硫酸或可溶性硫酸盐溶液,不生成沉淀(与锶盐区别)。

镁盐

1. 取供试品溶液,加氨试液,即生成白色沉淀;滴加氯化铵试液,沉淀溶解;再加磷酸氢二钠试液1滴,振摇,即生成白色沉淀。沉淀在氨试液中不溶。

2. 取供试品溶液,加氢氧化钠试液,即生成白色沉淀。分离,沉淀分成两份,一份中加过量的氢氧化钠试液,沉淀不溶;另一份中加碘试液,沉淀转成红棕色。

汞盐

1. 取供试品溶液,加氢氧化钠试液,即生成黄色沉淀。

2. 取供试品的中性试液,加碘化钾试液,即生成猩红色沉淀,能在过量的碘化钾试液中溶解;再以氢氧化钠试液碱化,加铵盐即生成红棕色的沉淀。

3. 取不含过量硝酸的供试品溶液,涂于光亮的铜箔表面,擦拭后即生成一层光亮似银的沉积物。

亚汞盐

1. 取供试品,加氨试液或氢氧化钠试液,即变黑色。

2. 取供试品,加碘化钾试液,振摇,即生成黄绿色沉淀,瞬即变为灰绿色,并逐渐转变为灰黑色。

钾盐

1. 取铂丝,用盐酸湿润后,蘸取供试品,在无色火焰中燃烧,火焰即显紫色;但有少量的钠盐混存时,须隔蓝色玻璃透视,方能辨认。

2. 取供试品,加热炽灼除去可能杂有的铵盐,放冷后,加水溶解,再加0.1%四苯硼钠溶液与醋酸,即生成白色沉淀。

钠盐

1. 取铂丝,用盐酸湿润后,蘸取供试品,在无色火焰中燃烧,火焰即显鲜黄色。

2. 取供试品的中性溶液,加醋酸氧铀锌试液,即生成黄色沉淀。

银盐

1. 取供试品溶液,加稀盐酸,即生成白色凝乳状沉淀;分离,沉淀能在氨试液中溶解,加硝酸,沉淀复生成。

2. 取供试品的中性溶液,加铬酸钾试液,即生成砖红色沉淀;分离,沉淀能在硝酸中溶解。

锌盐

1. 取供试品溶液,加亚铁氰化钾试液,即生成白色沉淀;分离,沉淀在稀盐酸中不溶解。

2. 取供试品溶液,以稀硫酸酸化,加0.1%硫酸铜溶液1滴及硫氰酸汞铵试液数滴,即生成紫色沉淀。

阴离子

醋酸盐

1. 取供试品,加硫酸和乙醇后,加热,即分解发生醋酸乙酯的香气。

2. 取供试品的中性溶液,加三氯化铁试液1滴,溶液呈深红色,加稀无机酸,红色即褪去。

Silver Salts

1. Add dilute hydrochloric acid to a solution of the substance being examined; a cruddy white precipitate is produced which is soluble in ammonia TS and reprecipitated on addition of nitric acid.

2. Add potassium chromate TS to a neutral solution of the substance being examined; a brick red precipitate is produced which is soluble in nitric acid.

Zinc Salts

1. Add potassium ferrocyanide TS to a solution of the substance being examined; a white precipitates is formed which is insoluble in dilute hydrochloric acid.

2. Acidify a solution of the substance being examined with dilute sulfuric acid, add 1 drop of 0.1% copper sulfate solution and a few drops off mercuric ammonium thiocyanate TS; a violet precipitate is produced.

Anion ions

Acetates

1. Heat a quantity of the substance being examined with sulfuric acid and ethanol; the characteristic odour of ethyl acetates is liberated.

2. To a neutral solution of the substance being examined add 1 drop of ferric chloride TS; a deep red colour is produced which disappears on addition of dilute mineral acid.

Borates

1. A solution of the substance being examined, acidified with hydrochloric acid, turns turmeric paper to brownish red. The colour deepens on drying and becomes greenish black when moistened with ammonia TS.

2. Mix a quantity of the substance being examined with sulfuric acid and add methanol, when the mixture is ignited, it burns with a green-bordered flame.

Bromides

1. Add silver nitrate TS to a solution of the substance being examined; a cruddy, pale yellow precipitate is formed which is slightly soluble in ammonia TS and practically insoluble in nitric acid.

2. Add chlorine TS dropwise to a solution of the substance being examined; bromide is liberated, add chloroform and shake, a yellow or reddish brown color is developed in chloroform layer.

Carbonates and Bicarbonates

1. Add dilute acid to a solution of the substance being examined, it effervesces with the evolution of carbon dioxide, producing a white precipitate when passed into calcium hydroxide TS.

2. Add magnesium sulfate TS to a solution of the substance being examined; a white precipitate is produced immediately (carbonates) or on boiling (bicarbonates).

3. A solution of the substance being examined is colorless or only slightly colored on the addition of phenolphthalein IS (bicarbonates), or an intense red color is produced (carbonates).

Chlorides

1. Acidify a solution of the substance being examined with nitric acid and add silver nitrate TS; a cruddy, white precipitate is formed which is soluble in ammonia TS and reprecipitated on addition of nitric acid. Organic bases should be removed by the addition of ammonia TS and filtration prior to the test.

2. Mix a small quantity of the substance being examined with an equal part of manganese

硼酸盐

1. 取供试品溶液,加盐酸呈酸性后,能使姜黄试纸变成棕红色;放置干燥,颜色即变深,用氨试液湿润,即变为绿黑色。

2. 取供试品,加硫酸,混合后,加甲醇,点火燃烧,即发生边缘带绿色的火焰。

溴化物

1. 取供试品溶液,加硝酸银试液,即生成淡黄色凝乳状沉淀;分离,沉淀能在氨试液中微溶,但在硝酸中几乎不溶。

2. 取供试品溶液,滴加氯试液,溴即游离,加氯仿振摇,氯仿层显黄色或红棕色。

碳酸盐与碳酸氢盐

1. 取供试品溶液,加稀酸,即泡沸,发生二氧化碳气体,导入氢氧化钙试液中,即生成白色沉淀。

2. 取供试品溶液,加硫酸镁试液,如为碳酸盐溶液,即生成白色沉淀;如为碳酸氢盐溶液,须煮沸,始生成白色沉淀。

3. 取供试品溶液,加酚酞指示液,如为碳酸盐溶液,即显深红色;如为碳酸氢盐溶液,不变色或仅显微红色。

氯化物

1. 取供试品溶液,加硝酸使呈酸性后,加硝酸银试液,即生成白色凝乳状沉淀;分离,沉淀加氨试液即溶解,再加硝酸,沉淀复生成。如供试品为生物碱或其他有机碱的盐酸盐,须先加氨试液使呈碱性,将析出的沉淀滤过除去,取滤液进行试验。

2. 取供试品少量,置试管中,加等量的二氧化锰,混匀,加硫酸湿润,缓缓加热,即发生氯气,能使湿润的碘化钾淀粉试纸显蓝色。

碘化物

1. 取供试品溶液,加硝酸银试液,即生成黄色凝乳状沉淀;分离,沉淀在硝酸或氨试液中均不溶解。

2. 取供试品溶液,加少量的氯试液,碘即游离;如加氯仿振摇,氯仿层显紫色;如加淀粉指示液,溶液显蓝色。

硝酸盐

1. 取供试品溶液,置试管中,加等量的硫酸,注意混合,冷后,沿管壁加硫酸亚铁试液,使成两液层,接界面显棕色。

2. 取供试品溶液,加硫酸与铜丝(或铜屑),加热,即发生红棕色的蒸气。

3. 取供试品溶液,滴加高锰酸钾试液,紫色不应褪去(与亚硝酸盐区别)。

磷酸盐

1. 取供试品的中性溶液,加硝酸银试液,即生成浅黄色沉淀;分离,沉淀在氨试液或稀硝酸中均易溶解。

2. 取供试品溶液,加氯化铵镁试液,即生成白色结晶性沉淀。

3. 取供试品溶液,加钼酸铵试液与硝酸后,加热即生成黄色沉淀;分离,沉淀能在氨试液中溶解。

亚硫酸盐或亚硫酸氢盐

1. 取供试品,加盐酸,即发生二氧化硫的气体,有刺激性臭味,并能使硝酸亚汞试液湿

dioxide, moisten with sulfuric acid and heat gently; chlorine is evolved which turns a strip of moistened starch-potassium iodide paper to blue.

Iodides

1. Add silver nitrate TS to a solution of the substance being examined; a cruddy, pale yellow precipitate is produced which is insoluble in nitric acid or ammonia TS.

2. Add chlorine TS dropwise to a solution of the substance being examined; iodine is liberated, add chloroform and shake, a violet colour is produced in the chloroform layer. If starch IS is added instead of chloroform, a blue colour is produced.

Nitrates

1. Mix cautiously a solution of the substance being examined with an equal volume of sulfuric acid and allow to cool. Add ferrous sulfate TS along the inner wall of the test tube; a brown ring is developed at the interface of the two layers.

2. To a solution of the substance being examined add cautiously sulfuric acid and metallic copper; a reddish brown fume is evolved on heating.

3. Add dropwise potassium permanganate TS to a solution of the substance being examined; the violet color does not disappear (distinction from nitrites).

Phosphates

1. Add silver nitrate TS to a neutral solution of the substance being examined; a light yellow precipitate is formed which is freely soluble in ammonia TS or dilute nitric acid.

2. Add magnesium ammonium chloride TS to a solution of the substance being examined; a white crystalline precipitate is produced.

3. To a solution of the substance being examined add ammonium molybdate TS and nitric acid; a yellow precipitate soluble in ammonia TS is produced on heating.

Sulfites and Bisulfites

1. Add hydrochloric acid to the substance being examined; the pungent odor of sulfur dioxide is perceived, the vapour blackens a strip of filter paper moistened with mercurous nitrate TS.

2. Add iodine TS dropwise to a solution of the substance being examined; the colour of iodine is discharged.

Sulfates

1. Add barium chloride TS to a solution of the substance being examined a white precipitate is formed which is insoluble in hydrochloric acid or nitric acid.

2. Add lead acetate TS to a solution of the substance being examined; a white precipitate is formed which is soluble ammonium acetate TS or sodium hydroxide TS.

3. Add hydrochloric acid to a solution of the substance being examined; no white precipitate is produced. (distinction from thiosulphates)

润的滤纸显黑色。

2. 取供试品溶液,滴加碘试液,碘的颜色即消褪。

硫酸盐

1. 取供试品溶液,加氯化钡试液,即生成白色沉淀;分离,沉淀在盐酸或硝酸中均不溶解。

2. 取供试品溶液,加醋酸铅试液,即生成白色沉淀;分离,沉淀在醋酸铵试液或氢氧化钠试液中溶解。

3. 取供试品溶液,加盐酸,不生成白色沉淀(与硫代硫酸盐区别)。

5.3 常见无机酸碱的解离常数

表 5-1 无机酸在水溶液中的解离常数(25 ℃)
Table 5-1 Dissociation Constants of Mineral Acids in Aqueous Solution(25 ℃)

序号(No.)	名称(Substance)	化学式(Chemical formula)	K_a	pK_a
1	偏铝酸	$HAlO_2$	6.3×10^{-13}	12.20
2	亚砷酸	H_3AsO_3	6.0×10^{-10}	9.22
3	砷 酸	H_3AsO_4	$6.3 \times 10^{-3}\,(K_1)$	2.20
			$1.05 \times 10^{-7}\,(K_2)$	6.98
			$3.2 \times 10^{-12}\,(K_3)$	11.50
4	硼 酸	H_3BO_3	$5.8 \times 10^{-10}\,(K_1)$	9.24
			$1.8 \times 10^{-13}\,(K_2)$	12.74
			$1.6 \times 10^{-14}\,(K_3)$	13.80
5	次溴酸	$HBrO$	2.4×10^{-9}	8.62
6	氢氰酸	HCN	6.2×10^{-10}	9.21
7	碳 酸	H_2CO_3	$4.2 \times 10^{-7}\,(K_1)$	6.38
			$5.6 \times 10^{-11}\,(K_2)$	10.25
8	次氯酸	$HClO$	3.2×10^{-8}	7.50
9	氢氟酸	HF	6.61×10^{-4}	3.18
10	锗 酸	H_2GeO_3	$1.7 \times 10^{-9}\,(K_1)$	8.78
			$1.9 \times 10^{-13}\,(K_2)$	12.72
11	高碘酸	HIO_4	2.8×10^{-2}	1.56
12	亚硝酸	HNO_2	5.1×10^{-4}	3.29
13	次磷酸	H_3PO_2	5.9×10^{-2}	1.23
14	亚磷酸	H_3PO_3	$5.0 \times 10^{-2}\,(K_1)$	1.30
			$2.5 \times 10^{-7}\,(K_2)$	6.60
15	磷 酸	H_3PO_4	$7.52 \times 10^{-3}\,(K_1)$	2.12
			$6.31 \times 10^{-8}\,(K_2)$	7.20
			$4.4 \times 10^{-13}\,(K_3)$	12.36
16	焦磷酸	$H_4P_2O_7$	$3.0 \times 10^{-2}\,(K_1)$	1.52
			$4.4 \times 10^{-3}\,(K_2)$	2.36
			$2.5 \times 10^{-7}\,(K_3)$	6.60
			$5.6 \times 10^{-10}\,(K_4)$	9.25

续表

序号(No.)	名称(Substance)	化学式(Chemical formula)	K_a	pK_a
17	氢硫酸	H_2S	$1.3\times10^{-7}(K_1)$	6.88
			$7.1\times10^{-15}(K_2)$	14.15
18	亚硫酸	H_2SO_3	$1.23\times10^{-2}(K_1)$	1.91
			$6.6\times10^{-8}(K_2)$	7.18
19	硫酸	H_2SO_4	$1.0\times10^{3}(K_1)$	−3.0
			$1.02\times10^{-2}(K_2)$	1.99
20	硫代硫酸	$H_2S_2O_3$	$2.52\times10^{-1}(K_1)$	0.60
			$1.9\times10^{-2}(K_2)$	1.72
21	氢硒酸	H_2Se	$1.3\times10^{-4}(K_1)$	3.89
			$1.0\times10^{-11}(K_2)$	11.0
22	亚硒酸	H_2SeO_3	$2.7\times10^{-3}(K_1)$	2.57
			$2.5\times10^{-7}(K_2)$	6.60
23	硒酸	H_2SeO_4	$1\times10^{3}(K_1)$	−3.0
			$1.2\times10^{-2}(K_2)$	1.92
24	硅酸	H_2SiO_3	$1.7\times10^{-10}(K_1)$	9.77
			$1.6\times10^{-12}(K_2)$	11.80
25	亚碲酸	H_2TeO_3	$2.7\times10^{-3}(K_1)$	2.57
			$1.8\times10^{-8}(K_2)$	7.74

表 5-2 无机碱在水溶液中的解离常数(25 ℃)

Table 5-2 Dissociation Constants of Mineral Bases in Aqueous Solution (25 ℃)

序号(No.)	名称(Substance)	化学式(Chemical formula)	K_b	pK_b
1	氢氧化铝	$Al(OH)_3$	$1.38\times10^{-9}(K_3)$	8.86
2	氢氧化银	$AgOH$	1.10×10^{-4}	3.96
3	氢氧化钙	$Ca(OH)_2$	$3.72\times10^{-3}(K_1)$	2.43
			$3.98\times10^{-2}(K_2)$	1.40
4	氨水	NH_3+H_2O	1.78×10^{-5}	4.75
5	肼(联氨)	$N_2H_4+H_2O$	$9.55\times10^{-7}(K_1)$	6.02
			$1.26\times10^{-15}(K_2)$	14.9
6	羟氨	NH_2OH+H_2O	9.12×10^{-9}	8.04
7	氢氧化铅	$Pb(OH)_2$	$9.55\times10^{-4}(K_1)$	3.02
			$3.0\times10^{-8}(K_2)$	7.52
8	氢氧化锌	$Zn(OH)_2$	9.55×10^{-4}	3.02

5.4 常用基准物质(Common Primary Standards)

标定实验用的基准物质必须符合以下 4 条要求。① 用作基准物的物质,应该非常纯净,纯度至少在 99.9% 以上;其组成应与其化学式完全相符。② 要稳定,不易被空气所氧化,也不易吸收空气中的水分和 CO_2 等;在进行干燥时组成不变;尽量避免使用带结晶水的物质。③ 被标定的物质之间的反应应该有确定的化学计量关系,反应速度要快。④ 最好能采用具有较大摩尔质量的物质,这样可以减小称量误差。

表 5-3 常用基准物质的使用
Table 5-3 Usages of Common Primary standards

滴定方法	标准溶液	基准物质	优缺点
酸碱滴定	HCl	Na_2CO_3	便宜,易得纯品,易吸湿
		$Na_2B_4O_7 \cdot 10H_2O$	易得纯品,不易吸湿,摩尔质量大,湿度小时会先结晶水
	NaOH	$C_6H_4 \cdot COOH \cdot COOK$	易得纯品,不吸湿,摩尔质量大
		$H_2C_2O_4 \cdot 2H_2O$	便宜,结晶水不稳定,纯度不理想
络合滴定	EDTA	金属 Zn 或 ZnO	纯度高,稳定,既可在 pH=5~6 又可在 pH=9~10 应用
氧化还原滴定	$KMnO_4$	$Na_2C_2O_4$	易得纯品,稳定,无显著吸湿
	$K_2Cr_2O_7$	$K_2Cr_2O_7$	易得纯品,非常稳定,可直接配制标准溶液
	$Na_2S_2O_3$	$K_2Cr_2O_7$	易得纯品,非常稳定,可直接配制标准溶液
	I_2	升华碘	纯度高,易挥发,水中溶解度很小
		As_2O_3	能得纯品,产品不吸湿,剧毒
	$KBrO_3$	$KBrO_3$	
	$KBrO_3$+过量 KBr	$KBrO_3$	易得纯品,稳定
沉淀滴定	$AgNO_3$	$AgNO_3$	易得纯品,防止光照及有机物玷污
		NaCl	易得纯品,易吸湿

5.5 常用 pH 缓冲溶液(Common pH Buffer Solution)

在化学中,有一类能够减缓因外加强酸或强碱以及稀释等而引起的 pH 急剧变化的作用的溶液,此种溶液被称为 pH 缓冲溶液。pH 缓冲溶液一般都是由浓度较大的弱酸及其共轭碱所组成,如 $HAc-Ac^-$,$NH_4^+-NH_3$ 等,此种缓冲溶液具有抗外加强酸强碱的作用,同时还有抗稀释的作用。在高浓度的强酸或强碱溶液中,由于 H^+ 或 OH^- 浓度本来就很高,外加少量酸或碱基本不会对溶液的酸度产生太大的影响。在这种情况下,强酸(pH<2)、强碱(pH>12)也是缓冲溶液,但此类缓冲溶液不具有抗稀释的作用。

在分析化学中用到的缓冲溶液,大多数是用于控制溶液的 pH,也有一部分是专门用于测量溶液的 pH 值时的参照标准,被称为标准缓冲溶液。

表 5-4 常用 pH 缓冲溶液的配制和 pH 值
Table 5-4 Preparation and pH Values of Common pH Buffer Solutions

序号	溶液名称	配制方法	pH值
1	氯化钾—盐酸	13.0 mL 0.2 mol·L^{-1} HCl 与 25.0 mL 0.2 mol·L^{-1} KCl 混合均匀后,加水稀释至 100 mL	1.7
2	氨基乙酸—盐酸	500 mL 水中溶解氨基乙酸 150 g,加 480 mL 浓盐酸,再加水稀释至 1 L	2.3
3	一氯乙酸—氢氧化钠	200 mL 水中溶解 2 g 一氯乙酸后,加 40 g NaOH,溶解完全后再加水稀释至 1 L	2.8
4	邻苯二甲酸氢钾—盐酸	25.0 mL 0.2 mol·L^{-1} 的邻苯二甲酸氢钾溶液与 6.0 mL 0.1 mol·L^{-1} HCl 混合均匀,加水稀释至 100 mL	3.6
5	邻苯二甲酸氢钾—氢氧化钠	25.0 mL 0.2 mol·L^{-1} 的邻苯二甲酸氢钾溶液与 17.5 mL 0.1 mol·L^{-1} NaOH 混合均匀,加水稀释至 100 mL	4.8
6	六亚甲基四胺—盐酸	200 mL 水中溶解六亚甲基四胺 40 g,加浓 HCl 10 mL,再加水稀释至 1 L	5.4
7	磷酸二氢钾—氢氧化钠	25.0 mL 0.2 mol·L^{-1} 的磷酸二氢钾与 23.6 mL 0.1 mol·L^{-1} NaOH 混合均匀,加水稀释至 100 mL	6.8
8	硼酸—氯化钾—氢氧化钠	25.0 mL 0.2 mol·L^{-1} 的硼酸—氯化钾与 4.0 mL 0.1 mol·L^{-1} NaOH 混合均匀,加水稀释至 100 mL	8.0
9	氯化铵—氨水	0.1 mol·L^{-1} 氯化铵与 0.1 mol·L^{-1} 氨水以 2:1 比例混合均匀	9.1
10	硼酸—氯化钾—氢氧化钠	25.0 mL 0.2 mol·L^{-1} 的硼酸—氯化钾与 43.9 mL 0.1 mol·L^{-1} NaOH 混合均匀,加水稀释至 100 mL	10.0
11	氨基乙酸—氯化钠—氢氧化钠	49.0 mL 0.1 mol·L^{-1} 氨基乙酸—氯化钠与 51.0 mL 0.1 mol·L^{-1} NaOH 混合均匀	11.6
12	磷酸氢二钠—氢氧化钠	50.0 mL 0.05 mol·L^{-1} Na$_2$HPO$_4$ 与 26.9 mL 0.1 mol·L^{-1} NaOH 混合均匀,加水稀释至 100 mL	12.0
13	氯化钾—氢氧化钠	25.0 mL 0.2 mol·L^{-1} KCl 与 66.0 mL 0.2 mol·L^{-1} NaOH 混合均匀,加水稀释至 100 mL	13.0

5.6 常用指示剂

表 5-5 酸碱指示剂 (Acid-base Indicators)

序号 (No.)	名 称 (Indicator)	pH变色范围 (Color range)	酸 色 (Acid color)	碱 色 (Base color)	pK_a	浓 度 (Concentration)
1	甲基紫(第一次变色)	0.13~0.5	黄	绿	0.8	0.1%水溶液
2	甲酚红(第一次变色)	0.2~1.8	红	黄	—	0.04%乙醇(50%)溶液
3	甲基紫(第二次变色)	1.0~1.5	绿	蓝	—	0.1%水溶液
4	百里酚蓝(第一次变色)	1.2~2.8	红	黄	1.65	0.1%乙醇(20%)溶液
5	茜素黄R(第一次变色)	1.9~3.3	红	黄	—	0.1%水溶液
6	甲基紫(第三次变色)	2.0~3.0	蓝	紫	—	0.1%水溶液
7	甲基黄	2.9~4.0	红	黄	3.3	0.1%乙醇(90%)溶液
8	溴酚蓝	3.0~4.6	黄	蓝	3.85	0.1%乙醇(20%)溶液
9	甲基橙	3.1~4.4	红	黄	3.40	0.1%水溶液
10	溴甲酚绿	3.8~5.4	黄	蓝	4.68	0.1%乙醇(20%)溶液
11	甲基红	4.4~6.2	红	黄	4.95	0.1%乙醇(60%)溶液
12	溴百里酚蓝	6.0~7.6	黄	蓝	7.1	0.1%乙醇(20%)
13	中性红	6.8~8.0	红	黄	7.4	0.1%乙醇(60%)溶液
14	酚红	6.8~8.0	黄	红	7.9	0.1%乙醇(20%)溶液
15	甲酚红(第二次变色)	7.2~8.8	黄	红	8.2	0.04%乙醇(50%)溶液
16	百里酚蓝(第二次变色)	8.0~9.6	黄	蓝	8.9	0.1%乙醇(20%)溶液
17	酚酞	8.2~10.0	无色	紫红	9.4	0.1%乙醇(60%)溶液
18	百里酚酞	9.4~10.6	无色	蓝	10.0	0.1%乙醇(90%)溶液
19	茜素黄R(第二次变色)	10.1~12.1	黄	紫	11.16	0.1%水溶液
20	靛胭脂红	11.6~14.0	蓝	黄	12.2	25%乙醇(50%)溶液

表 5-6 氧化还原指示剂 (Redox Indicators)

序号 (No.)	名 称 (Indicator)	氧化型颜色 (Oxidized color)	还原型颜色 (Reduced color)	E_{ind}/V	浓度 (Concentration)
1	二苯胺	紫	无色	+0.76	1%浓硫酸溶液
2	二苯胺磺酸钠	紫红	无色	+0.84	0.2%水溶液
3	亚甲基蓝	蓝	无色	+0.532	0.1%水溶液
4	中性红	红	无色	+0.24	0.1%乙醇溶液
5	喹啉黄	无色	黄	—	0.1%水溶液
6	淀粉	蓝	无色	+0.53	0.1%水溶液
7	孔雀绿	棕	蓝	—	0.05%水溶液
8	劳氏紫	紫	无色	+0.06	0.1%水溶液
9	邻二氮菲-亚铁	浅蓝	红	+1.06	(1.485 g 邻二氮菲＋0.695 g 硫酸亚铁)溶于 100 mL 水
10	酸性绿	橘红	黄绿	+0.96	0.1%水溶液
11	专利蓝 V	红	黄	+0.95	0.1%水溶液

表 5-7 元素的原子量(1999)

(按照原子序数排列，以 $A_r(^{12}C)=12$ 为基准)

元素			原子序	原子量	元素			原子序	原子量
符号	名称	英文名			符号	名称	英文名		
H	氢	Hydrogen	1	1.00794(7)	Y	钇	Yttrium	39	88.90585(2)
He	氦	Helium	2	4.002602(2)	Zr	锆	Zirconium	40	91.224(2)
Li	锂	Lithium	3	6.941(2)	Nb	铌	Niobium	41	92.90638(2)
Be	铍	Beryllium	4	9.012182(3)	Mo	钼	Molybdenum	42	95.94(1)
B	硼	Boron	5	10.811(7)	Tc	锝	Technetium	43	[98]
C	碳	Carbon	6	12.0107(8)	Ru	钌	Ruthenium	44	101.07(2)
N	氮	Nitrogen	7	14.0067(2)	Rh	铑	Rhodium	45	102.90550(2)
O	氧	Oxygen	8	15.9994(3)	Pd	钯	Palladium	46	106.42(1)
F	氟	Fluorine	9	18.9984032(5)	Ag	银	Silver	47	107.8682(2)
Ne	氖	Neon	10	20.1797(6)	Cd	镉	Cadmium	48	112.411(8)
Na	钠	Sodium	11	22.989770(2)	In	铟	Indium	49	114.818(3)
Mg	镁	Magnesium	12	24.3050(6)	Sn	锡	Tin	50	118.710(7)
Al	铝	Aluminium	13	26.981538(2)	Sb	锑	Antimony	51	121.760(1)
Si	硅	Silicon	14	28.0855(3)	Te	碲	Tellurium	52	127.60(3)
P	磷	Phosphorus	15	30.973761(2)	I	碘	Iodine	53	126.90447(3)
S	硫	Sulfur	16	32.065(5)	Xe	氙	Xenon	54	131.293(6)
Cl	氯	Chlorine	17	35.453(2)	Cs	铯	Cesium	55	132.90545(2)
Ar	氩	Argon	18	39.948(1)	Ba	钡	Barium	56	137.327(7)
K	钾	Potassium	19	39.0983(1)	La	镧	Lanthanum	57	138.9055(2)
Ca	钙	Calcium	20	40.078(4)	Ce	铈	Cerium	58	140.116(1)
Sc	钪	Scandium	21	44.955910(8)	Pr	镨	Praseodymium	59	140.90765(2)
Ti	钛	Titanium	22	47.867(1)	Nd	钕	Neodymium	60	144.24(3)
V	钒	Vanadium	23	50.9415(1)	Pm	钷	Promethium	61	[145]
Cr	铬	Chromium	24	51.9961(6)	Sm	钐	Samarium	62	150.36(3)
Mn	锰	Manganese	25	54.938049(9)	Eu	铕	Europium	63	151.964(1)
Fe	铁	Iron	26	55.845(2)	Gd	钆	Gadolinium	64	157.25(3)
Co	钴	Cobalt	27	58.933200(9)	Tb	铽	Terbium	65	158.92534(2)
Ni	镍	Nickel	28	58.6934(2)	Dy	镝	Dysprosium	66	162.50(3)
Cu	铜	Copper	29	63.546(3)	Ho	钬	Holmium	67	164.93032(2)
Zn	锌	Zinc	30	65.39(2)	Er	铒	Erbium	68	167.259(3)
Ga	镓	Gallium	31	69.723(1)	Tm	铥	Thulium	69	168.93421(2)
Ge	锗	Germanium	32	72.64(1)	Yb	镱	Ytterbium	70	173.04(3)
As	砷	Arsenic	33	74.92160(2)	Lu	镥	Lutetium	71	174.967(1)
Se	硒	Selenium	34	78.96(3)	Hf	铪	Hafnium	72	178.49(2)
Br	溴	Bromine	35	79.904(1)	Ta	钽	Tantalum	73	180.9479(1)
Kr	氪	Krypton	36	83.30(1)	W	钨	Tungsten	74	183.84(1)
Rb	铷	Rubidium	37	85.4678(3)	Re	铼	Rhenium	75	186.207(1)
Sr	锶	Strontium	38	87.62(1)	Os	锇	Osmium	76	190.23(3)

续表

元素			原子序	原子量	元素			原子序	原子量
符号	名称	英文名			符号	名称	英文名		
Ir	铱	Iridium	77	192.217(3)	Pa	镤	Protactinium	91	231.03588(2)
Pt	铂	Platinum	78	195.078(2)	U	铀	Uranium	92	238.02891(3)
Au	金	Gold	79	196.96655(2)	Np	镎	Neptunium	93	[237]
Hg	汞	Mercury	80	200.59(2)	Pu	钚	Plutonium	94	[244]
Tl	铊	Thallium	81	204.3833(2)	Am	镅	Americium	95	[243]
Pb	铅	Lead	82	207.2(1)	Cm	锔	Curium	96	[247]
Bi	铋	Bismuth	83	208.98038(2)	Bk	锫	Berkelium	97	[247]
Po	钋	Polonium	84	[209]	Cf	锎	Californium	98	[251]
At	砹	Astatine	85	[210]	Es	锿	Einsteinium	99	[252]
Rn	氡	Radon	86	[222]	Fm	镄	Fermium	100	[257]
Fr	钫	Francium	87	[223]	Md	钔	Mendelevium	101	[258]
Ra	镭	Radium	88	[226]	No	锘	Nobelium	102	[259]
Ac	锕	Actinium	89	[227]	Lr	铹	Lawrencium	103	[262]
Th	钍	Thorium	90	232.0381(1)					

注：录自1999年国际原子量表(IUPAC Commission of Atomic Weights and Isotopic Abundances. Atomic Weights of the Elements 1999. Pure Appl. Chem.,2001,73:667—683).( )表示原子量最后一位的不确定性,[ ]中的数值为没有稳定同位素元素的半衰期最长同位素的质量数。